Basics of Electromagnetics and Transmission Lines

G. Jagadeeswar Reddy

T. Jayachandra Prasad

BSP BS Publications

CRC Press
Taylor & Francis Group
Boca Raton London New York

CRC Press is an imprint of the
Taylor & Francis Group, an **informa** business

CRC Press
Taylor & Francis Group
6000 Broken Sound Parkway NW, Suite 300
Boca Raton, FL 33487-2742

CRC Press is an imprint of the Taylor & Francis Group, an informa business

International Standard Book Number-13: 978-0-367-36330-7 (Hardback)

Library of Congress Cataloging-in-Publication Data
A catalog record has been requested

Visit the Taylor & Francis Web site at
http://www.taylorandfrancis.com

and the CRC Press Web site at
http://www.crcpress.com

BSP **BS Publications**

MIX
Paper from responsible sources
FSC™ C013985
FSC www.fsc.org

Printed in the United Kingdom
by Henry Ling Limited

Contents

Chapter 1

Static Electric Fields

Chapter 2

Static Magnetic Fields

Chapter 3

Maxwell's Equations for Time Varying Fields

Chapter 4

EM Wave Characteristics

Chapter 5
Transmission Lines

Dedicated to,

To my Father Late Sri. G.C. Subba Reddy

Preface

The applications of Physics play a very important role in our daily life. Electromagnetism is one of the main branches of Physics, which has a wide range of such significance. It is the study of the electromagnetic force; a type of force exerted among electrically charged particles and magnetic lines. The electromagnetic force can be generated by means of three types of fields known as electric fields, magnetic fields and optics. Electromagnetism is the force, which is behind all the practical phenomena encountered by man in daily life. The basics of the well-known technologies like Microwave, RADAR (weather, aircraft, ground penetrating), Optics, Satellite Imaging, Wireless transport of Electricity, Medical Imaging etc., are always related with the concepts of electromagnetic field theory.

After the second great unification in physics; it was clear that, "Magnetism and electricity are not separable by nature. It means without the knowledge of magnetic induction, it is not possible to explain magnetism; without the knowledge of electric charge, it is not possible to explain the electricity; and without the knowledge of electricity and magnetism, it is not possible to explain the electromagnetic fields."

Maxwell is an eminent physicist, who gives the complete mathematical relationships between the electric and magnetic fields. Maxwell demonstrated that electric and magnetic fields together travel through free space as waves moving with a speed which is equal to the speed of light, in his publication '*A Dynamical Theory of the Electromagnetic Field*' in 1865. Maxwell anticipated the nature of light as an electromagnetic wave, which is the cause of electric and magnetic phenomena. As a result of the combination of light and electrical phenomena, the existence of radio waves is estimated.

To comprehend the cutting edge technologies, the electronics engineer must acquire the concepts of the field theory by studying "Basics of Electromagnetics and Transmission Lines". It is a toughest subject among all the other subjects that the students come across. Hence, this book is designed by targeting below average students to make them more comfortable with the subject, though it is tough.

In order to guide the students in an effective manner, this book has been written in a very simple and clear language. The entire curriculum of this subject was divided into 5 units.

- UNIT I: Understanding of electric fields.
- UNIT II: Understanding of magnetic fields.
- UNIT III: Understanding of electromagnetic fields (Maxwell equations).
- UNIT IV: Understanding the Characteristics and propagation of EM waves.
- UNIT V: Understanding the concepts of Transmission lines.

Salient Features:

- This book has been written in such a straightforward language that a slow learner can understand the concepts very clearly.
- Each chapter includes a sufficient content of problems with solutions which includes questions from previous papers. The solutions for the problems are given clearly step by step, immediately after the related theory.
- At the end of each chapter, multiple choice questions, short answer questions with answers and exercises are also given, which will help the students to face the examinations confidently. The impedance matching problems using smith chart are also explained in a very easy manner.

Chapter

1

Static Electric Fields

1.1 Introduction

Electrostatic in the sense static or rest or time in-varying electric fields. Electrostatic field can be obtained by the distribution of static charges.

The two fundamental laws which describe electrostatic fields are Coulomb's law and Gauss's law:

They are independent laws. i.e., one law does not depend on the other law.

Coulomb's law can be used to find electric field when the charge distribution is of any type, but it is easy to use Gauss's law to find electric field when the charge distribution is symmetrical.

1.2 Coulomb's Law

This law is formulated in the year 1785 by Coulomb. It deals with the force a point charge exerts on another point charge; generally a charge can be expressed in terms of coulombs.

$$1 \text{ coulomb} = 6 \times 10^{18} \text{ electrons}$$

$$1 \text{ electron charge} = -1.6 \times 10^{-19} \text{ Coulombs}$$

Coulomb's law states that the force between two point charges Q_1 and Q_2 is along the line joining between them, directly proportional to the product of two point charges, and inversely proportional to the square of the distance between them

$$\therefore \qquad F = \frac{KQ_1Q_2}{R^2}$$

where K is proportional constant

In SI, a unit for Q_1 and Q_2 is coulombs(C), for R meters(m) and for F newtons(N).

$$K = \frac{1}{4\pi \in_0}$$

where $\in_0$ = permittivity of free space (or) vacuum

$= 8.854 \times 10^{-12}$ farads/meter

$= \frac{10^{-9}}{36\pi}$ farads/m

$$K = \frac{36\pi}{4\pi \times 10^{-9}} = 9 \times 10^9 \text{ m/farads}$$

$$F = \frac{Q_1Q_2}{4\pi \in_0 R^2} \qquad \text{.....(1.2.1)}$$

Assume that the point charges Q_1 and Q_2 are located at (x_1, y_1, z_1) and (x_2, y_2, z_2) with the position vectors $\overline{r}_1$ and $\overline{r}_2$ respectively. Let the force on Q_2 due to Q_1 be $\overline{F}_{12}$ which can be written as

$$\overline{F}_{12} = \frac{Q_1Q_2}{4\pi \in_0 R^2}\overline{a}_{R12} \qquad \text{.....(1.2.2)}$$

where $\overline{a}_{R_{12}}$ is unit vector along the vector $\overline{R}_{12}$. Graphical representation of the vectors in rectangular coordinate system is shown in Fig.1.1

Where $\overline{a}_x$ is the unit vector along X-axis and $\overline{a}_y$ is the unit vector along Y-axis and $\overline{a}_z$ is the unit vector along Z-axis.

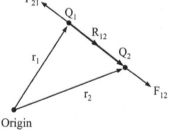

Fig. 1.1 Graphical representation of the vectors

From Fig.1.1, we can write $\overline{r}_1 + \overline{R}_{12} = \overline{r}_2$

i.e., $\overline{R}_{12} = \overline{r}_2 - \overline{r}_1$

where $\overline{r}_1 = x_1\overline{a}_x + y_1\overline{a}_y + z_1\overline{a}_z$

$\overline{r}_2 = x_2\overline{a}_x + y_2\overline{a}_y + z_2\overline{a}_z$

Now $$\bar{F}_{12} = \frac{Q_1 Q_2}{4\pi \in_0 R^2} \frac{\bar{R}_{12}}{|\bar{R}_{12}|}$$

$$\because \quad \bar{a}_{R_{12}} = \frac{\bar{R}_{12}}{|\bar{R}_{12}|}$$

$$= \frac{Q_1 Q_2}{4\pi \in_0 R^2} \frac{\bar{R}_{12}}{R} \qquad \because \quad |R_{12}| = R$$

$$= \frac{Q_1 Q_2}{4\pi \in_0} \frac{\bar{R}_{12}}{R^3}$$

$$= \frac{Q_1 Q_2}{4\pi \in_0} \frac{\bar{r}_2 - \bar{r}_1}{|\bar{r}_2 - \bar{r}_1|^3} \qquad \text{.....(1.2.3)}$$

and force on Q_1 due to Q_2 is $\bar{F}_{21} = -\bar{F}_{12}$

If we have more than two point charges i.e., Q_1, Q_2,... Q_N with the position vectors $\bar{r}_1, \bar{r}_2, ...\bar{r}_N$ respectively, then the force on a point charge Q, whose position vector is $\bar{r}$, can be written as

$$F = \frac{QQ_1}{4\pi \in_0} \frac{\bar{r} - \bar{r}_1}{|\bar{r} - \bar{r}_1|^3} + \frac{QQ_2}{4\pi \in_0} \frac{\bar{r} - \bar{r}_2}{|\bar{r} - \bar{r}_2|^3} + ... + \frac{QQ_N}{4\pi \in_0} \frac{\bar{r} - \bar{r}_N}{|\bar{r} - \bar{r}_N|^3}$$

$$= \frac{Q}{4\pi \in_0} \sum_{K=1}^{N} Q_K \frac{\bar{r} - \bar{r}_K}{|\bar{r} - \bar{r}_K|^3} \qquad \text{.....(1.2.4)}$$

1.3 Electric Field Intensity

Electric field intensity is defined as force per unit charge in an electric field. The other name of electric field intensity is electric field strength and it is denoted by $\bar{E}$.

$$\therefore \quad \bar{E} = \frac{\bar{F}}{Q} \text{ N/C or Volts/meter}$$

i.e., $$\bar{E} = \frac{Q\,Q}{Q 4\pi \in_0 R^2} = \frac{Q}{4\pi \in_0 R^2} \qquad \text{.....(1.3.1)}$$

Consider a point charge Q with position vector $\bar{r}$, then the electric field intensity $\bar{E}$ at some point with position vector $\bar{r}_1$ due to point charge Q is

$$\bar{E} = \frac{Q}{4\pi \in_0 R^2} \bar{a}_R \qquad \text{.....(1.3.2)}$$

where $\bar{a}_R$ is the unit vector along $\bar{R}$. Graphical representation of vector is shown in Fig.1.2

From Fig. 1.2, $\bar{R} = \bar{r}_1 - \bar{r}$

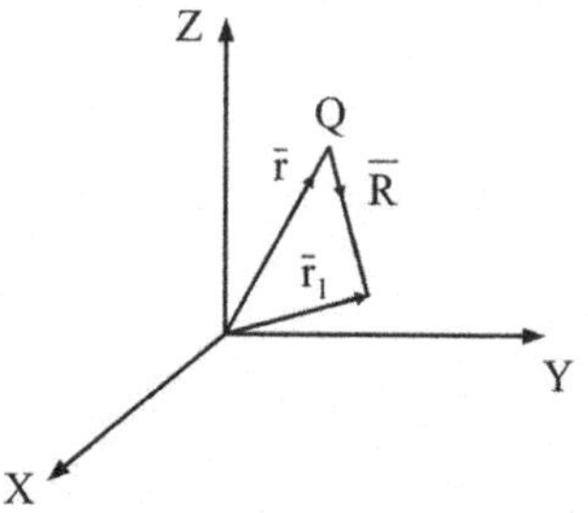

Fig. 1.2 Graphical representation

$$\bar{E} = \frac{Q}{4\pi \in_0 R^2} \frac{\bar{R}}{|\bar{R}|}$$

$$= \frac{Q}{4\pi \in_0 R^2} \frac{\bar{R}}{R}$$

$$= \frac{Q}{4\pi \in_0} \frac{\bar{R}}{R^3} = \frac{Q}{4\pi \in_0} \frac{\bar{r}_1 - \bar{r}}{|\bar{r}_1 - \bar{r}|^3}$$

If we have more than one point charge i.e., Q_1, Q_2,… Q_N with the position vectors $\bar{r}_1, \bar{r}_2, \ldots \bar{r}_N$ respectively. Then the electric field intensity $\bar{E}$ at some point with position vector $\bar{r}$ can be written as

$$\bar{E} = \frac{Q_1}{4\pi \in_0} \frac{\bar{r} - \bar{r}_1}{|\bar{r} - \bar{r}_1|^3} + \frac{Q_2}{4\pi \in_0} \frac{\bar{r} - \bar{r}_2}{|\bar{r} - \bar{r}_2|^3} + \ldots + \frac{Q_N}{4\pi \in_0} \frac{\bar{r} - \bar{r}_N}{|\bar{r} - \bar{r}_N|^3}$$

$$= \frac{1}{4\pi \in_0} \sum_{K=1}^{N} Q_K \frac{\bar{r} - \bar{r}_K}{|\bar{r} - \bar{r}_K|^3} \qquad \text{.....(1.3.3)}$$

Problem 1.1

Point charges 1 mC and –2 mC are located at (3, 2, –1) and (–1, –1, 4) respectively. Calculate the electric force on a 10 nC charge located at (0, 3, 1) and the electric field intensity at that point.

Solution

We know

$$\bar{F} = \frac{Q}{4\pi \in_0} \sum_{K=1}^{2} Q_K \frac{\bar{r} - \bar{r}_K}{|\bar{r} - \bar{r}_K|^3}$$

$$= \frac{10 \times 10^{-9}}{4\pi \in_0} \left[1 \times 10^{-3} \frac{(-3\bar{a}_x + \bar{a}_y + 2\bar{a}_z)}{(\sqrt{9+1+4})^3} - 2 \times 10^{-3} \frac{(\bar{a}_x + 4\bar{a}_y - 3\bar{a}_z)}{(\sqrt{1+16+9})^3} \right]$$

$\because \quad \epsilon_0 = 8.854\times10^{-12}$ and $\pi = 3.14$

$$= 90\left[\frac{\left(-3\bar{a}_x+\bar{a}_y+2\bar{a}_z\right)\times10^{-3}}{52.38}-10^{-3}\frac{\left(2\bar{a}_x+8\bar{a}_y-6\bar{a}_z\right)}{132.57}\right]$$

$$= 90\times10^{-3}\left[\bar{a}_x\left(\frac{-3}{52.38}-\frac{2}{132.57}\right)+\bar{a}_y\left(\frac{1}{52.38}-\frac{8}{132.57}\right)+\bar{a}_z\left(\frac{2}{52.38}+\frac{6}{132.57}\right)\right]$$

$$= 90\times10^{-3}\left[-0.0723\bar{a}_x-0.0413\bar{a}_y+0.0834\bar{a}_z\right]$$

$$= -0.0065\,\bar{a}_x-0.0037\,\bar{a}_y+0.0075\,\bar{a}_z \text{ N.}$$

Also we know $\bar{E}=\dfrac{\bar{F}}{Q}$

$$= -\frac{0.0065}{10\times10^{-9}}\bar{a}_x-\frac{0.037}{10\times10^{-9}}\bar{a}_y+\frac{0.0075}{10\times10^{-9}}\bar{a}_z$$

$$= -650\bar{a}_x-370\bar{a}_y+750\bar{a}_z \text{ kV/m.}$$

Problem 1.2

Point charges 5 nC and –2 nC are located at $2\bar{a}_x+4a_z$ and $-3\bar{a}_x+5\bar{a}_z$ respectively. (a) Determine the force on a 1 nC point charge located at $\bar{a}_x-3\bar{a}_y+7\bar{a}_z$. (b) Find the electric field $\bar{E}$ *at* $\bar{a}_x-3\bar{a}_y+7\bar{a}_z$.

Solution

(a) We know

$$\bar{F}=\frac{Q}{4\pi\epsilon_0}\sum_{K=1}^{2}Q_K\frac{\bar{r}-\bar{r}_K}{\left|\bar{r}-\bar{r}_K\right|^3}$$

$$= 10^{-9}\times9\times10^{9}\times10^{-9}\left[5\frac{\left(-\bar{a}_x-3\bar{a}_y+3\bar{a}_z\right)}{\left(\sqrt{1+9+9}\right)^3}-\frac{2\left(4\bar{a}_x-3\bar{a}_y+2\bar{a}_z\right)}{\left(\sqrt{16+9+4}\right)^3}\right]$$

$$=9\times10^{-9}\left[\bar{a}_x\left(\frac{-5}{82.81}-\frac{8}{156.169}\right)+\bar{a}_y\left(\frac{-15}{82.81}+\frac{6}{156.169}\right)+\bar{a}_z\left(\frac{15}{82.81}-\frac{4}{156.169}\right)\right]$$

$$= 9\times10^{-9}\left[\bar{a}_x(-0.112)+\bar{a}_y(-0.143)+\bar{a}_z(0.155)\right]$$

$$= -1.008\bar{a}_x-1.287\bar{a}_y+1.395\bar{a}_z \text{ nN}$$

(b) $\bar{E} = \dfrac{\bar{F}}{Q}$, here Q = 1 nC

$$\therefore \quad \bar{E} = -1.008\bar{a}_x - 1.287\bar{a}_y + 1.395\bar{a}_z \text{ V/m}$$

Problem 1.3

Point charges Q_1 and Q_2 are respectively located at (4, 0, –3) and (2, 0, 1). If Q_2 = 4 nC, Find Q_1 such that (a) The $\bar{E}$ at (5, 0, 6) has no Z-component. (b) The force on a test charge at (5, 0, 6) has no X-component.

Solution

We have $\bar{F} = \dfrac{Q}{4\pi \in_0} \sum_{K=1}^{2} Q_K \dfrac{\bar{r} - \bar{r}_K}{|\bar{r} - \bar{r}_K|^3}$

$$\text{(a)} \quad \bar{E} = \frac{\bar{F}}{Q} = \frac{1}{4\pi \in_0}\left[\frac{Q_1|(5,0,6)-(4,0,-3)|}{\left(\sqrt{1+81}\right)^3} + \frac{4\times10^{-9}|(5,0,6)-(2,0,1)|}{\left(\sqrt{9+25}\right)^3}\right]$$

Given $\bar{E}$ has no Z – component, considering only Z components on both sides

$$0 = \frac{1}{4\pi \in_0}\left[\frac{Q_1 \times 9}{\left(\sqrt{82}\right)^3} + \frac{4\times10^{-9}\times5}{\left(\sqrt{34}\right)^3}\right]$$

$$\frac{Q_1 \times 9}{\left(\sqrt{82}\right)^3} = -\frac{4\times10^{-9}\times5}{\left(\sqrt{34}\right)^3}$$

$$Q_1 = -\frac{20}{9}\left(\sqrt{\frac{41}{17}}\right)^3 nC = -8.3\text{nC}$$

(b) Given the force on test charge has no X-component

$$0 = \frac{Q}{4\pi \in_0}\left[\frac{Q_1}{\left(\sqrt{82}\right)^3} + \frac{4\times10^{-9}\times3}{\left(\sqrt{34}\right)^3}\right]$$

$$= \frac{Q_1}{\left(\sqrt{82}\right)^3} = -\frac{4\times10^{-9}\times3}{\left(\sqrt{34}\right)^3}$$

$$Q_1 = -12\left(\sqrt{\frac{41}{17}}\right)^3 \text{nC} = -44.95\text{nC}$$

Problem 1.4

Two point charges of equal mass '*m*', charge '*Q*' are suspended at a common point by two threads of negligible mass and length '*l*'. Show that at equilibrium the inclination angle 'α' of each thread to the vertical is given by $Q^2 = 16\ \pi\epsilon_0\ mgl^2\ \sin^2\alpha\ \tan\alpha$, (or)

$$\frac{\tan^3\alpha}{1+\tan^2\alpha} = \frac{Q^2}{16\pi\in_0 mgl^2},$$

if 'α' is very small

Show that $$\alpha = \sqrt[3]{\frac{Q^2}{16\pi\in_0 mgl^2}}$$

Solution:

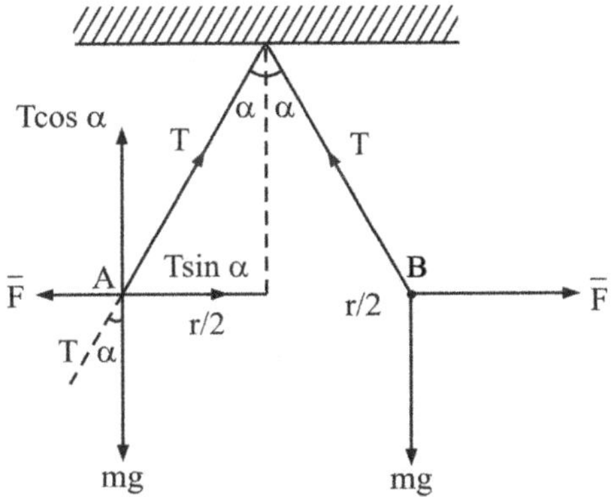

Fig. 1.3 Suspended charge particles

When two charges are suspended from a common point with threads of length '*l*', we can represent graphically as sown in Fig.1.3, where *T* is the tension in thread '*mg*' is the weight of charge towards ground due to gravitational force and $\bar{F}$ is force on charge at 'A'(B) due to charge at 'B'(A). $T\cos\alpha$ is the vertical component of '*T*' which is upwards and $T\sin\alpha$ is the horizontal component of '*T*' which is opposite to $\bar{F}$. To form equilibrium either at 'A' or 'B'

$$T\cos\alpha = mg \qquad \text{.....(1.3.4)}$$

$$T\sin\alpha = \bar{F} \qquad \text{.....(1.3.5)}$$

$$\frac{(1.3.4)}{(1.3.5)} = \frac{T\sin\alpha}{T\cos\alpha} = \frac{\bar{F}}{mg}$$

$$\Rightarrow \qquad \tan\alpha = \frac{\bar{F}}{mg}$$

where $$\bar{F} = \frac{Q^2}{4\pi \in_0 r^2}$$

From Fig.1.3 $$\sin\alpha = \frac{r/2}{l}$$

$$\Rightarrow r = 2l\sin\alpha$$

$$\tan\alpha = \frac{Q^2}{4mg\pi \in_0 r^2}$$

$$= \frac{Q^2}{4mg\pi \in_0 4l^2 \sin^2\alpha}$$

$$\tan\alpha = \frac{Q^2}{16mgl^2\pi \in_0 \sin^2\alpha}$$

$$\sin^2\alpha \tan\alpha = \frac{Q^2}{16mgl^2\pi \in_0} \qquad .....(1.3.6)$$

$$\Rightarrow \quad Q^2 = 16\pi \in_0 mgl^2 \sin^2\alpha \tan\alpha \qquad .....(1.3.7)$$

From (1.3.6)

$$\cos^2\alpha \frac{\sin^2\alpha}{\cos^2\alpha} \tan\alpha = \frac{Q^2}{16\pi \in_0 mgl^2}$$

$$\frac{\tan^3\alpha}{\sec^2\alpha} = \frac{Q^2}{16\pi \in_0 mgl^2}$$

$$\frac{\tan^3\alpha}{1+\tan^2\alpha} = \frac{Q^2}{16\pi \in_0 mgl^2}$$

If α is very small, $\sin\alpha = \tan\alpha = \alpha$

From (1.3.4) $Q^2 = 16\pi\in_0 mg\, l^2\, \alpha^3$

$$\alpha^3 = \frac{Q^2}{16\pi \in_0 mg\, l^2}$$

$$\alpha = \sqrt[3]{\frac{Q^2}{16\pi \in_0 mg\, l^2}}$$

Problem 1.5

Two small identical conducting spheres have charges of 2×10^{-9} and -0.5×10^{-9} C respectively. (a) When they are placed 4 cm apart what is the force between them? (b) If they are brought into contact and then separated by 4 cm. What is the force between them?

Solution

(a) We know

$$\bar{F} = \frac{Q_1 Q_2}{4\pi \in_0 R^2}$$

$$\because \frac{1}{4\pi \in_0} = 9 \times 10^9$$

$$\bar{F} = \frac{-2 \times 10^{-9} \times 0.5 \times 10^{-9} \times 9 \times 10^9}{16 \times 10^{-4}}$$

$$= -5.625\ \mu\text{N}$$

(b) When they are brought into contact, charges will be added and again when they are separated charge will be distributed equally

$Q_1 = 0.758 \times 10^{-9}$ C

$Q_2 = 0.75 \times 10^{-9}$ C

$\bar{F} = 3.164\ \mu$N

Problem 1.6

If the charges in the above problem are separated with the same distance in a kerosene ($\in_r = 2$), then find (a) and (b) as in the previous problem.

Solution

(a) $\bar{F}_k = \frac{-5.625}{2}\ \mu\text{N}$

$= -2.8125\ \mu$N

(b) $\bar{F}_k = \frac{3.164}{2} = 1.582\ \mu\text{N}$

Problem 1.7

Three equal +Ve charges of 4×10^{-9} C each are located at 3 corners of a square, side 20 cm. Determine the magnitude and direction of the electric field at the vacant corner point of the square.

Solution

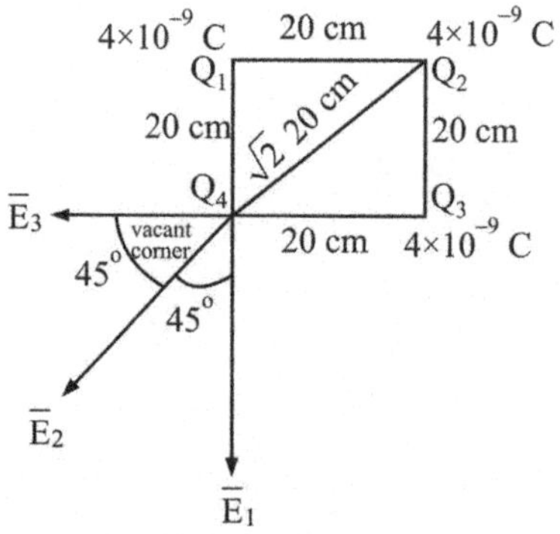

Fig. 1.4

$\bar{E}_1$ = Electric field intensity at Q_4 due to Q_1

$$= \frac{Q_1}{4\pi \in_0 R^2}$$

$$= 900 \text{ V/m}$$

$\bar{E}_2 = 450$ V/m

$\bar{E}_3 = 900$ V/m

The electric field intensity at vacant point is

$$\bar{E} = \bar{E}_2 + \bar{E}_1 \cos 45^\circ + \bar{E}_3 \cos 45^\circ$$

$$= 450 + \frac{900}{\sqrt{2}} + \frac{900}{\sqrt{2}}$$

$$= 450 + 900\sqrt{2}$$

$$= 1722.792206 \text{ V/m}$$

1.4 Coordinate Systems

The most widely used coordinate systems are Cartesian or rectangular co-ordinate system, Circular or cylindrical co-ordinate system, and Spherical co-ordinate system.

1.4.1 Cartesian Co-ordinate System

In this system the co-ordinates are X, Y, Z in which three are mutually perpendicular to each other. This system is shown in Fig. 1.5, where $\bar{a}_x, \bar{a}_y$ & $\bar{a}_z$ are unit vectors along X, Y and Z respectively. In Cartesian co-ordinate system the dot product of any unit vector with itself gives '1'.

i.e., $\bar{a}_x \cdot \bar{a}_x = 1$ $\quad = 1$ $\quad \bar{a}_z \cdot \bar{a}_z = 1$

and the dot product of one unit vector with the other one gives '0'.

i.e., $\bar{a}_x \cdot \bar{a}_y = 0 \quad \bar{a}_y \cdot \bar{a}_z = 0 \quad \bar{a}_z \cdot \bar{a}_x = 0$

The cross product of one unit vector with the other unit vector, which is next to the first one in anticlockwise direction, results the last unit vector in anticlockwise direction.

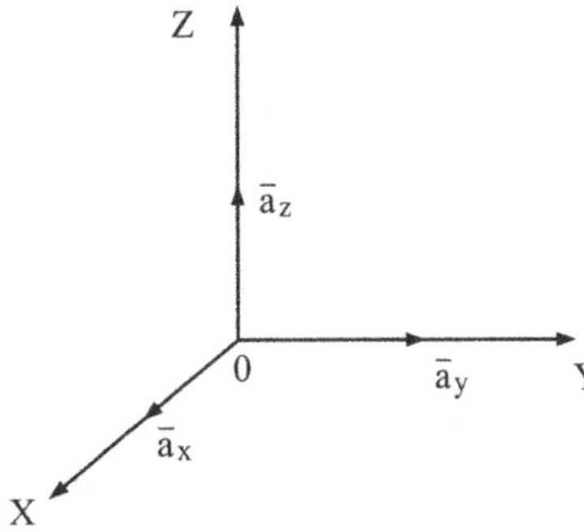

Fig. 1.5 Cartesian co-ordinate system

i.e., $\bar{a}_x \times \bar{a}_y = \bar{a}_z \quad \bar{a}_y \times \bar{a}_z = \bar{a}_x \quad \bar{a}_z \times \bar{a}_x = \bar{a}_y$

Consider a general vector $\bar{A}$ with components A_x, A_y, A_z along X, Y, Z respectively, then it can be represented in Cartesian coordinate system as

$$\bar{A} = A_x\bar{a}_x + A_y\bar{a}_y + A_z\bar{a}_z$$

Here X ranges from $-\infty$ to ∞, Y from $-\infty$ to ∞, and Z from $-\infty$ to ∞.

Note:

1. Differential displacement or elemental length is
$$d\bar{l} = dx\bar{a}_x + dy\bar{a}_y + dz\bar{a}_z$$
2. Differential or elemental normal area is $d\bar{S} = dy\,dz\,\bar{a}_x$
$$= dx\,dz\,\bar{a}_y$$
$$= dx\,dy\,\bar{a}_z$$
3. Differential or elemental volume is $dv = dx\;dy\;dz$

1.4.2 Cylindrical Co-ordinate System

In this system ρ, ϕ and z are coordinates in which all are mutually orthogonal to each other.

Note: If the given problem is of circular symmetry, then it would be better to use cylindrical coordinates rather than Cartesian coordinates.

Where ρ is the radial distance from origin, ϕ is the azimuthal angle from X-axis to the radial distance and Z is same as in Cartesian coordinate system. The cylindrical coordinate system is shown in Fig. 1.6.

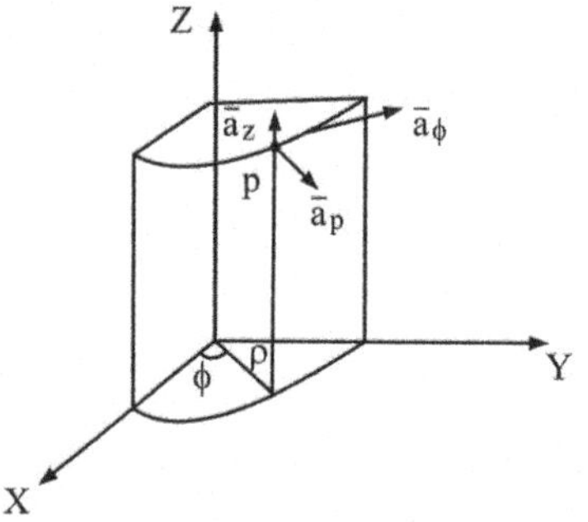

Fig. 1.6 Cylindrical coordinate system

Where $\bar{a}_\rho, \bar{a}_\phi$ and $\bar{a}_z$ are unit vectors along radial axis, azimuthal angle and z-direction respectively.

The dot product of any unit vector with itself gives '1'.

i.e. $\bar{a}_\rho \cdot \bar{a}_\rho = 1$ $\bar{a}_\phi \cdot \bar{a}_\phi = 1$ $\bar{a}_z \cdot \bar{a}_z = 1$

The dot product of any unit vector with the other unit vector gives '0'

i.e. $\bar{a}_\rho \cdot \bar{a}_\phi = 0$ $\bar{a}_\phi . \bar{a}_z = 0$ $\bar{a}_z \cdot \bar{a}_\rho = 0$

The cross product of any unit vector with the other unit vector, which is next to the first one in anticlockwise direction, results last unit vector in the anticlockwise direction.

i.e., $\bar{a}_\rho \times \bar{a}_\phi = \bar{a}_z$ $\bar{a}_\phi \times \bar{a}_z = \bar{a}_\rho$ $\bar{a}_z \times \bar{a}_\rho = \bar{a}_\phi$

Consider a general vector $\bar{A}$ with components A_ρ, A_ϕ, A_z along the three axes, then it can be represented as

$$\bar{A} = A_\rho \bar{a}_\rho + A_\phi \bar{a}_\phi + A_z \bar{a}_z$$

In this system $0 \le \rho < \infty$, $0 \le \phi < 2\pi$, and $-\infty < z < \infty$

The relation between Cylindrical and Cartesian coordinate system is shown in Fig.1.7.

The component of ρ on X-axis is ρ cosϕ and the component of ρ on Y-axis is ρ sin ϕ.

$$\therefore \quad X = \rho \cos\phi, \quad Y = \rho \sin\phi, \quad Z = z$$

from Fig.1.7 $\tan\phi = \dfrac{Y}{X} \Rightarrow \phi = \tan^{-1}\left(\dfrac{Y}{X}\right)$

$$X^2 + Y^2 = \rho^2 \Rightarrow \rho = \sqrt{X^2 + Y^2}$$

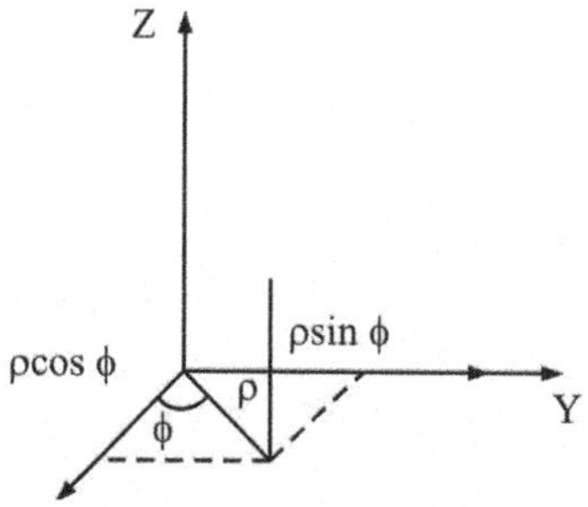

Fig.1.7 Relation between cylindrical and cartesian coordinate system

To find the relation among $\bar{a}_x$ and $\bar{a}_\rho, \bar{a}_\phi$ consider the Fig.1.8. A component of $\bar{a}_\rho$ on $\bar{a}_x$ is $\bar{a}_\rho \cos\phi$ and the component of $-\bar{a}_\phi$ on $\bar{a}_x$ is $-\bar{a}_\phi \sin\phi$.

$\therefore$ $\quad \bar{a}_x$ can be written as $\bar{a}_x = \bar{a}_\rho \cos\phi - \bar{a}_\phi \sin\phi$

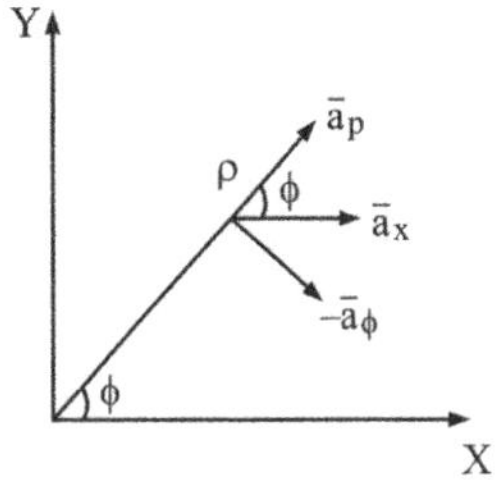

Fig. 1.8

To find the relation among $\bar{a}_y$ and $\bar{a}_\rho$, $\bar{a}_\phi$ consider the Fig.1.9. The component of $\bar{a}_\rho$ on $\bar{a}_y$ is $\bar{a}_\rho \sin\phi$ and the component of $\bar{a}_\phi$ on $\bar{a}_y$ is $\bar{a}_\phi \cos\phi$.

$\therefore$ $\quad \bar{a}_y = \bar{a}_\rho \sin\phi + \bar{a}_\phi \cos\phi$

The unit vector $\bar{a}_z$ of Cartesian coordinate system and cylindrical coordinate system is same $\therefore \bar{a}_z = \bar{a}_z$

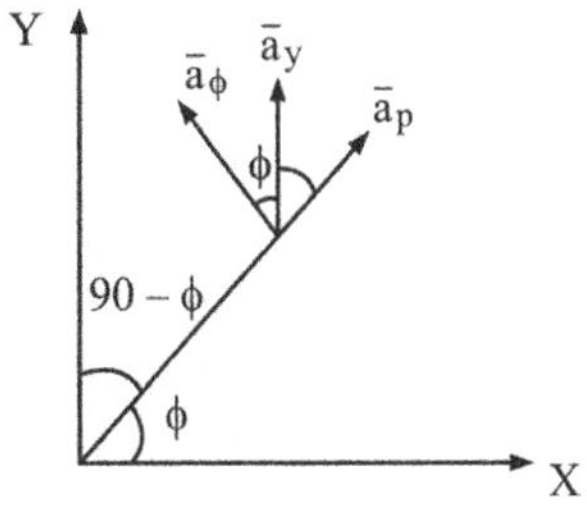

Fig. 1.9

We know that in Cartesian co-ordinate system

$$\bar{A} = A_x\bar{a}_x + A_y\bar{a}_y + A_z\bar{a}_z\,.$$

Substituting unit vectors,

$$\bar{A} = A_x\left(\bar{a}_\rho \cos\phi - \bar{a}_\phi \sin\phi\right) + A_y\left(\bar{a}_\rho \sin\phi + \bar{a}_\phi \cos\phi\right) + A_z\bar{a}_z$$

$$\bar{A} = \left(A_x \cos\phi + A_y \sin\phi\right)\bar{a}_\rho + \left(A_y \cos\phi - A_x \sin\phi\right)\bar{a}_\phi + A_z\bar{a}_z$$

$$\bar{A} = A_\rho\bar{a}_\rho + A_\phi\bar{a}_\phi + A_z\bar{a}_z$$

where

$$A_\rho = A_x \cos\phi + A_y \sin\phi$$

$$A_\phi = -A_x \sin\phi + A_y \cos\phi$$

$$A_z = A_z$$

i.e., in matrix form

$$\begin{bmatrix} A_\rho \\ A_\phi \\ A_z \end{bmatrix} = \begin{bmatrix} \cos\phi & \sin\phi & 0 \\ -\sin\phi & \cos\phi & 0 \\ 0 & 0 & 1 \end{bmatrix} \begin{bmatrix} A_x \\ A_y \\ A_z \end{bmatrix}$$

The above matrix in terms of unit vectors is given by

$$\begin{bmatrix} A_\rho \\ A_\phi \\ A_z \end{bmatrix} = \begin{bmatrix} \bar{a}_\rho\cdot\bar{a}_x & \bar{a}_\rho\cdot\bar{a}_y & \bar{a}_\rho\cdot\bar{a}_z \\ \bar{a}_\phi\cdot\bar{a}_x & \bar{a}_\phi\cdot\bar{a}_y & \bar{a}_\phi\cdot\bar{a}_z \\ \bar{a}_z\cdot\bar{a}_x & \bar{a}_z\cdot\bar{a}_y & \bar{a}_z\cdot\bar{a}_z \end{bmatrix} \begin{bmatrix} A_x \\ A_y \\ A_z \end{bmatrix}$$

or
$$\begin{bmatrix} A_x \\ A_y \\ A_z \end{bmatrix} = \begin{bmatrix} \bar{a}_x \cdot \bar{a}_\rho & \bar{a}_x \cdot \bar{a}_\phi & \bar{a}_x \cdot \bar{a}_z \\ \bar{a}_y \cdot \bar{a}_\rho & \bar{a}_y \cdot \bar{a}_\phi & \bar{a}_y \cdot \bar{a}_z \\ \bar{a}_z \cdot \bar{a}_\rho & \bar{a}_z \cdot \bar{a}_\phi & \bar{a}_z \cdot \bar{a}_z \end{bmatrix} \begin{bmatrix} A_\rho \\ A_\phi \\ A_z \end{bmatrix}$$

Note:

1. Differential displacement or elemental length is
$$d\bar{l} = d\rho \bar{a}_\rho + \rho d\phi \bar{a}_\phi + dz \bar{a}_z$$
2. Differential or elemental normal area is $d\bar{S} = \rho d\phi\, dz\, \bar{a}_\rho$
$$= d\rho dz \bar{a}_\phi$$
$$= \rho d\phi d\rho \bar{a}_z$$
3. Differential or elemental volume is dv = $\rho d\rho\, d\phi\, dz$

1.4.3 Spherical Coordinate System

When the given problem is of spherical symmetry, it is better to use spherical coordinate system to solve the problem instead of either Cartesian or cylindrical coordinate system.

In this system r, θ, ϕ are coordinates in which all are mutually orthogonal to each other. Where 'r' is the distance from origin to the point (where the vector is located). θ is the co-latitude angle which is taken from z axis to the radial distance and ϕ is same as in cylindrical coordinate system.

The spherical coordinate system is shown in Fig.1.10. Where $\bar{a}_r$ is the unit vector along r, $\bar{a}_\theta$ is the unit vector in increasing direction of θ and $\bar{a}_\phi$ is the unit vector in increasing direction of ϕ.

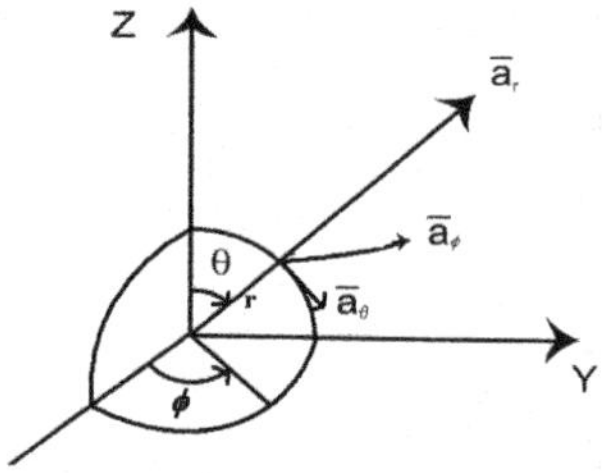

Fig. 1.10 Spherical coordinate system

The dot product of any unit vector with itself gives unity

i.e., $\bar{a}_r \cdot \bar{a}_r = 1 \quad \bar{a}_\theta \cdot \bar{a}_\theta = 1 \quad \bar{a}_\phi \cdot \bar{a}_\phi = 1$

The dot product of any unit vector with the other unit vector gives '0'

i.e., $\bar{a}_r \cdot \bar{a}_\theta = 0 \quad \bar{a}_\theta \cdot \bar{a}_\phi = 0 \quad \bar{a}_\phi \cdot \bar{a}_r = 0$

The cross product of unit vectors is: $\bar{a}_r \times \bar{a}_\theta = \bar{a}_\phi$, $\bar{a}_\theta \times \bar{a}_\phi = \bar{a}_r$, $\bar{a}_\phi \times \bar{a}_r = \bar{a}_\theta$

Here $0 \le r < \infty$, $0 \le \theta \le \pi$, and $0 \le \phi \le 2\pi$.

To convert from Cartesian to cylindrical or spherical co-ordinate system consider Fig. 1.11.

The component of r on z-axis is $r \cos \theta$, and the component of r on ρ is $r \sin\theta$.

$$\therefore \quad z = r\cos\theta$$

$$\rho = r\sin\theta \text{ and}$$

we know $x = \rho\cos\phi$ & $y = \rho\sin\phi$

From Cartesian to cylindrical, the conversion is $x = \rho\cos\phi$, $y = \rho\sin\phi$, and $z = z$

To get conversion from Cartesian to spherical co-ordinate system, substitute $\rho = r \sin \theta$ in the above equations.

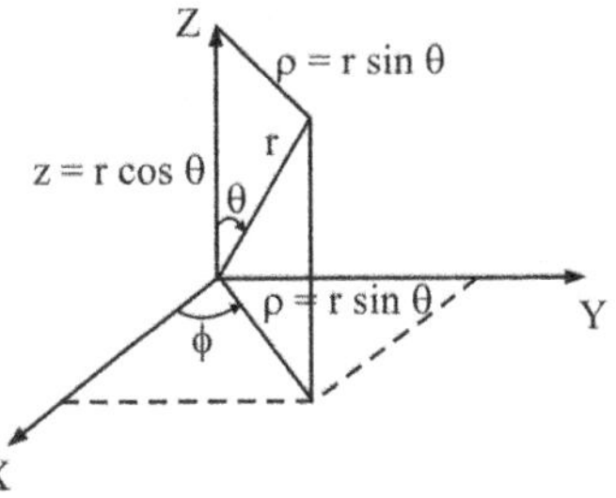

Fig. 1.11

$$x = r\sin\theta\cos\phi,$$

$$y = r\sin\theta\sin\phi, \text{ and}$$

$$z = r\cos\theta$$

From the above equations $r = \sqrt{x^2 + y^2 + z^2}$

From Fig.1.11 $\qquad \phi = \tan^{-1}\left(\dfrac{y}{x}\right)$

and $\qquad \tan\theta = \left(\dfrac{\rho}{z}\right) = \dfrac{\sqrt{x^2+y^2}}{z}$

$$\theta = \tan^{-1}\frac{\sqrt{x^2+y^2}}{z}$$

relation between unit vectors of Cartesian and spherical co-ordinate systems is as follows:

$$\bar{a}_x = \sin\theta\cos\phi\bar{a}_r + \cos\theta\cos\phi\bar{a}_\theta - \sin\phi\,\bar{a}_\phi$$

$$\bar{a}_y = \sin\theta\sin\phi\bar{a}_r + \cos\theta\sin\phi\bar{a}_\theta + \cos\phi\,\bar{a}_\phi$$

$$\bar{a}_z = \cos\theta\,\bar{a}_r - \sin\theta\,\bar{a}_\theta$$

***Note*:**

1. Differential displacement or elemental length is

$$d\bar{l} = dr\,\bar{a}_r + rd\theta\,\bar{a}_\theta + r\sin\theta d\phi\,\bar{a}_\phi$$

2. Differential or elemental normal area is

$$d\bar{S} = r^2 \sin\theta \, d\theta \, d\phi \, \bar{a}_r$$

$$= r \sin\theta dr d\phi \bar{a}_\theta$$

$$= r dr d\theta \bar{a}_\phi$$

3. Differential or elemental volume is $dv = r^2 \sin\theta \, dr d\theta d\phi$

1.5 Electric Fields due to Continuous Charge Distributions

So far we have discussed the electric field or force due to point charges. Let us see the electric field due to continuous charge distribution along a line, on a surface and in a volume. If the charge is distributed along a line the distribution can be represented with the line charge density ρ_L(C/m), which is shown in Fig.1.12(a). If the charge is distributed on a surface it's distribution can be represented with the surface charge density ρ_s(C/m^2), which is shown in Fig. 1.12(b). If the charge is distributed in a volume it's distribution can be represented with the volume charge density ρ_v(C/m^3), which is shown in Fig. 1.12(c).

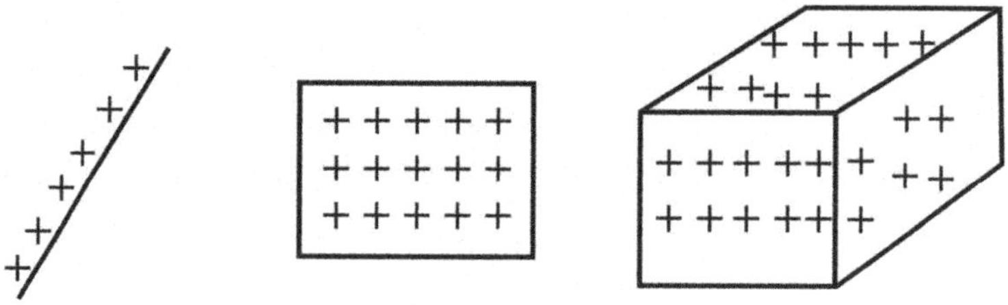

(a) a line charge (b) surface charge (c) volume charge

Fig.1.12 Charge distribution

The elemental charge dQ along a line can be written as

$dQ = \rho_l dl$, where dl is the elemental length.

So $$Q = \int_l \rho_l \, dl$$

∴ Electric field intensity due to line charge distribution is

$$\bar{E} = \int_l \frac{\rho_l \, dl}{4\pi \epsilon_0 R^2} \bar{a}_R \quad .....(1.5.1)$$

The elemental charge dQ on a surface can be written as $dQ = \rho_s ds$

$$\Rightarrow \qquad Q = \int_S \rho_s ds$$

∴ Electric field intensity due to surface charge distribution is

$$\bar{E} = \int_s \frac{\rho_s \, ds}{4\pi \in_0 R^2} \bar{a}_R \qquad .....(1.5.2)$$

The elemental charge dQ in a volume can be written as $dQ = \rho_v dv$

$$\Rightarrow \qquad Q = \int_v \rho_v dv$$

∴ Electric field intensity due to volume charge distribution is

$$\bar{E} = \int_v \frac{\rho_v \, dv}{4\pi \in_0 R^2} \bar{a}_R \qquad .....(1.5.3)$$

1.5.1 Line Charge Distribution

Consider a line charge distribution from A to B along Z-axis as shown in Fig.1.13.

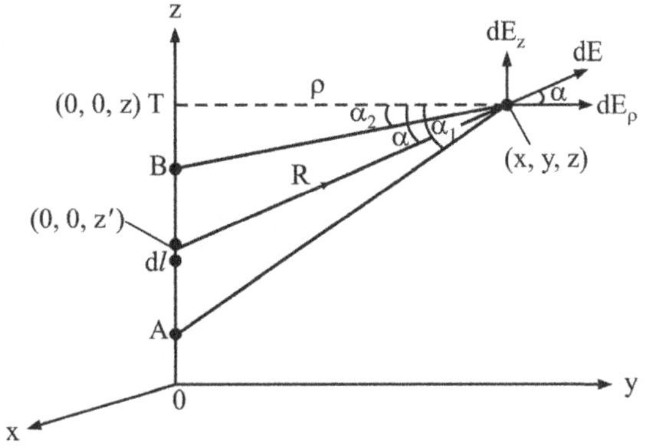

Fig.1.13 Finding $\bar{E}$ due to line charge distribution

Let us find the electric field at point (x, y, z) due to line charge distribution along Z- axis. We know electric field intensity due to line charge distribution as

$$\bar{E} = \int_l \frac{\rho_l \, dl}{4\pi \in_0 R^2} \bar{a}_R$$

where $\qquad \bar{a}_R = \dfrac{\bar{R}}{|\bar{R}|}$, $dl = dz'$,

Since the charge distribution has cylindrical symmetry, we use cylindrical coordinate system to obtain Electric field intensity.

From the Fig. 1.13

$$\bar{R} = \rho \bar{a}_\rho + (z - z')\bar{a}_z$$

$$\bar{E} = \int_l \frac{\rho_L \, dz'}{4\pi \in_0 R^2} \frac{\bar{R}}{|\bar{R}|}$$

$$= \int_l \frac{\rho_L \, dz'}{4\pi \in_0} \frac{\bar{R}}{R^3} = \int_l \frac{\rho_L \, dz'}{4\pi \in_0} \frac{\left[\rho \bar{a}_\rho + (z - z')\bar{a}_z\right]}{\left(\rho^2 + (z - z')^2\right)^{3/2}}$$

$$= \int_l \frac{\rho_L \, dz'}{4\pi \in_0} \frac{\left[\rho \bar{a}_\rho + (z - z')\bar{a}_z\right]}{\left(\rho^2 + (z - z')^2\right)^{3/2}}$$

From the Fig. 1.13

$$\tan\alpha = \frac{z - z'}{\rho} \Rightarrow z - z' = \rho \tan\alpha$$

$$\cos\alpha = \frac{\rho}{R} \Rightarrow R = \rho \sec\alpha \Rightarrow \sqrt{\rho^2 + \left(z - z'\right)^2} = \rho \sec\alpha$$

$$z' = \mathrm{OT} - (z - z') = \mathrm{OT} - \rho \tan\alpha$$

$$dz' = 0 - \rho \sec^2\alpha \, d\alpha$$

$$\bar{E} = \frac{-\rho_L}{4\pi \in_0} \int_l \frac{\rho \sec^2\alpha \, d\alpha \left(\rho \bar{a}_\rho + \rho \tan\alpha \, \bar{a}_z\right)}{\rho^3 \sec^3\alpha}$$

$$= \frac{-\rho_L}{4\pi \in_0} \int_{\alpha_1}^{\alpha_2} \frac{\rho \sec^2\alpha \, d\alpha \; \rho \sec\alpha \left(\bar{a}_\rho \cos\alpha + \bar{a}_z \sin\alpha\right)}{\rho^3 \sec^3\alpha}$$

$$= \frac{-\rho_L}{4\pi \in_0 \rho} \int_{\alpha_1}^{\alpha_2} (\cos\alpha \, \bar{a}_\rho + \sin\alpha \, \bar{a}_z) \, d\alpha$$

$$= \frac{-\rho_L}{4\pi \in_0 \rho} \left[\left[\sin\alpha \, \bar{a}_\rho\right]_{\alpha_1}^{\alpha_2} - \left[\cos\alpha \, \bar{a}_z\right]_{\alpha_1}^{\alpha_2}\right]$$

$$= \frac{-\rho_L}{4\pi \in_0 \rho}\left[(\sin\alpha_2 - \sin\alpha_1)\bar{a}_\rho + (-\cos\alpha_2 + \cos\alpha_1)\bar{a}_z\right]$$

which is electric field at point (x, y, z) due to line charge distribution from 'A' to 'B' along Z-axis. If 'A' is tending to $-\infty$ then α_1 becomes $\pi/2$ and 'B' is tending to ∞ then α_2 becomes $-\pi/2$.

$$\bar{E} = \frac{-\rho_L}{4\pi \in_0 \rho}\left[\left(\sin\left(-\frac{\pi}{2}\right) - \sin\left(\frac{\pi}{2}\right)\right)\bar{a}_\rho + \left(-\cos\left(-\frac{\pi}{2}\right) + \cos\left(\frac{\pi}{2}\right)\right)\bar{a}_z\right]$$

$$= \frac{2\bar{a}_\rho \rho_L}{4\pi \in_0 \rho}$$

$$\bar{E} = \frac{\rho_L}{2\pi \in_0 \rho}\bar{a}_\rho \qquad \text{.....(1.5.4)}$$

which is the electric field at point (*x, y, z*) due to infinite line charge distribution along Z-axis.

1.5.2 Surface Charge Distribution

Consider an infinite sheet lying on XY plane which is perpendicular to Z-axis as shown in the Fig. 1.14.

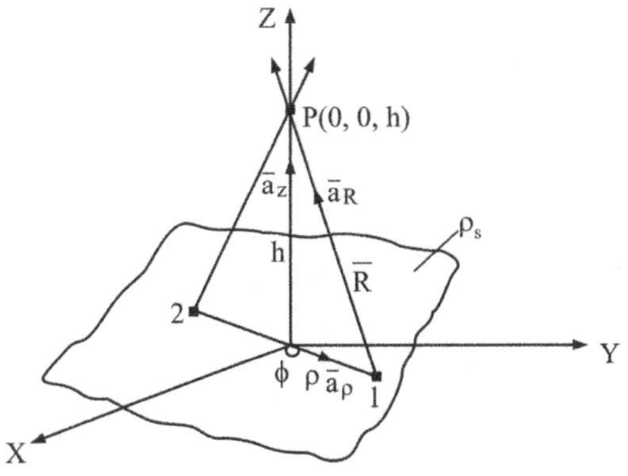

Fig. 1.14 Finding $\bar{E}$ due to infinite sheet of charge

Assume that the elemental surfaces are located on the sheet at '1' and '2'.

Then the elemental charge dQ on elemental surface ds is $dQ = \rho_s ds$.

∴ The elemental electric field at point (0, 0, h) due to the elemental surface ds is

$$d\bar{E} = \frac{dQ}{4\pi \in_0 R^2}\bar{a}_R$$

where

$$dQ = \rho_s ds \text{ and } \bar{a}_R = \frac{\bar{R}}{|\bar{R}|}$$

Since the surface is infinite it has circular symmetry, hence we can use cylindrical coordinate system to obtain electric field intensity.

Here ds lies on ρ and ϕ axises, Hence $ds = d\rho\, \rho\, d\phi$

From Fig.1.14

$$\rho\bar{a}_\rho + \bar{R} = h\bar{a}_z$$

$$\Rightarrow \quad \bar{R} = h\bar{a}_z - \rho\bar{a}_\rho$$

$$\therefore \quad d\bar{E} = \frac{dQ}{4\pi \in_0}\frac{\bar{R}}{|\bar{R}|^3}$$

$$= \frac{dQ}{4\pi \in_0}\frac{-\rho\bar{a}_\rho + h\bar{a}_z}{\left(\rho^2 + h^2\right)^{3/2}}$$

Since the sheet is symmetry with respect to origin on XY plane, for every electric field due to elemental surface (for example elemental surface located at '1') there will be an equal and opposite electric field due to the elemental surface on the other side(for example elemental surface located at '2') in the direction of 'ρ' (radial length), so finally when we add up the electric fields due to all the elemental surfaces on the sheet the electric field in the 'ρ' direction will get cancelled. We will have only the electric field perpendicular to the sheet i.e., along Z-direction.

By integrating the above equation, $\bar{E} = \dfrac{Q}{4\pi \in_0}\dfrac{h\bar{a}_z}{\left(\rho^2 + h^2\right)^{3/2}}$

Where

$$Q = \int_\phi \int_\rho \rho_s\, \rho\, d\rho\, d\phi$$

$$\therefore \quad \bar{E} = \int_0^{2\pi}\int_0^{\infty} \frac{h\bar{a}_z}{\left(\rho^2 + h^2\right)^{3/2}} \rho_s\, \rho\, d\rho\, d\phi \frac{1}{4\pi \in_0}$$

$$\bar{E} = \frac{\rho_s}{4\pi \in_0} \int_0^{2\pi} d\phi \int_0^{\infty} \frac{h\rho}{\left(\rho^2 + h^2\right)^{3/2}} d\rho\, \bar{a}_z$$

$$= \frac{\rho_s h}{4\pi \in_0} (2\pi) \int_0^{\infty} \left(\rho^2 + h^2\right)^{-3/2} \frac{1}{2} d\left(\rho^2\right) \bar{a}_z$$

$$= \frac{\rho_s h}{2 \in_0} \frac{1}{2} \left[\frac{\left(\rho^2 + h^2\right)^{-\frac{3}{2}+1}}{-\frac{3}{2}+1} \bar{a}_z \right]_0^{\infty}$$

$$\bar{E} = \frac{\rho_s h}{2 \in_0} \frac{1}{2} \left[\frac{-\left(h^2\right)^{-1/2}}{-1/2} \right] \bar{a}_z$$

$$\bar{E} = \frac{\rho_s}{2 \in_0} \bar{a}_z \qquad \text{.....(1.5.5)}$$

If we observe the above equation, the electric field is independent of the height 'h' i.e., the point can be considered at anywhere on the Z-axis.

The above equation can be generalized as

$$\bar{E} = \frac{\rho_s}{2 \in_0} \bar{a}_n \qquad \text{.....(1.5.6)}$$

Where $\bar{a}_n$ is the unit vector which is perpendicular to the sheet.

Consider a parallel plate capacitor of equal and opposite charge on each plate, the electric field due to these parallel plates can be written as

$$\bar{E} = \frac{\rho_s}{2 \in_0} \bar{a}_n + \frac{(-\rho_s)}{2 \in_0} \left(-\bar{a}_n\right) = \frac{\rho_s}{\in_0} \bar{a}_n \qquad \text{.....(1.5.7)}$$

Problem 1.8

A circular ring of radius 'a' carries a uniform charge ρ_L C/m and is placed on the XY plane with axis the same as the Z-axis.

(a) Show that $\bar{E}(0,0,h) = \dfrac{\rho_L a h}{2 \in_0 \left(h^2 + a^2\right)^{3/2}} \bar{a}_z$.

(b) What values of h gives the maximum value of $\bar{E}$

(c) If the total charge on the ring is Q. Find $\bar{E}$ as 'a' tends to zero.

Solution

(a) Here

$$dl = a\, d\phi$$

$$dQ = \rho_L\, dl$$

$$= \rho_L\, a d\phi$$

$$\therefore \quad d\overline{E} = \frac{dQ}{4\pi \in_0 R^2}\overline{a}_r$$

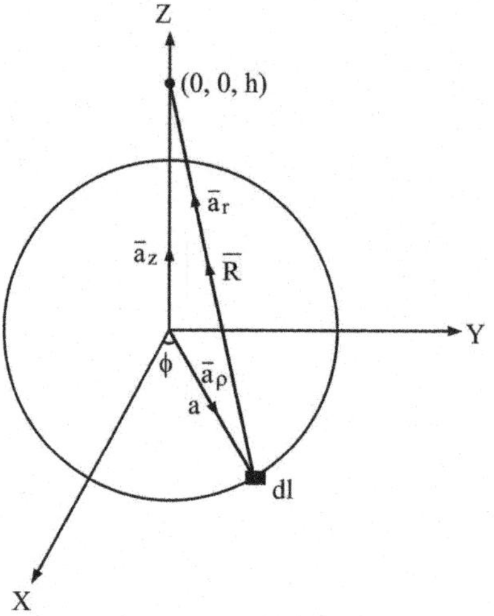

Fig. 1.15

$$\overline{a}_r = \frac{\overline{R}}{|R|};\ \frac{\overline{a}_r}{R^2} = \frac{\overline{R}}{R^3}$$

$$\therefore \quad d\overline{E} = \frac{dQ}{4\pi \in_0}\frac{\left[-a\,\overline{a}_\rho + h\overline{a}_z\right]}{\left(a^2+h^2\right)^{3/2}}$$

$$dQ = \rho_L\, a\, d\phi$$

$$Q = \int \rho_L\, a\, d\phi$$

when we add up electric fields, the electric field in ρ direction gets cancelled.

$$\therefore \quad \overline{E} = \frac{dQ}{4\pi \in_0}\frac{h\overline{a}_z}{\left(a^2+h^2\right)^{3/2}}$$

$$= \int \frac{\rho_L a \, d\phi}{4\pi \in_0} \frac{h\bar{a}_z}{\left(a^2 + h^2\right)^{3/2}}$$

$$= \frac{\rho_L a}{4\pi \in_0} \frac{h\bar{a}_z}{\left(a^2 + h^2\right)^{3/2}} \int_0^{2\pi} d\phi = \frac{\rho_L a h}{2 \in_0 \left(a^2 + h^2\right)^{3/2}} \bar{a}_z$$

(b) $\frac{d\bar{E}}{dh} = 0$

$$\frac{\rho_L a}{2 \in_0} \bar{a}_z \frac{\left(a^2 + h^2\right)^{3/2} .1 - h\frac{3}{2}(a^2 + h^2)^{1/2} 2h}{\left(a^2 + h^2\right)^3} = 0$$

$$(a^2 + h^2) - 3h^2 = 0$$

$$a^2 - 2h^2 = 0$$

$$2h^2 = a^2$$

$$h = \pm \frac{a}{\sqrt{2}}$$

(c) When 'a' tends to zero, it becomes a point charge 'Q' located at origin and we have to find electric field at (0, 0, h) due to point charge 'Q' located at origin.

$$\therefore \qquad \bar{E} = \frac{Q}{4\pi \in_0 h^2} \bar{a}_z$$

Problem 1.9

Derive an expression for the electric field strength due to a circular ring of radius 'a' and uniform charge density ρ_L C/m. Obtain the value of height 'h' along Z-axis at which the net electric field becomes zero. Assume the ring to be placed in X-Y plane.

Solution

Derivation is as in Problem. 1.8.

$$\bar{E} = \frac{\rho_L a h}{2 \in_0 \left(a^2 + h^2\right)^{3/2}} \bar{a}_z$$

Which can be written as

$$\bar{E} = \frac{\rho_L a}{2 \in_0 h^2 \left(\frac{a^2}{h^2} + 1 \right)^{3/2}} \bar{a}_z$$

From the above equation we can say that for $h = \infty$, the net electric field becomes zero.

Problem 1.10

A circular ring of radius 'a' carries uniform charge ρ_L C/m and is in XY-plane. Find the Electric field at point (0, 0, 2) along its axis.

Solution

Replacing 'h' in problem.1.8 with '2' and solving, we get

$$\bar{E} = \frac{\rho_L a^2}{2 \in_0 \left(a^2 + 4\right)^{3/2}} \bar{a}_z$$

1.5.3 Volume Charge Distribution

Consider a sphere of radius '*a*' as shown in the Fig.1.16.

Assume elemental volume *dv* is placed at point (r', θ', ϕ'). The elemental charge *dQ* due to the elemental volume *dv*, whose volume charge density ρ_v is

$$dQ = \rho_v dv$$

$$Q = \rho_v \int_v dv$$

$$= \rho_v \frac{4}{3} \pi a^3$$

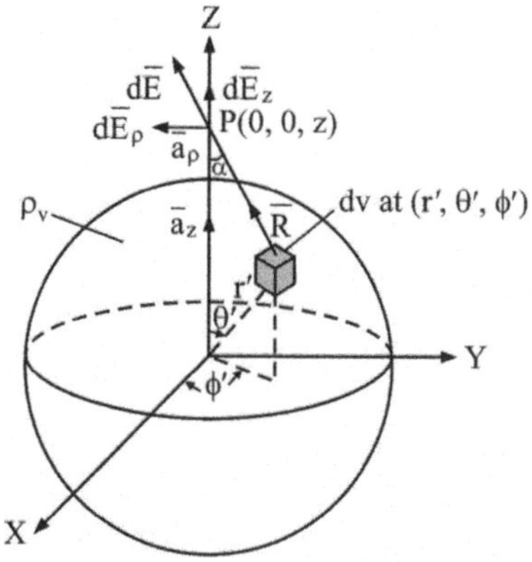

Fig. 1.16 Finding $\bar{E}$ due to volume charge distribution

The elemental electric field $d\bar{E}$ due to elemental volume dv is

$$d\bar{E} = \frac{dQ}{4\pi \in_0 R^2} \bar{a}_R$$

$$= \frac{\rho_v dv}{4\pi \in_0 R^2} \bar{a}_R$$

where $$\bar{a}_R = \cos\alpha\, \bar{a}_z + \sin\alpha\, \bar{a}_\rho$$

Due to symmetry, the electric field in 'ρ' direction will be zero. Finally total electric field will be in Z-direction.

$$\bar{E}_z = \bar{E} \cdot \bar{a}_z = \int_v \frac{\rho_v dv}{4\pi \in_0 R^2} \cos\alpha$$

In spherical coordinate system

$$dv = dr'\, r'\, d\theta'\, r' \sin\theta' d\phi'$$

$$dv = (r')^2 \sin\theta' dr' d\theta' d\phi'$$

$$\bar{E}_z = \int_v \frac{\rho_v (r')^2 \sin\theta' dr' d\theta' d\phi' \cos\alpha}{4\pi \in_0 R^2}$$

By applying cosine rule in the Fig.1.16

$$(r')^2 = z^2 + R^2 - 2zR\cos\alpha$$

$$\cos\alpha = \frac{-(r')^2 + z^2 + R^2}{2zR}$$

Similarly

$$R^2 = z^2 + (r')^2 - 2zr'\cos\theta'$$

$$\Rightarrow \qquad \cos\theta' = \frac{z^2 + (r')^2 - R^2}{2zr'} \qquad .....(1.5.8)$$

On differentiating equation (1.5.8), we get

$$-\sin\theta'\, d\theta' = \frac{-2R}{2zr'} dR$$

$$\sin\theta'\, d\theta' = \frac{R}{zr'} dR$$

Here as θ' varies from 0 to π, R changes from $z-r'$ to $z+r'$ respectively

Substituting $\cos\alpha$ and $\sin\theta' d\theta'$ in $\bar{E}_z$ equation, we get

$$\bar{E}_z = \frac{\rho_V}{4\pi \epsilon_0} \int_{\phi'=0}^{2\pi} d\phi' \int_{r'=0}^{a} \int_{R=z-r'}^{z+r'} r'^2 \frac{RdR}{zr'} dr' \frac{z^2+R^2-r'^2}{2zR} \frac{1}{R^2}$$

$$\bar{E}_z = \frac{\rho_v 2\pi}{8\pi \epsilon_0 z^2} \int_{r'=0}^{a} \int_{R=z-r'}^{z+r'} r' \left[1+\frac{z^2 r'^2}{R^2}\right] dR\, dr'$$

$$\bar{E}_z = \frac{\rho_v \pi}{4\pi \epsilon_0 z^2} \int_{r'=0}^{a} r' \left[R - \frac{z^2-r'^2}{R}\right]_{z-r'}^{z+r'} dr'$$

$$\bar{E}_z = \frac{\rho_v \pi}{4\pi \epsilon_0 z^2} \int_{r'=0}^{a} 4r'^2 dr'$$

$$\bar{E}_z = \frac{\rho_v}{\epsilon_0 z^2} \frac{a^3}{3} = \frac{\rho_v}{4\pi \epsilon_0 z^2} \frac{4}{3}\pi a^3$$

$$\bar{E} = \frac{Q}{4\pi \epsilon_0 z^2} \bar{a}_z \qquad \text{.....(1.5.9)}$$

The electric field due to a sphere of radius 'a' with volume charge density ρ_v is similar to the electric field due to a point charge which is placed at origin.

Problem 1.11

A circular disk of radius 'a' is uniformly charged with ρ_s C/m^2. If the disk lies on the Z = 0 plane with it's axis along the Z-axis

(a) Show that at point (0, 0, h), $\bar{E} = \frac{\rho_s}{2\epsilon_0}\left[1 - \frac{h}{\left(h^2+a^2\right)^{1/2}}\right]\bar{a}_z$

(b) From this derive the $\bar{E}$ due to an infinite sheet of charge on the Z = 0 plane.

(c) If $a << h$, Show that $\bar{E}$ is similar to the field due to a point charge.

Solution

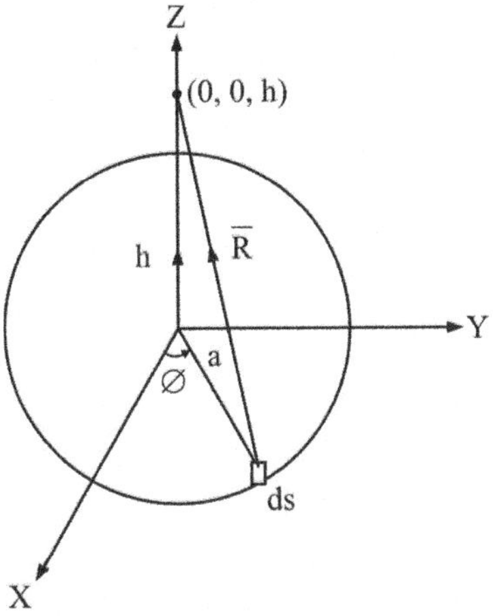

Fig. 1.17

(a)

$$d\overline{E} = \frac{dQ}{4\pi \in_0 R^2}\overline{a}_r$$

$$dQ = \rho_s\, ds; \quad ds = d\rho.\rho\, d\phi,$$

$$= \rho_s\, \rho d\rho\, d\phi$$

$$\rho\overline{a}_\rho + \overline{R} = h\overline{a}_z$$

$$\overline{R} = h\overline{a}_z - \rho\overline{a}_\rho$$

$$\overline{E} = \int_s \frac{\rho_s \rho\, d\rho\, d\phi}{4\pi \in_0} \frac{\left(h\overline{a}_z - \rho\overline{a}_\rho\right)}{\left(h^2 + \rho^2\right)^{3/2}}$$

$$\overline{E} = \frac{\rho_s}{4\pi \in_0}\overline{a}_z \int_0^{2\pi} d\phi \int_0^a \frac{\rho h}{\left(h^2 + \rho^2\right)^{3/2}} d\rho$$

$$= \frac{\rho_s}{4\pi \in_0}\overline{a}_z 2\pi h \int_0^a \frac{1}{2}\left(h^2 + \rho^2\right)^{-3/2} d\left(\rho^2\right)$$

$$= \frac{\rho_s h}{2 \in_0} \bar{a}_z \frac{1}{2} \left[\frac{\left(h^2 + \rho^2\right)^{\frac{-3}{2}+1}}{\frac{-3}{2}+1} \right]_0^a$$

$$= \frac{\rho_s h}{4 \in_0} \bar{a}_z \left\{ -2 \left[\left(h^2 + a^2\right)^{-1/2} - \left(h^2\right)^{-1/2} \right] \right\}$$

$$= \frac{-\rho_s h \bar{a}_z}{2 \in_0} \left[\frac{1}{\sqrt{\left(h^2 + a^2\right)}} - \frac{1}{h} \right]$$

$$\bar{E} = \frac{\rho_s}{2 \in_0} \left[1 - \frac{h}{\left(h^2 + a^2\right)^{1/2}} \right] \bar{a}_z$$

(b) $a \rightarrow \infty$;

$$\therefore \quad \bar{E} = \frac{\rho_s}{2 \in_0} \bar{a}_z$$

(c) when a<< h, the volume charge density becomes a point charge located at origin,

$$\therefore \quad \bar{E} = \frac{Q}{4\pi \in_0 h^2} \bar{a}_z$$

Problem 1.12

The finite sheet $0 < x < 1$, $0 < y < 1$ on the $Z = 0$ plane has a charge density $\rho_s = xy\,(x^2 + y^2 + 25)^{3/2}$ nC/m^2.

Find

(a) the total charge on the sheet

(b) the electric field at (0, 0, 5)

(c) the force experienced by a – 1 nC charge located at (0, 0, 5)

Solution

(a) $dQ = \rho_s\, ds$

$$Q = \int_s \rho_s ds$$

$$= \int_{x=0}^{1}\int_{y=0}^{1} xy\left(x^2+y^2+25\right)^{3/2} n\,dx\,dy$$

$$= n\int_{x=0}^{1} x\int_{y=0}^{1}\left(x^2+y^2+25\right)^{3/2}\frac{1}{2}d(y^2)dx$$

$$= n\int_{x=0}^{1} x\left[\left(x^2+y^2+25\right)^{5/2}\right]_0^1\frac{2}{5}\frac{1}{2}dx$$

$$= \frac{n}{5}\int_{x=0}^{1}\left[\left(x^2+26\right)^{5/2}-\left(x^2+25\right)^{5/2}\right]\frac{1}{2}d(x^2)$$

$$= \frac{n}{5}\left[\left(x^2+26\right)^{7/2}-\left(x^2+25\right)^{7/2}\right]_0^1\frac{1}{7}$$

$$= \frac{n}{35}\left[(27)^{7/2}-2(26)^{7/2}+(25)^{7/2}\right]$$

$$= \frac{n}{35}\left[102275.868136-179240.733942+78125\right]$$

$Q = 33.15$ nC

(b) Electric field at (0, 0, 5)

$$d\bar{E} = \frac{\rho_s ds}{4\pi \in_0 R^2}\bar{a}_R; \qquad \text{on Z-plane point is } (x, y, 0)$$

$$\therefore \quad \bar{R} = (0,0,5)-(x,y,0) = -x\bar{a}_x - y\bar{a}_y + 5\bar{a}_z$$

$$\frac{\bar{a}_R}{R^2} = \frac{\bar{R}}{|\bar{R}|^3} = \frac{-x\bar{a}_x - y\bar{a}_y + 5\bar{a}_z}{\left(\sqrt{x^2+y^2+25}\right)^3}$$

$$\bar{E} = \int_S \frac{\rho_s ds}{4\pi \in_0}\frac{\bar{R}}{|\bar{R}|^3}$$

$$= \int_{x=0}^{1}\int_{y=0}^{1}\frac{xy\left(x^2+y^2+25\right)^{3/2}\times 10^{-9}}{4\pi \in_0}\left(\frac{-x\bar{a}_x - y\bar{a}_y + 5\bar{a}_z}{\left(\sqrt{x^2+y^2+25}\right)^3}\right)dx\,dy$$

$$=\frac{1}{4\pi\in_0}\int_{x=0}^{1}\int_{y=0}^{1}-x^2y\bar{a}_x-xy^2\bar{a}_y+5xy\bar{a}_z dx\,dy\times10^{-9}$$

$$=\frac{1}{4\pi\in_0}\int_{x=0}^{1}-x^2\left[\frac{y^2}{2}\right]_0^1\bar{a}_x-x\left[\frac{y^3}{3}\right]_0^1\bar{a}_y+5x\left[\frac{y^2}{2}\right]_0^1\bar{a}_z\,dx\times10^{-9}$$

$$=\frac{1}{4\pi\in_0}\int_{x=0}^{1}-\frac{x^2}{2}\bar{a}_x-\frac{x}{3}\bar{a}_y+\frac{5}{2}x\bar{a}_z\,dx\times10^{-9}$$

$$=\frac{1}{4\pi\in_0}\left[\left[-\frac{x^3}{6}\right]_0^1\bar{a}_x-\left[\frac{x^2}{6}\right]_0^1\bar{a}_y+\frac{5}{2}\left[\frac{x^2}{2}\right]_0^1\bar{a}_z\right]\times10^{-9}$$

$$=\frac{1}{4\pi\in_0}\left[-\frac{1}{6}\bar{a}_x-\frac{1}{6}\bar{a}_y+\frac{5}{4}\bar{a}_z\right]\times10^{-9}$$

$$=9\times10^9\left[-\frac{1}{6}\bar{a}_x-\frac{1}{6}\bar{a}_y+\frac{5}{4}\bar{a}_z\right]\times10^{-9}$$

$$=-1.5\bar{a}_x-1.5\bar{a}_y+11.25\bar{a}_z\ \text{V/m}$$

(c) $\bar{F}=q\bar{E}$

$$=(-1\,nC)\left[-1.5\bar{a}_x-1.5\bar{a}_y+11.25\bar{a}_z\right]$$

$$=1.5\bar{a}_x+1.5\bar{a}_y-11.25\bar{a}_z\,\text{nN}$$

Problem 1.13

A square plane described by $-2 < x < 2$, $-2 < y < 2$, $z = 0$ carries a charge density $12|y|$ mC/m^2. Find the total charge on the plate and the electric field intensity at (0, 0, 10)

Solution

$$dQ=\rho_s\,ds$$

$$Q=\int_s\rho_s ds$$

$$=\int_{x=-2}^{2}\int_{y=-2}^{2}12\,|\,y\,|\times10^{-3}dxdy$$

$$= 10^{-3} \int_{x=-2}^{2} \left[\int_{y=-2}^{0} -12y\,dy + \int_{y=0}^{2} 12y\,dy \right] dx$$

$$= 10^{-3} \int_{x=-2}^{2} -12\left[\frac{y^2}{2}\right]_{-2}^{0} + 12\left[\frac{y^2}{2}\right]_{0}^{2} dx$$

$$= 10^{-3} \int_{x=-2}^{2} 12(2) + 12(2)\,dx$$

$$= 48\times10^{-3} \int_{x=-2}^{2} dx = 48\times10^{-3}\times4 = 192 \text{ mC}$$

$$d\bar{E} = \frac{\rho_s ds}{4\pi \in_0 R^2}\bar{a}_R\,; \quad \bar{R} = (0, 0, 10) - (x, y, 0) = -x\bar{a}_x - y\bar{a}_y + 10\bar{a}_z$$

$$d\bar{E} = \frac{\rho_s ds}{4\pi \in_0} \frac{\bar{R}}{R^3}$$

$$\bar{E} = \int_s \frac{\rho_s ds}{4\pi \in_0} \frac{\bar{R}}{R^3}$$

$$= \int_{x=-2}^{2} \int_{y=-2}^{2} \frac{12|y|\times10^{-3}}{4\pi \in_0} \left(\frac{-x\bar{a}_x - y\bar{a}_y + 10\bar{a}_z}{\left(\sqrt{x^2 + y^2 + 100}\right)^3} \right) dx\,dy$$

$$= 9\times10^6\times12 \int_{x=-2}^{2} \left[\int_{y=-2}^{0} \frac{xy\bar{a}_x + y^2\bar{a}_y - 10y\bar{a}_z}{\left(x^2 + y^2 + 100\right)^{3/2}} dy + \int_{y=0}^{2} \frac{-xy\bar{a}_x - y^2\bar{a}_y + 10y\bar{a}_z}{\left(x^2 + y^2 + 100\right)^{3/2}} dy \right] dx$$

Replacing y with –y in the first integral and simplifying

$$\bar{E} = 108\times10^6 \int_{x=-2}^{2} \left[\int_{y=0}^{2} \frac{-2xy\bar{a}_x + 20y\bar{a}_z}{\left(x^2 + y^2 + 100\right)^{3/2}} dy \right] dx$$

$$=108\times10^6\int_{x=-2}^{2}\left[-x\int_{y=0}^{2}2y\bar{a}_x\left(x^2+y^2+100\right)^{-3/2}dy+10\int_{y=0}^{2}2y\bar{a}_z\left(x^2+y^2+100\right)^{-3/2}dy\right]dx$$

$$=108\times10^6\int_{x=-2}^{2}\left[-x\int_{y=0}^{2}\bar{a}_x\left(x^2+y^2+100\right)^{-3/2}d(y^2)+10\int_{y=0}^{2}\bar{a}_z\left(x^2+y^2+100\right)^{-3/2}d(y^2)\right]dx$$

$$=108\times10^6\int_{x=-2}^{2}\left[-x\left[\frac{\left(x^2+y^2+100\right)^{-1/2}}{-1/2}\right]_0^2\bar{a}_x+10\left[\frac{\left(x^2+y^2+100\right)^{-1/2}}{-1/2}\right]_0^2\bar{a}_z\right]dx$$

$$=108\times10^6\int_{x=-2}^{2}\left\{\left[2x\left(x^2+104\right)^{-1/2}-2x\left(x^2+100\right)^{-1/2}\right]\bar{a}_x-20\left[\left(x^2+104\right)^{-1/2}-\left(x^2+100\right)^{-}\right.\right.$$

$\because$ $x\left(x^2+104\right)^{-1/2}$ & $x\left(x^2+100\right)^{-1/2}$ are odd functions

and $\left(x^2+104\right)^{-1/2}$ & $\left(x^2+100\right)^{-1/2}$ are even functions

$$\int_{-a}^{a}f(x)dx=\begin{cases}0 & \text{if } f \text{ is odd}\\ 2\int_0^a f(x)dx & \text{if } f \text{ is even}\end{cases}$$

$$\therefore\quad \bar{E}=-20\times108\times10^6\times2\int_{x=0}^{2}\left[\frac{1}{\sqrt{x^2+\left(\sqrt{104}\right)^2}}-\frac{1}{\sqrt{x^2+10^2}}\right]\bar{a}_z dx$$

$$=-40\times108\times10^6\left[\sinh^{-1}\left(\frac{x}{\sqrt{104}}\right)-\sinh^{-1}\left(\frac{x}{10}\right)\right]_0^2\bar{a}_z$$

$$=-40\times108\times10^6\left[\sinh^{-1}\left(\frac{2}{\sqrt{104}}\right)-\sinh^{-1}\left(\frac{1}{5}\right)\right]\bar{a}_z$$

$$=-40\times108\times10^6\left[0.19488-0.19869\right]\bar{a}_z$$

$\bar{E}=16.46\ \bar{a}_z$ MV/m.

1.6 Electric Flux Density or Displacement Density

It is also called Electric displacement and to understand the concept of Electric flux density, one needs to know about line integral, surface integral and electric flux, which are explained as follows.

1.6.1 Line Integral

If a vector $\bar{A}$ is passing through a line as shown in the Fig.1.18. The line integral can be defined as the tangential component of vector $\bar{A}$ along the line, which can be written as

$$\int_L |\bar{A}| \cos\theta \, dL = \int_L \bar{A}.d\bar{L}$$

If a line is closed curve then the above integral can be written as $\oint_L \bar{A}.d\bar{l}$ which is called as contour line integral.

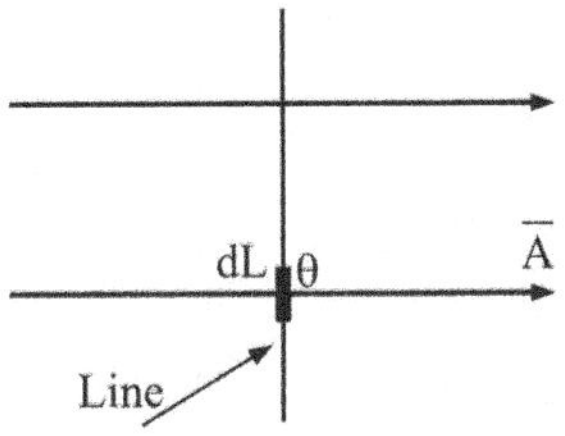

Fig. 1.18 Evaluation of line integral

1.6.2 Surface Integral

Similarly, if a vector $\bar{A}$ is passing through a surface as shown in Fig. 1.19

The flux (ψ) of a vector $\bar{A}$ or surface integral can be written as

$$\psi = \int_s |\bar{A}| \cos\theta \, ds$$

$$= \int_s \bar{A}.d\bar{s}$$

If it is closed surface then the above integral can be be written as $\oint_s \bar{A}.d\bar{s}$ which is called as contour surface integral.

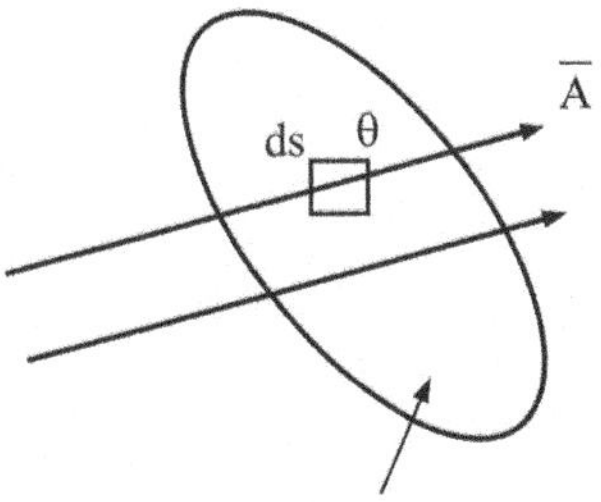

Fig. 1.19 Evaluation of surface integral

1.6.3 Electric Flux

We know that electric field intensity depends upon the medium in which it passes. Let us define a new vector $\bar{D}$ such that it is independent of medium i.e.,

$\bar{D} = \epsilon_0 \bar{E}$. Then the flux of $\bar{D}$, i.e., $\psi = \oint_S \bar{D}.d\bar{s}$, where ψ is the electric flux. Which can be defined according to SI units as one line of flux originates from +1 Coloumb and terminates at –1 Coloumb. So the unit of Electric flux is also Coloumb and $\bar{D}$ is the electric flux density whose unit is columb/m^2.

The formulae for $\bar{D}$ can be obtained by multiplying the formulae of $\bar{E}$ with ϵ_0.

∴ Electric flux density due to a point charge $\bar{D}_Q = \frac{Q}{4\pi}\frac{\bar{a}_R}{R^2}$ .....(1.6.1)

and Electric flux density due to an infinite line with line chare density

ρ_L *is* $\bar{D}_L = \frac{\rho_L}{2\pi\rho}\bar{a}_\rho$ (1.6.2)

Problem 1.14

Determine $\bar{D}$ at (4, 0, 3) if there is a point charge –5π mC at (4, 0, 0) and a line charge 3π mC/m along the Y-axis

Solution

$$\bar{D}_Q = \frac{Q}{4\pi}\frac{\bar{a}_R}{R^2}$$

where, $\bar{R} = (4,0,3) - (4,0,0) = (0,0,3)$

$$= \frac{-5\pi}{4\pi}\frac{3\bar{a}_z \times 10^{-3}}{(9)^{3/2}}$$

$$= \frac{-5}{4}\frac{3\bar{a}_z \times 10^{-3}}{27} = \frac{-5\bar{a}_z \times 10^{-3}}{36} = -0.139\bar{a}_z \times 10^{-3} \text{ C/m}^2$$

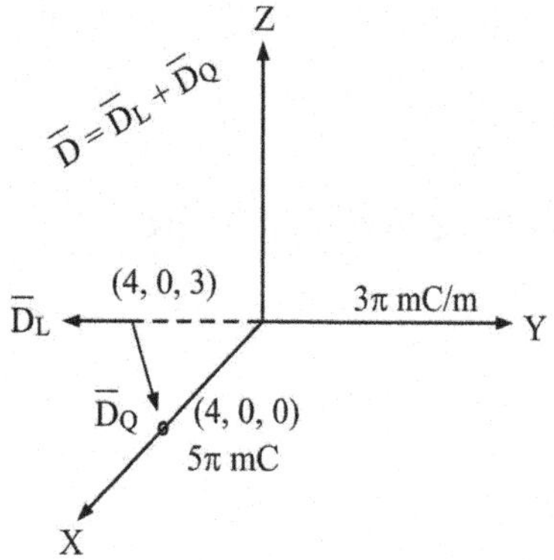

Fig. 1.20

$$\bar{a}_\rho = \frac{\bar{\rho}}{|\bar{\rho}|}$$

$$\bar{\rho} = (4,0,3) - (0,0,0) = 4\bar{a}_x + 3\bar{a}_z$$

$$\bar{D}_L = \frac{\rho_L}{2\pi\rho}\bar{a}_\rho$$

$$= \frac{3\pi}{2\pi} \times 10^{-3}\frac{4\bar{a}_x + 3\bar{a}_z}{25}$$

$$= 0.24\bar{a}_x + 0.18\bar{a}_z \text{ mC/m}^2$$

$$\bar{D} = \bar{D}_Q + \bar{D}_L = 240\bar{a}_x + 42\bar{a}_z \ \mu\text{C/m}^2$$

1.7 Divergence of a Vector

Divergence: The divergence of a vector $\overline{A}$ at a given point is the outward flux in a volume as volume shrinks about the point. It can be represented as

$$div\,\overline{A} = \nabla \cdot \overline{A} = \lim_{\Delta v \to 0} \frac{\oint_S \overline{A} \cdot \overline{ds}}{\Delta v} \quad .....(1.7.1)$$

Where ∇ is the del operator or gradient operator. ∇ can be operated on a vector or scalar. It has got different meanings when it is operating on a vector and scalar. If it is operating on a scalar V then it can be written as ∇V which is called as scalar gradient. If it is operating on a vector $\overline{A}$ with dot product then it is $\nabla.\overline{A}$ and it is called as divergence of vector $\overline{A}$ and If it is operating on a vector $\overline{A}$ with cross product then it is $\nabla \times \overline{A}$ and it is called as curl of vector $\overline{A}$.

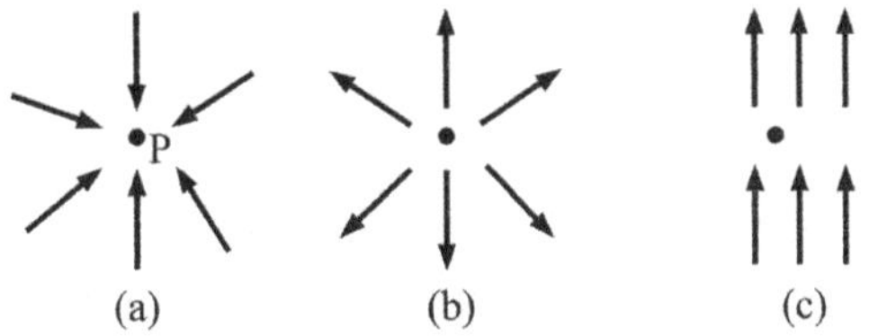

Fig. 1.21 Flux lines

Physically divergence can be interpreted as the measure of how much field diverges or emanates from a point. Let us consider the Fig.1.21(a) in which field is reaching to the point. Divergence at that point is –Ve or it is also called as convergence. In Fig.1.21(b) the field is going away from the point, therefore divergence is +Ve. In Fig.1.21(c) some of the flux lines or field lines are reaching to the point and same number of field lines are leaving from the point hence the divergence is zero.

To determine $\nabla.\overline{A}$ let us consider the volume in Cartesian co-ordinate systems as shown in the Fig.1.22. In Cartesian co-ordinate system, the vector $\overline{A}$ with it's unit vectors and components along X, Y, Z is

$$\overline{A} = A_x\overline{a}_x + A_y\overline{a}_y + A_z\overline{a}_z$$

Assume the elemental volume $\Delta V = \Delta x \Delta y \Delta z$. The flux of a vector $\overline{A}$ on Y-axis that enters in to the left side of the volume is $A_y \Delta x \Delta z$. The flux which is leaving from right side of the volume on Y-axis can be written as $(A_y + \Delta A_y)\,\Delta x \Delta z$. This equation can be modified as

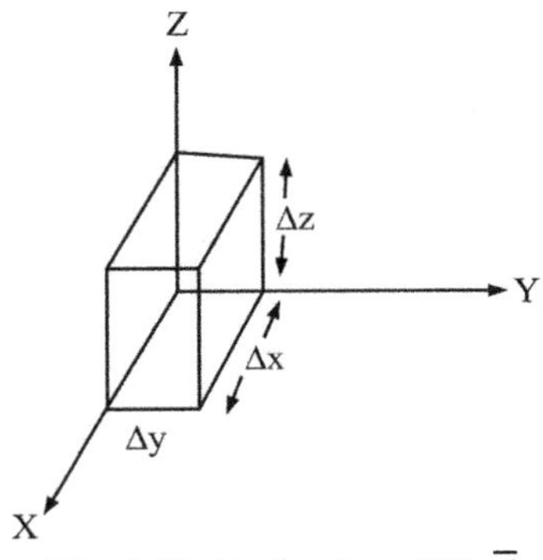

Fig. 1.22 Evaluation of $\nabla.\overline{A}$

$\left(A_y + \frac{\Delta A_y}{\Delta_y}\Delta_y\right)\Delta x\,\Delta z$. So the total flux on Y-axis is $A_y\Delta x\Delta z + \frac{\partial A_y}{\partial y}\Delta x\Delta y\Delta z - A_y\Delta x\Delta z$

$$= \frac{\partial A_y}{\partial y}\Delta x\Delta y\Delta z$$

Similarly on X and Z-axises also.

The entire flux in all the directions is $\psi = \left(\frac{\partial A_x}{\partial x} + \frac{\partial A_y}{\partial y} + \frac{\partial A_z}{\partial z}\right)\Delta x\Delta y\Delta z$. We know $\psi = \oint_s \bar{A}\cdot \overline{ds}$

$$\frac{\oint_s \bar{A}\cdot\overline{ds}}{\Delta v} = \frac{\partial A_x}{\partial x} + \frac{\partial A_y}{\partial y} + \frac{\partial A_z}{\partial z}$$

Applying Limit on both sides

$$\lim_{\Delta v\to 0}\frac{\oint_s \bar{A}\cdot\overline{ds}}{\Delta v} = \lim_{\Delta v\to 0}\frac{\partial A_x}{\partial x} + \frac{\partial A_y}{\partial y} + \frac{\partial A_z}{\partial z}$$

$$\nabla\cdot\bar{A} = \frac{\partial A_x}{\partial x} + \frac{\partial A_y}{\partial y} + \frac{\partial A_z}{\partial z}$$

Conclusion

The divergence of a vector results a scalar. The divergence of a scalar has no meaning

$$\nabla\cdot\left(\bar{A} + \bar{B}\right) = \nabla.\bar{A} + \nabla.\bar{B}$$

$$\nabla\cdot\left(V\bar{A}\right) = V\nabla.\bar{A} + \bar{A}\cdot\nabla V$$

$$\nabla\cdot\bar{A} = \left(\frac{\partial\bar{a}_x}{\partial x} + \frac{\partial\bar{a}_y}{\partial y} + \frac{\partial\bar{a}_z}{\partial z}\right).\left(A_x\bar{a}_x + A_y\bar{a}_y + A_z\bar{a}_z\right)$$

$$= \frac{\partial A_x}{\partial x} + \frac{\partial A_y}{\partial y} + \frac{\partial A_z}{\partial z}$$

So from the above equation, the gradient operator is

$$\nabla = \frac{\partial a_x}{\partial x} + \frac{\partial a_y}{\partial y} + \frac{\partial a_z}{\partial z} \qquad .....(1.7.2)$$

and the scalar gradient is

$$\nabla V = \frac{\partial V}{\partial x}\bar{a}_x + \frac{\partial V}{\partial y}\bar{a}_y + \frac{\partial V}{\partial z}\bar{a}_z$$

1.7.1 Divergence Theorem

Statement

This theorem states that the outward flux flows through a closed surface is same as the volume integral of divergence of a vector.

$$\oint_s \bar{A} \cdot d\bar{s} = \int_v \nabla \cdot \bar{A}\, dv \qquad \text{..... (1.7.3)}$$

Proof:

Consider a vector $\bar{A} = A_x\bar{a}_x + A_y\bar{a}_y + A_z\bar{a}_z$.

Similarly $d\bar{s} = ds_x\bar{a}_x + ds_y\bar{a}_y + ds_z\bar{a}_z$ and we know that divergence of vector $\bar{A}$ i.e.,

$$\nabla \cdot \bar{A} = \frac{\partial A_x}{\partial x} + \frac{\partial A_y}{\partial y} + \frac{\partial A_z}{\partial z}$$

Assume dv = dx dy dz

consider the volume integral

$$\int_v \nabla \cdot \bar{A}\, dv = \iiint_v \left(\frac{\partial A_x}{\partial x} + \frac{\partial A_y}{\partial y} + \frac{\partial A_z}{\partial z} \right) dx\, dy\, dz$$

The second term in the above integral can be written as

$$\iiint_v \frac{\partial A_y}{\partial y} dx\, dy\, dz = \oiint_s \left[\int \frac{dA_y}{dy} dy \right] dx\, dz = \oiint_s A_y ds_y$$

where ds_y = The elemental surface on XZ plane.

Similarly the first and third terms can be written as

$$\oiint_s A_x ds_x \text{ and } \oiint_s A_z ds_z$$

$$\therefore \quad \int_v \nabla \cdot \bar{A} dv = \oint\!\!\oint_s (A_x ds_x + A_y ds_y + A_z ds_z)$$

$$= \oint\!\!\oint_s \left(A_x \bar{a}_x + A_y \bar{a}_y + A_z \bar{a}_z\right) \cdot \left(ds_x \bar{a}_x + ds_y \bar{a}_y + ds_z \bar{a}_z\right) = \oint\!\!\oint_s \bar{A} \cdot \overline{ds}$$

Hence proved

Formulae for Gradient

in Cartesian co-ordinate system

$$\nabla V = \frac{\partial V}{\partial x}\bar{a}_x + \frac{\partial V}{\partial y}\bar{a}_y + \frac{\partial V}{\partial z}\bar{a}_z \qquad \text{.....(1.7.4)}$$

in cylindrical co-ordinate system

$$\nabla V = \frac{\partial V}{\partial \rho}\bar{a}_\rho + \frac{1}{\rho}\frac{\partial V}{\partial \phi}\bar{a}_\phi + \frac{\partial V}{\partial z}\bar{a}_z \qquad \text{.....(1.7.5)}$$

in spherical co-ordinate system

$$\nabla V = \frac{\partial V}{\partial r}\bar{a}_r + \frac{1}{r}\frac{\partial V}{\partial \theta}\bar{a}_\theta + \frac{1}{r\sin\theta}\frac{\partial V}{\partial \phi}\bar{a}_\phi \qquad \text{.....(1.7.6)}$$

Problem 1.15

Find the gradient of the following scalar fields

(a) $V = e^{-z} \sin 2x \cos hy$

(b) $U = \rho^2 z \cos 2\phi$

(c) $W = 10r \sin^2\theta \cos\phi$

Solution

(a) Since given V is in x and y, consider gradient in Cartesian co-ordinate system

$$\nabla V = \frac{\partial V}{\partial x}\bar{a}_x + \frac{\partial V}{\partial y}\bar{a}_y + \frac{\partial V}{\partial z}\bar{a}_z$$

$$= e^{-z} \cos h\, y \cos 2x\, 2\bar{a}_x + e^{-z} \sin 2x \sin h\, y\, \bar{a}_y + \sin 2x \cos h\, y\, e^{-z}(-1)\bar{a}_z$$

$$= 2\cos 2x \cos h\, y e^{-z}\bar{a}_x + \sin 2x \sin h\, y e^{-z}\,\bar{a}_y - \sin 2x \cos h\, y\, e^{-z}\bar{a}_z$$

(b) Since given U is in ρ, z and ϕ, consider gradient in cylindrical co-ordinate system

$$\nabla U = \frac{\partial U}{\partial \rho}\bar{a}_\rho + \frac{1}{\rho}\frac{\partial U}{\partial \phi}\bar{a}_\phi + \frac{\partial U}{\partial z}\bar{a}_z$$

$$= Z\cos 2\phi\, 2\rho\,\bar{a}_\rho + \rho z(-\sin 2\phi)2\bar{a}_\phi + \rho^2\cos 2\phi\,\bar{a}_z$$

(c) Since given W is in r, θ and ϕ, consider gradient in spherical co-ordinate system

$$\nabla W = \frac{\partial W}{\partial r}\bar{a}_r + \frac{1}{r}\frac{\partial W}{\partial \theta}\bar{a}_\theta + \frac{1}{r\sin\theta}\frac{\partial W}{\partial \phi}\bar{a}_\phi$$

$$= 10\sin^2\theta\cos\phi\,\bar{a}_r + \left(\frac{10r}{r}\right)2\sin\theta\cos\theta\cos\phi\,\bar{a}_\theta + 10r\sin^2\theta(-\sin\phi)\bar{a}_\phi\cdot\frac{1}{r\sin\theta}$$

Formulae for Divergence of a Vector

in Cartesian co-ordinate system

$$\nabla\cdot\bar{A} = \frac{\partial A_x}{\partial x} + \frac{\partial A_y}{\partial y} + \frac{\partial A_z}{\partial z} \qquad \text{.....(1.7.7)}$$

in cylindrical co-ordinate system

$$\nabla\cdot\bar{A} = \frac{1}{\rho}\frac{\partial(\rho A_\rho)}{\partial \rho} + \frac{1}{\rho}\frac{\partial(A_\phi)}{\partial \phi} + \frac{\partial A_z}{\partial z} \qquad \text{.....(1.7.8)}$$

in spherical co-ordinate system

$$\nabla\cdot\bar{A} = \frac{1}{r^2}\frac{\partial(r^2 A_r)}{\partial r} + \frac{1}{r\sin\theta}\frac{\partial(\sin\theta\, A_\theta)}{\partial \theta} + \frac{1}{r\sin\theta}\frac{\partial A_\phi}{\partial \phi} \qquad \text{.....(1.7.9)}$$

Problem 1.16

Determine the divergence of the following vector fields.

(a) $\bar{P} = x^2yz\,\bar{a}_x + x^3zy\,\bar{a}_y + xy^2z^3\bar{a}_z$

(b) $\bar{Q} = \rho\sin\phi\,\bar{a}_\rho + \rho^2z\,\bar{a}_\phi + z\cos\phi\,\bar{a}_z$

(c) $\bar{T} = \frac{1}{r^2}\cos\theta\,\bar{a}_r + r\sin\theta\cos\phi\,\bar{a}_\theta + \cos\theta\,\bar{a}_\phi$

(d) $\bar{N} = r^3\sin\theta\,\bar{a}_r + \sin 2\theta\cos^2\phi\,\bar{a}_\theta + \cos\theta\, r^2\,\bar{a}_\phi$

Solution

(a) Given $\bar{P} = x^2yz\,\bar{a}_x + x^3zy\,\bar{a}_y + xy^2z^3\bar{a}_z$

$$\nabla\cdot\bar{P}=\frac{\partial P_x}{\partial x}+\frac{\partial P_y}{\partial y}+\frac{\partial P_z}{\partial z}$$

$$=2xyz+x^3z+3xy^2z^2$$

(b) Given $\bar{Q}=\rho\sin\phi\,\bar{a}_\rho+\rho^2 z\,\bar{a}_\phi+z\cos\phi\,\bar{a}_z$

$$\nabla\cdot\bar{Q}=\frac{1}{\rho}\frac{\partial(\rho Q_\rho)}{\partial\rho}+\frac{1}{\rho}\frac{\partial(Q_\phi)}{\partial\phi}+\frac{\partial Q_z}{\partial z}$$

$$=\frac{1}{\rho}2\rho\sin\phi+\frac{1}{\rho}(0)+\cos\phi$$

$$=2\sin\phi+\cos\phi$$

(c) Given $\bar{T}=\frac{1}{r^2}\cos\theta\,\bar{a}_r+r\sin\theta\cos\phi\,\bar{a}_\theta+\cos\theta\,\bar{a}_\phi$

$$\nabla\cdot\bar{T}=\frac{1}{r^2}\frac{\partial(r^2T_r)}{\partial r}+\frac{1}{r\sin\theta}\frac{\partial(\sin\theta\,T_\theta)}{\partial\theta}+\frac{1}{r\sin\theta}\frac{\partial T_\phi}{\partial\phi}$$

$$=\frac{1}{r^2}(0)+\frac{1}{r\sin\theta}r\,2\sin\theta\cos\theta\cos\phi+\frac{1}{r\sin\theta}(0)$$

$$=2\cos\theta\cos\phi$$

(d) Given $\bar{N}=r^3\sin\theta\,\bar{a}_r+\sin 2\theta\cos^2\phi\,\bar{a}_\theta+\cos\theta\, r^2\,\bar{a}_\phi$

$$\nabla\cdot\bar{N}=\frac{1}{r^2}\frac{\partial(r^2N_r)}{\partial r}+\frac{1}{r\sin\theta}\frac{\partial(\sin\theta\,N_\theta)}{\partial\theta}+\frac{1}{r\sin\theta}\frac{\partial N_\phi}{\partial\phi}$$

$$=\frac{1}{r^2}5r^4\sin\theta+\frac{1}{r\sin\theta}\frac{1}{2}\left(-\sin\theta+\frac{\sin 3\theta}{3}\right)\cos^2\phi+\frac{1}{r\sin\theta}(0)$$

$$=5r^2\sin\theta-\frac{1}{2r}\cos^2\phi+\frac{\sin 3\theta}{6r\sin\theta}\cos^2\phi$$

1.8 Gauss's Law and Applications

1.8.1 Gauss Law

Gauss law states that the flux flowing through a closed surface is equivalent to the charge enclosed by that surface.

According to the statement $\psi = Q_{enc}$ .....(1.8.1)

Where ψ is the flux flowing through a closed surface. Q_{enc} is the charge enclosed by the closed surface.

We know $$\psi = \oint_S \bar{D} \cdot \bar{ds}$$

The charge enclosed within a volume or closed surface whose volume charge density ρ_v is

$$Q = \int_v \rho_v \, dv$$

According to Gauss's law we can write as

$$\psi = \oint_S \bar{D} \cdot \bar{ds} = \int_v \rho_v \, dv \quad .....(1.8.1a)$$

According to divergence theorem we can write

$$\oint_S \bar{D} \cdot \bar{ds} = \int_V \nabla \cdot \bar{D} \, dv \quad .....(1.8.1b)$$

By comparing the volume integrals in equations (1.8.1a) and (1.8.1b) we can write as

$$\rho_V = \nabla \cdot \bar{D} \quad .....(1.8.2)$$

which is the Maxwell's first equation for electrostatics (time in-varying fields)

Consider unsymmetrical distribution as shown in Fig. 1.23a. The flux flowing through the closed surface shown in Fig. 1.23a is $\psi = 5 - 2 = 3$ nC. The charge enclosed by the surface is $Q = 3$ nC.

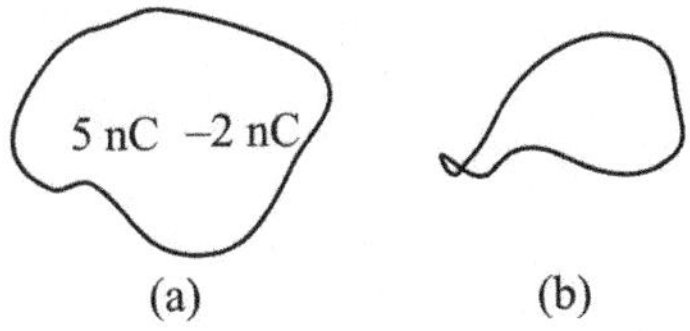

Fig. 1.23 Closed surface

Consider an empty closed surface as shown in Fig. 1.23b. Flux flowing through the closed surface shown in Fig. 1.23b is $\psi = 0$ and hence charge enclosed by the closed surface is zero.

Conclusion

Gauss law holds good even if the charge distribution is unsymmetrical as shown in Figs.1.23a & b. But to find either $\bar{E}$ or $\bar{D}$, the charge distribution must be symmetrical. It can be rectangular symmetry or cylindrical symmetry or spherical symmetry.

If the continuous charge distribution depends on either 'x' or 'y' or 'z', then the distribution will have rectangular symmetry. So to find either $\bar{E}$ or $\bar{D}$, we can use rectangular co-ordinates.

If the continuous charge distribution depends only on ρ and is independent of ϕ and z then the distribution will have cylindrical symmetry. So, to find either $\bar{E}$ or $\bar{D}$, we can use cylindrical co-ordinates.

If the continuous charge distribution depends on 'r' and is independent of θ and ϕ then the symmetry it will have is spherical. So to find either $\bar{E}$ or $\bar{D}$, we can use spherical co-ordinates.

1.8.2 Applications of Gauss's Law – Point Charge

We need to find $\bar{D}$ at any point surrounded by Q. Assume that the point charge is located at origin, then a sphere can be assumed, that surrounds the point charge as shown in Fig.1.24, which shows the problem has spherical symmetry and spherical coordinate system can be used to obtain $\bar{D}$. Let us find out $\bar{D}$ at point 'P' due to a point charge.

The electric flux density $\bar{D}$ is normal or perpendicular to the spherical surface. i.e., $\bar{D} = D_r \bar{a}_r$.

The elemental surface *ds* lies on θ and ϕ axises. i.e., *ds* is normal to r axis.

$$\therefore \quad d\bar{s} = r^2 \sin\theta \, d\theta \, d\phi \, \bar{a}_r$$

Flux flowing through the sphere is

$$\psi = \oint_S \bar{D}.d\bar{s}$$

$$\therefore \quad \psi = \int_{\phi=0}^{2\pi} \int_{\theta=0}^{\pi} \bar{D}_r \bar{a}_r . r^2 \sin\theta \, d\theta \, d\phi \, \bar{a}_r$$

$$\psi = D_r \int_{\phi=0}^{2\pi} \int_{\theta=0}^{\pi} r^2 \sin\theta \, d\theta \, d\phi$$

$$= D_r \int_{\phi=0}^{2\pi} r^2 [-cos\,\theta]_0^{\pi} \, d\phi$$

$$\psi = 2D_r r^2 [2\pi]$$

$$\psi = 4\pi r^2 D_r$$

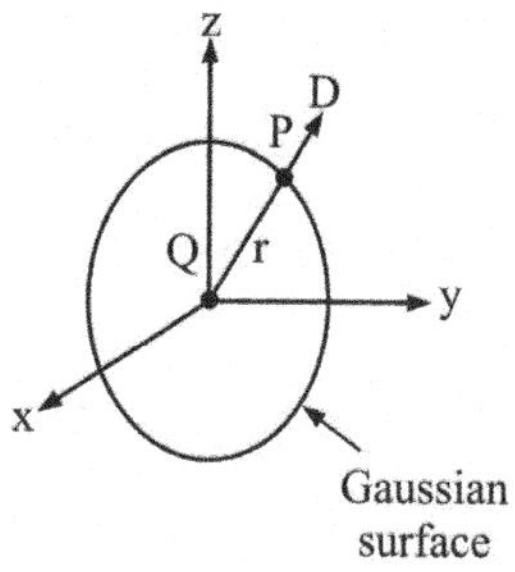

Fig. 1.24 Gaussian surface about a point charge

The charge enclosed by the sphere is

$$Q_{enc} = Q$$

According to Gauss's law

$$\psi = Q_{enc}$$

$\therefore$ $$Q = 4\pi r^2 D_r$$

$\Rightarrow$ $$D_r = \frac{Q}{4\pi r^2}$$

or $$\bar{D} = \frac{Q}{4\pi r^2}\bar{a}_r$$

and $$\bar{E} = \frac{Q}{4\pi \in_0 r^2} a_r$$

Which is similar to the formula derived by using Coulomb's law

1.8.3 Applications of Gauss's Law - Infinite Line Charge

Let us consider that charge is distributed along Z-axis with the charge density ρ_L C/m. Since the charge distribution is along a line, a cylinder of length 'l' can be assumed that surrounds the line charge distribution as shown in Fig.1.25. Hence it is better to consider cylindrical coordinate system to find either $\bar{E}$ or $\bar{D}$ at a point 'p' on the surface of the cylinder.

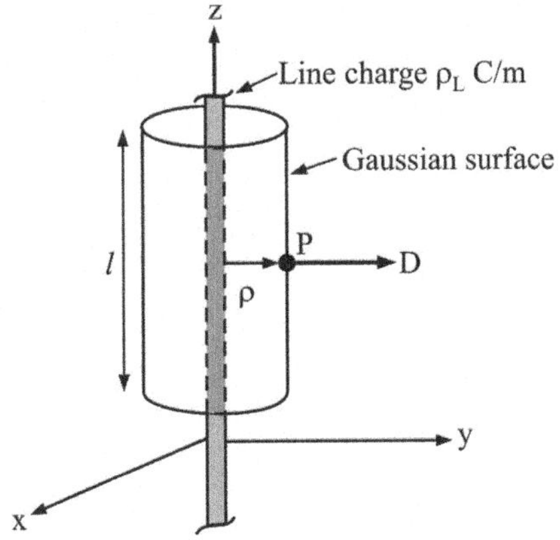

Fig. 1.25 Gaussian surface about an infinite line charge

Here $\bar{D}$ the electric flux density is perpendicular to the surface of the cylinder i.e., it will be in 'ρ' direction in cylindrical co-ordinate systems.

$$\therefore \quad \bar{D} = D_\rho \bar{a}_\rho$$

The elemental surface ds lies on ϕ and Z axises

$$\therefore \quad d\bar{s} = \rho \, d\phi \, dz \, \bar{a}_\rho$$

Flux flowing through the cylinder can be written as

$$\psi = \oint_S \bar{D} \cdot d\bar{s}$$

$$\psi = \int_{z=0}^{l} \int_{\phi=0}^{2\pi} D\rho \bar{a}_\rho . \rho d\phi dz \bar{a}_\rho$$

$$\psi = D_\rho \rho \int_{z=0}^{l} dz \int_{\phi=0}^{2\pi} d\phi$$

$$\psi = D_\rho 2\pi\rho l \qquad .....(1.8.3)$$

The charge enclosed by the cylinder is

$$Q_{enc} = \rho_L l \qquad .....(1.8.4)$$

According to Gauss's law

$$\psi = Q_{enc}$$

Substitute ψ and Q_{enc} from equations (1.8.3) and (1.8.4) in the above equation

$$\rho_L l = D_\rho 2\pi\rho l$$

$$\Rightarrow \quad D_\rho = \frac{\rho_L}{2\pi\rho}$$

$$\bar{D} = \frac{\rho_L}{2\pi\rho} \bar{a}_\rho \text{ and}$$

$$\bar{E} = \frac{\bar{D}}{\epsilon_0} = \frac{\rho_L}{2\pi \epsilon_0 \rho} \bar{a}_\rho \qquad .....(1.8.5)$$

Which is similar to the formula derived by using Coulomb's law.

Problem 1.17

Given $\bar{D} = z\rho\cos^2\phi \bar{a}_z$ C/m^2. Calculate the charge density at (1, π/4, 3) and the total charge enclosed by the cylinder of radius 1m with $-2 \leq z \leq 2\ m$.

Solution

We know

$$\rho_v = \nabla \cdot \bar{D}$$

in cylindrical co-ordinate system the divergence can be written as

$$\rho_v = \frac{1}{\rho}\frac{\partial(\rho D_\rho)}{\partial \rho} + \frac{1}{\rho}\frac{\partial(D_\phi)}{\partial \phi} + \frac{\partial D_z}{\partial z}$$

$$\rho_v = \frac{\partial D_z}{\partial z} \quad \text{since } \bar{D} \text{ has only Z- component}$$

$$\rho_v = \rho \cos^2 \phi$$

$$(\rho_v)\left(1,\ \frac{\pi}{4},\ 3\right) = (1)\cos^2\left(\frac{\pi}{4}\right) = \frac{1}{2} \ \text{ C/m}^3$$

change enclosed = $Q_{enc} = \int_v \rho_v dv$ where $dv = \rho\, d\rho\, d\phi\, dz$

$$Q_{enc} = \int_{\rho=0}^{1} \int_{\phi=0}^{2\pi} \int_{z=-2}^{2} \rho \cos^2 \phi \rho\, d\rho\, d\phi\, dz$$

$$= \int_{\rho=0}^{1} \int_{\phi=0}^{2\pi} \rho^2 \cos^2 \phi (4)\, d\rho\, d\phi$$

$$= 4 \int_{\rho=0}^{1} \rho^2 \left[\frac{1}{2}(2\pi) + \frac{1}{2}\sin 4\phi\right] d\rho$$

$$= 4\pi \int_{\rho=0}^{1} \rho^2 d\rho = 4\pi \left[\frac{\rho^3}{3}\right]_0^1 = \frac{4\pi}{3} \text{C}$$

Problem 1.18

If $\bar{D} = (2y^2 + z)\bar{a}_x + 4xy\,\bar{a}_y + x\,\bar{a}_z$ C/m^2. Find

(a) the volume charge density at (–1, 0, 3)

(b) the flux through the cube defined by $0 \le x \le 1,\ 0 \le y \le 1,\ 0 \le z \le 1$

(c) the total charge enclosed by the cube

Solution

According to Maxwell's I equation

$$\rho_v = \nabla \cdot \bar{D}$$

$$\rho_v = \frac{\partial D_x}{\partial x} + \frac{\partial D_y}{\partial y} + \frac{\partial D_z}{\partial z}$$

$$= 0 + 4\text{x} + 0$$

$$= 4\text{x C/m}^3$$

(a) $\left(\rho_v\right)_{(-1,0,3)} = 4(-1) = -4\ \text{C/m}^2$

(b) & (c) $\psi = \int\limits_v \rho_v dv = Q_{enc}$

$$= \int\limits_{x=0}^{1} \int\limits_{y=0}^{1} \int\limits_{z=0}^{1} 4x\, dx\, dy\, dz$$

$$= \int\limits_{x=0}^{1} \int\limits_{y=0}^{1} 4x(1)\, dx\, dy$$

$$= \int\limits_{x=0}^{1} 4x(1)\, dx$$

$$= 4\left[\frac{x^2}{2}\right]_0^1 = \frac{4}{2} = 2C$$

Problem 1.19

Given the electric flux density $\bar{D} = 0.3r^2\bar{a}_r$ nC/m^2, in free space. Find

(a) $\bar{E}$ at point (2, 25°, 90°)

(b) the total charge within the sphere r = 3

(c) the total electric flux leaving the sphere r = 4

Solution

(a) Given $\bar{D} = 0.3r^2\bar{a}_r$ nC/m^2

$$\therefore \qquad \bar{E} = \frac{\bar{D}}{\epsilon_0} = \frac{0.3r^2\bar{a}_r}{8.854\times 10^{-12}}$$

$$\left(\bar{E}\right)_{\left(2,\ 25^{o},90^{o}\right)}=\frac{0.3(4)}{8.854\times10^{-12}}\bar{a}_r=1.355\times10^{11}\bar{a}_r\times10^{-9}=135.5\,\bar{a}_r\ \text{V/m}$$

(b) we know $\rho_v=\nabla\cdot\bar{D}$

$$=\frac{1}{r^2}\frac{\partial\left(r^2D_r\right)}{\partial r}+\frac{1}{r\sin\theta}\frac{\partial\left(\sin\theta\ D_\theta\right)}{\partial\theta}+\frac{1}{r\sin\theta}\frac{\partial D_\phi}{\partial\phi}=\frac{1}{r^2}0.3(4)r^3\,\text{n}=1.2\text{r n}$$

Also known $Q=\int_v\rho_v dv$ where $dv=r\sin\theta\,d\phi\,r\,d\theta\,dr$

$$=r^2\sin\theta\,d\theta\,d\phi\,dr$$

$$\therefore\quad Q=\int_{r=0}^{3}\int_{\theta=0}^{\pi}\int_{\phi=0}^{2\pi}1.2r\,n\,r^2\sin\theta\,d\theta\,d\phi\,dr$$

$$=n\int_{r=0}^{3}\int_{\theta=0}^{\pi}1.2r^3\sin\theta(2\pi)d\theta\,dr$$

$$=2.4\pi n\int_{r=0}^{3}r^3\left[-\cos\theta\right]_0^\pi dr$$

$$=2.4\pi n(2)\left[\frac{r^4}{4}\right]_0^3=305.4\ \text{nC}$$

(c) $Q=\int_{r=0}^{4}\int_{\theta=0}^{\pi}\int_{\phi=0}^{2\pi}1.2\,nr^3\sin\theta\,d\theta\,d\phi\,dr$

Upon simplifying, we get

$Q=965.09$ nC

1.8.4 Applications of Gauss's Law - Infinite Sheet of Charge

Consider an infinite sheet with surface charge density ρ_s C/m^2 is lying on XY plane as shown in the Fig.1.26. Since Electric flux density $\bar{D}$ is always normal to the surface, we need to find Electric flux density at any point on either side of the sheet. Since the charge distribution depends on X and Y axeses, rectangular coordinate system can be used to find $\bar{D}$ at any point on either side of the sheet.

Hence Consider a rectangular box that is cut symmetrically by the sheet as shown in the Fig.1.26. As $\bar{D}$ is perpendicular to the sheet it will have components only in the Z-direction i.e., components on X and Y-directions are zero. Let us find out $\bar{D}$ as

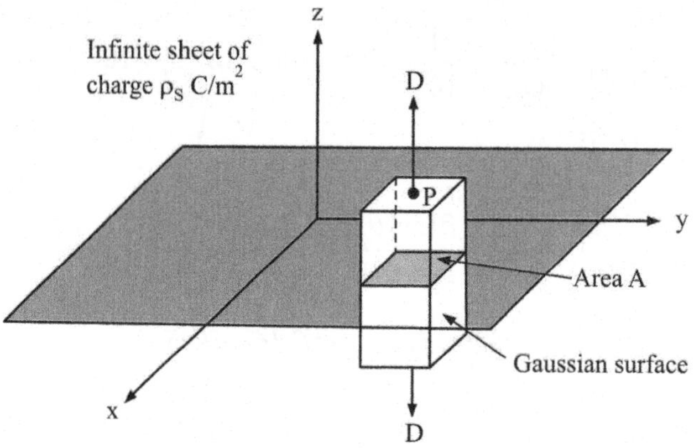

Fig.1.26 Gaussian surface about an infinite sheet of charge

Flux flowing through the rectangular box is

$$\psi = \oint_s \bar{D} \cdot \overline{ds}$$

Here $\bar{D} = D_z \bar{a}_z$ & $\overline{ds} = ds\bar{a}_z$

The flux due to bottom and top surfaces of rectangular box exists, but the flux due to the other surfaces of box is zero.

∴ above equation becomes

$$\psi = \oint_s D_z \bar{a}_z \cdot ds\bar{a}_z$$

$$\psi = D_z \left[\int_{top} ds + \int_{bottom} ds \right]$$

Assume that the area of the elemental surface as A, then

$$\psi = D_z [A + A]$$

$$\psi = 2AD_z$$

Charge enclosed by the rectangular box is

$$Q_{enc} = \int \rho_s ds$$

$$Q_{enc} = \rho_s \int ds$$

$$Q_{enc} = \rho_s A$$

According to Gauss's law

$$\psi = Q_{enc}$$

$$\rho_s A = 2AD_Z$$

$$D_z = \frac{\rho_s}{2}$$

$$\bar{D} = \frac{\rho_s}{2}\bar{a}_z$$

and $$\bar{E} = \frac{\bar{D}}{\epsilon_0} = \frac{\rho_s}{2\epsilon_0}\bar{a}_z \qquad \text{.....(1.8.6)}$$

which is similar to the formula derived by using Coulomb's law.

1.8.5 Applications of Gauss's Law - Uniformly Charged Sphere

Case I: (r < a)

Consider a sphere of radius 'a', which has uniform charge distribution with volume charge density ρ_v C/m^3 as shown in Fig.1.27. Since it is sphere, to find $\bar{D}$ at any point in side the sphere, consider a sphere of radius 'r' where r < a and is assumed as Gaussian surface. Hence spherical co-ordinate system can be used to find $\bar{D}$.

The charge enclosed by the sphere of radius 'r' is

$$Q_{enc} = \int_v \rho_v dv$$

$$Q_{enc} = \rho_v \int_v dv$$

$$= \rho_v \int_{\phi=0}^{2\pi} \int_{\theta=0}^{\pi} \int_{r=0}^{r} r^2 \sin\theta \, d\theta \, d\phi \, dr$$

$$= \rho_v \frac{4}{3}\pi r^3$$

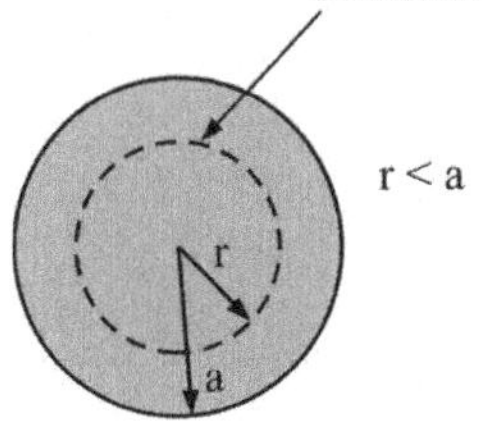

Fig. 1.27 Gaussian surface for uniformly charged sphere

The flux flowing through the spherical surface

$$\psi = \oint_s \bar{D}.\overline{ds}$$

As the flux density is normal to the surface it will have components only in 'r' direction.

$$= D_r \int_{\theta=0}^{\pi} \int_{\phi=0}^{2\pi} r^2 \sin\theta \, d\theta \, d\phi$$

$$= D_r \, 4\pi r^2$$

According to Gauss's law charge enclosed = flux flowing through the surface i.e., $Q_{enc} = \psi$

$$\rho_v \frac{4}{3}\pi r^3 = D_r \, 4\pi r^2$$

$$D_r = \frac{\rho_v}{3} r$$

$$\bar{D} = \frac{\rho_v}{3} r \, \bar{a}_r$$

and $$\bar{E} = \frac{\bar{D}}{\epsilon_0} = \frac{\rho_v}{3\,\epsilon_0} r \, \bar{a}_r \qquad .....(1.8.7)$$

Case II (r > a)

To find the electric flux density out side the sphere of radius 'a', consider a sphere of radius 'r', which is treated as Gaussian surface as shown in Fig.1.28.

Charge enclosed by the sphere of radius 'r' is

$$Q_{enc} = \int_v \rho_v dv$$

$$Q_{enc} = \rho_v \int_v dv$$

$$= \rho_v \int_{\theta=0}^{\pi} \int_{\phi=0}^{2\pi} \int_{r=0}^{a} r^2 \sin\theta \, d\theta \, d\phi \, dr$$

$$= \rho_v \frac{4}{3}\pi a^3$$

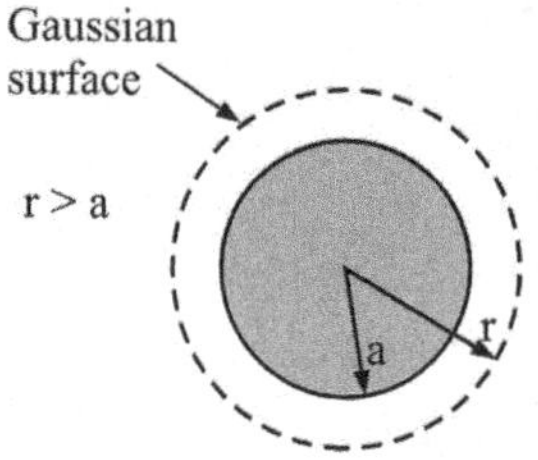

Fig. 1.28 Gaussian surface for uniformly charged sphere

Flux flowing through the surface

$$\psi = D_r \int_{\theta=0}^{\pi} \int_{\phi=0}^{2\pi} r^2 \sin\theta \, d\theta \, d\phi$$

$$= D_r \, 4\pi r^2$$

$Q_{enc} = \psi$ according to Gauss's law

$$\rho_v \frac{4}{3}\pi a^3 = D_r \, 4\pi r^2$$

$$D_r = \frac{\rho_v}{3r^2} a^3$$

$$\bar{D} = \frac{\rho_v a^3}{3r^2} \bar{a}_r$$

and $$\bar{E} = \frac{\rho_v a^3}{3r^2 \in_0} \bar{a}_r$$

$$\bar{E} = \frac{\rho_v a^3 4\pi}{3r^2 \in_0 4\pi} \bar{a}_r$$

$$\bar{E} = \frac{Q}{4\pi \in_0 r^2} \bar{a}_r \qquad .....(1.8.8)$$

which is similar to the formula derived by using Coulomb's law.

Problem 1.20

A charge distribution with spherical symmetry has density

$$\rho_v = \begin{cases} \rho_0 \dfrac{r}{R}, & 0 \le r \le R \\ 0, & r > R \end{cases}$$

Determine $\bar{E}$ everywhere

Solution:

Replace 'a' with 'R' in Fig.1.27, Then

Case I: Inside the sphere of radius 'R'

The charge enclosed by the sphere of radius 'r' is $Q_{enc} = \int_v \rho_v dv$

$$Q_{enc} = \int_v \rho_0 \frac{r}{R} dv$$

$$= \rho_0 \int_{\phi=0}^{2\pi} \int_{\theta=0}^{\pi} \int_{r=0}^{r} \frac{r^3}{R} \sin\theta \, d\theta \, d\phi \, dr$$

$$= \frac{\rho_0}{R} \int_{\phi=0}^{2\pi} d\phi \int_{\theta=0}^{\pi} \sin\theta \, d\theta \int_{r=0}^{r} r^3 dr$$

$$= \frac{4\pi r^4 \rho_0}{4R}$$

$$Q_{enc} = \frac{\rho_0}{R} \pi r^4$$

The flux flowing through the spherical surface

$$\psi = \oint_s \bar{D}.\overline{ds}$$

As the flux density is normal to the surface it will have components only in 'r' direction.

$$= D_r \int_{\theta=0}^{\pi} \int_{\varphi=0}^{2\pi} r^2 \sin\theta \, d\theta \, d\varphi$$

$$\psi = D_r 4\pi r^2$$

According to Gauss's law charge enclosed = flux flowing through the surface i.e., $Q_{enc} = \psi$

$$\frac{\rho_0}{R} \pi r^4 = D_r 4\pi r^2$$

$$D_r = \frac{\rho_0}{4R} r^2$$

$$\bar{D} = \frac{\rho_0}{4R} r^2 \bar{a}_r$$

and $$\bar{E} = \frac{\bar{D}}{\epsilon_0} = \frac{\rho_0}{4R \epsilon_0} r^2 \bar{a}_r$$

Case II: Outside the sphere of radius 'R'

Charge enclosed by the sphere of radius 'r' is

$$Q_{enc} = \int_v \rho_v dv$$

$$Q_{enc} = \int_v \rho_0 \frac{r}{R} dv$$

$$= \frac{\rho_0}{R} \int_{\theta=0}^{\pi} \int_{\phi=0}^{2\pi} \int_{r=0}^{R} r^3 \sin\theta \, d\theta \, d\phi \, dr$$

$$= \rho_0 \pi R^3$$

Flux flowing through the surface

$$\psi = D_r \int_{\theta=0}^{\pi} \int_{\phi=0}^{2\pi} r^2 \sin\theta \, d\theta \, d\phi$$

$$= D_r \, 4\pi r^2$$

$Q_{enc} = \psi$ according to Gauss's law

$$\rho_0 \, \pi R^3 = D_r \, 4\pi r^2$$

$$D_r = \frac{\rho_0 R^3}{4r^2}$$

$$\bar{D} = \frac{\rho_0 R^3}{4r^2} \bar{a}_r$$

and $$\bar{E} = \frac{\rho_0 R^3}{4r^2 \in_0} \bar{a}_r$$

Problem 1.21

A sphere of radius 'a' is filled with a uniform charge density of ρ_v C/m^3. Determine the electric field inside and outside the sphere.

Solution

The answer is as derived in section 1.8.5 case-I(inside the sphere) and case-II(outside the sphere).

Problem 1.22

A charge distribution in free space has $\rho_v = 2r$ nC/m^3 for $0 < r < 10$ m and '0' otherwise. Determine $\bar{E}$ at $r = 2\ m$ and $r = 12$ m

Solution

Replace 'a' with '10 m' in Fig.1.27, Then

At r = 2 m

The charge enclosed by the sphere of radius '2m' is $Q_{enc} = \int_v \rho_v dv$

$$Q_{enc} = \int_v 2rn dv$$

$$= 2n \int_{\phi=0}^{2\pi} \int_{\theta=0}^{\pi} \int_{r=0}^{2} r^3 \sin\theta \, d\theta \, d\phi \, dr$$

$$= 32\pi \ nC$$

The flux flowing through the spherical surface

$$\psi = \oint_s \bar{D}.\overline{ds}$$

As the flux density is normal to the surface it will have components only in 'r' direction.

$$= D_r \int_{\theta=0}^{\pi} \int_{\phi=0}^{2\pi} r^2 \sin\theta \, d\theta \, d\phi$$

$$= D_r \, 16\pi$$

According to Gauss's law charge enclosed = flux flowing through the surface i.e., $Q_{enc} = \psi$

$$32\pi \, n = D_r \, 16\pi$$

$$D_r = 2n$$

$$\bar{D} = 2n\,\bar{a}_r \text{ and}$$

$$\bar{E} = \frac{\bar{D}}{\epsilon_0} = 226\bar{a}_r \text{ V/m}$$

At $r = 12$ m

Charge enclosed by the sphere of radius '12 m' is

$$Q_{enc} = \int_v \rho_v dv$$

$$Q_{enc} = \int_v 2rn dv$$

$$= 2n \int_{\theta=0}^{\pi} \int_{\phi=0}^{2\pi} \int_{r=0}^{10} r^3 \sin\theta \, d\theta \, d\phi \, dr$$

$$= 20\pi \, \mu C$$

Flux flowing through the surface

$$\psi = D_r \int_{\theta=0}^{\pi} \int_{\phi=0}^{2\pi} r^2 \sin\theta \, d\theta \, d\phi$$

$$= D_r \, 4\pi \, 12^2$$

$Q_{enc} = \psi$ according to Gauss's law

$$20\pi\,\mu = D_r\,4\pi\,12^2$$

$$D_r = 0.0347\,\mu$$

$$\bar{D} = 0.0347\,\mu\,\bar{a}_r \text{ and}$$

$$\bar{E} = 3.92\,\bar{a}_r\,\text{kV/m}$$

1.9 Electric Potential

To find electric field intensity $\bar{E}$, so far we have used Coulomb's law if the charge distribution is of any type and Gauss's law if the charge distribution has symmetry. Another method to find electric field intensity is by using electric potential which is a scalar. So obviously this method is easier when compared with the other two methods.

If we move a point charge from A to B in an electric field having electric field intensity $\bar{E}$ as shown in Fig.1.29.

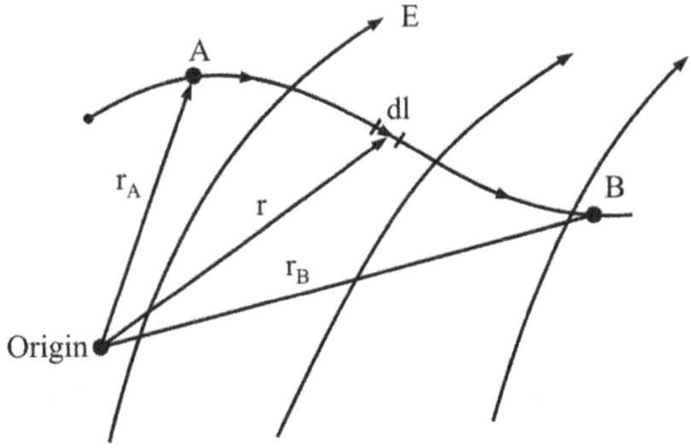

Fig. 1.29 Displacement of point charge in an electrostatic field

The elemental work done to move a point charge by an elemental distance dL is

$$dW = -\bar{F}\cdot d\bar{L}$$

The total work done in moving a point charge from A to B is

$$W = -\int_A^B \bar{F}\cdot d\bar{L}$$

-ve sign indicates work is being done by an external agent

We have $\bar{F} = Q\bar{E}$

then $$W = -\int_A^B Q\bar{E}\cdot d\bar{L}$$

$$W = -Q\int_A^B \bar{E} \cdot d\bar{L}$$

$$\Rightarrow \qquad \frac{W}{Q} = -\int_A^B \bar{E} \cdot d\bar{L}$$

which is work done per unit charge and it is also called potential difference V_{AB}.

We know that electric field intensity $\bar{E}$ due to a point charge is $\dfrac{Q}{4\pi \epsilon_0 r^2}\bar{a}_r$

and elemental length $d\bar{L} = dr\bar{a}_r$

then $$V_{AB} = \int_{r_A}^{r_B} \frac{Q}{4\pi \epsilon_0 r^2}\bar{a}_r \cdot dr\,\bar{a}_r$$

Where r_A and r_B are position vectors of point A and point B from origin

$$V_{AB} = \frac{Q}{4\pi \epsilon_0}\left[\frac{1}{r_B} - \frac{1}{r_A}\right]$$

$$= \frac{Q}{4\pi \epsilon_0 r_B} - \frac{Q}{4\pi \epsilon_0 r_A} = V_B - V_A \qquad \text{.....(1.9.1)}$$

Where V_B and V_A are absolute potentials at point B and A respectively. From the above equation V_{AB} is the potential at B with reference to the potential at A.

If A is at ∞ then $V_A = 0$.

The above equation can be generalized for a potential (V) at any point having distance 'r' as

$$V = \frac{Q}{4\pi \epsilon_0 r} \quad \text{(Here Q is located at origin)} \qquad \text{.....(1.9.2)}$$

If the point charge is placed at a distance r', then the electric potential at point 'r' can be written as

$$V = \frac{Q}{4\pi \epsilon_0 |r - r'|} \qquad \text{.....(1.9.3)}$$

If we have 'n' number of point charges $Q_1, Q_2, \ldots, Q_n$ with position vectors $r_1, r_2, \ldots,$ r_n respectively, then the potential at 'r' is

$$V = \frac{Q_1}{4\pi \epsilon_0 |r - r_1|} + \frac{Q_2}{4\pi \epsilon_0 |r - r_2|} + \ldots + \frac{Q_n}{4\pi \epsilon_0 |r - r_n|} \qquad \text{.....(1.9.4)}$$

For line charge distribution with charge density ρ_L, in the above equation Q can be replaced by $\int \rho_L dL$.

For surface charge distribution with charge density ρ_s, in equation (1.9.4), Q can be replaced by $\int \rho_S ds$.

Similarly Q can be replaced by $\int \rho_v dv$, For volume charge distribution with charge density ρ_v.

Problem 1.23

Two point charges – 4 μC and 5 μC are located at (2, –1, 3) and (0, 4, –2) respectively. Find the potential at (1, 0, 1). Assuming '0' potential at infinity.

Solution

$$V = \frac{Q_1}{4\pi \in_0 |r - r_1|} + \frac{Q_2}{4\pi \in_0 |r - r_2|}$$

$$V = \frac{-4\times10^{-6}}{4\pi \in_0 |(1,0,1)-(2,-1,3)|} + \frac{5\times10^{-6}}{4\pi \in_0 |(1,0,1)-(0,4,-2)|}$$

Simplifying, we get

$$V = -5.872 \text{ kV}$$

Problem: 1.24

A point charge 3 μC is located at the origin in addition to the two charges of previous problem. Find the potential at (–1, 5, 2). Assuming V(∞) = 0.

Solution:

$$r - r_1 = \sqrt{1+25+4} = 5.478$$

$$r - r_2 = \sqrt{9+36+1} = 6.782$$

$$r - r_3 = \sqrt{16+1+1} = 4.243$$

$$V = \left[\frac{3\times10^3}{5.478} + \frac{-4\times10^3}{6.782} + \frac{5\times10^3}{4.243}\right]\times 9$$

$$= 10.23 \text{ kV}$$

Problem 1.25

A point charge of 5 nC is located at the origin if $V = 2\ V$ at (0, 6, –8) find

(a) the potential at A (–3, 2, 6)

(b) the potential at B (1, 5, 7)

(c) the potential difference V_{AB}

Solution

(a) $$V_A - V = \frac{Q}{4\pi \epsilon_0}\left(\frac{1}{r_A} - \frac{1}{r}\right)$$

$$r_A = (-3, 2, 6) - (0, 0, 0) = \sqrt{3^2 + 2^2 + 6^2} = 7$$

$$r = (0, 6, -8) - (0, 0, 0) = \sqrt{0 + 6^2 + 8^2} = 10$$

$$V_A - 2 = \frac{5\times 10^{-9}}{4\pi \times \frac{10^{-9}}{36\pi}}\left(\frac{1}{7} - \frac{1}{10}\right)$$

$V_A = 3.929$ V

(b) $$V_B - V = \frac{Q}{4\pi \epsilon_0}\left(\frac{1}{r_B} - \frac{1}{r}\right)$$

$$r_B = (1, 5, 7) - (0, 0, 0) = \sqrt{1 + 5^2 + 7^2} = \sqrt{75}$$

$$V_B - 2 = \frac{5\times 10^{-9}}{4\pi \times \frac{10^{-9}}{36\pi}}\left(\frac{1}{\sqrt{75}} - \frac{1}{10}\right)$$

$V_B = 2.696$ V.

(c) $V_{AB} = V_B - V_A = -1.233$ V

***Problem 1.26**

A point of 5 nC is located at (–3, 4, 0), while line y = 1, z = 1 carries uniform charge 2 nC/m.

(a) If V = 0 V at O(0, 0, 0), find V at A(5, 0, 1).

(b) If V= 100 V at B(1, 2, 1), find V at C(–2, 5, 3).

(c) If V = – 5 V at O, find V_{BC}.

Solution:

Let the potential at any point be

$$V = V_Q + V_L$$

Where V_Q is potential due to point charge

i.e., $$V_Q = \frac{Q}{4\pi \in_0 r}$$

by neglecting constant of integration

and V_L is potential due to line charge distribution,

for infinite line, we have

$$\bar{E} = \frac{\rho_L}{2\pi \in_0 \rho} \bar{a}_\rho$$

$$\therefore \quad V_L = -\int \bar{E}.d\bar{l} = -\int \frac{\rho_L}{2\pi \in_0 \rho} \bar{a}_\rho . d\rho \bar{a}_\rho$$

$$\therefore \quad V_L = -\frac{\rho_L}{2\pi \in_0} \ln \rho$$

by neglecting constant of integration.

Here ρ is the perpendicular distance from the line y = 1, z = 1(which is parallel to the x-axis) to the field point.

Let the field point be (x, y, z), then

$$\rho = |(x, y, z) - (x, 1, 1)| = \sqrt{(y-1)^2 + (z-1)^2}$$

$$\therefore \quad V = -\frac{\rho_L}{2\pi \in_0} \ln \rho + \frac{Q}{4\pi \in_0 r}$$

by neglecting constant of integration.

(a) $\rho_O = |(0, 0, 0) - (0,1,1)| = \sqrt{2}$

$\rho_A = |(5,0,1) - (5,1,1)| = 1$

$r_O = |(0,0,0) - (-3,4,0)| = 5$

$r_A = |(5,0,1) - (-3,4,0)| = 9$

$$\therefore \quad V_O - V_A = -\frac{\rho_L}{2\pi \in_0} \ln \rho_O + \frac{\rho_L}{2\pi \in_0} \ln \rho_A + \frac{Q}{4\pi \in_0 r_O} - \frac{Q}{4\pi \in_0 r_A}$$

$$V_O - V_A = -\frac{\rho_L}{2\pi \in_0} \ln \frac{\rho_O}{\rho_A} + \frac{Q}{4\pi \in_0} \left[\frac{1}{r_O} - \frac{1}{r_A} \right]$$

$$0 - V_A = -\frac{2\times 10^{-9}}{2\pi \times \frac{10^{-9}}{36\pi}} \ln \frac{\sqrt{2}}{1} + \frac{5\times 10^{-9}}{4\pi \times \frac{10^{-9}}{36\pi}} \left[\frac{1}{5} - \frac{1}{9}\right]$$

$$-V_A = -36 \ln \sqrt{2} + 45\left[\frac{1}{5} - \frac{1}{9}\right]$$

$$V_A = 36 \ln \sqrt{2} - 4 = 8.477\ V$$

(b) $\rho_B = |(1,2,1) - (1,1,1)| = 1$

$\rho_C = |(-2,5,3) - (-2,1,1)| = \sqrt{20}$

$r_B = |(1,2,1) - (-3,4,0)| = \sqrt{21}$

$r_C = |(-2,5,3) - (-3,4,0)| = \sqrt{11}$

$$\therefore \quad V_C - V_B = -\frac{\rho_L}{2\pi \in_0} \ln \frac{\rho_C}{\rho_B} + \frac{Q}{4\pi \in_0}\left[\frac{1}{r_C} - \frac{1}{r_B}\right]$$

$$V_C - 100 = -36 \ln \frac{\sqrt{21}}{1} + 45\left[\frac{1}{\sqrt{11}} - \frac{1}{\sqrt{21}}\right]$$

$$V_C - 100 = -51.052$$

$$V_C = 48.94\ \text{V}$$

(c) $V_{BC} = V_C - V_B = 48.94 - 100 = -51.052$ V

1.10 Conservative and Non-Conservative Fields

1.10.1 Conservative Field

If the field is parallel to a straight line as shown in Fig. 1.30. Let $\overline{A}$ be a vector field. Choose a path P to Q as shown in Fig.1.30. $\overline{A}.d\overline{L}$ in moving from P to Q will be 'M' (scalar) and $\overline{A}.d\overline{L}$ in moving from Q to P is (–M).

$\therefore$ The $\oint \overline{A}.d\overline{L} = M - M = 0$. Chosen path may be of any shape, the contour line integral of $\overline{A}.d\overline{L}$ becomes '0'. The field whose contour line integral gives 'zero' is called conservative (or) irrotational field.

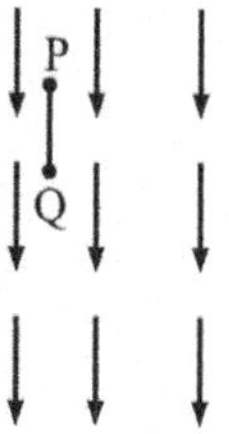

Fig. 1.30 Evaluation of conservative field

1.10.2 Non Conservative Field

In the conservative field, the filed vector is parallel to a straight line. Let us consider a field in circular fashion as shown in Fig.1.31(a). In this case $\bar{A}.d\bar{L}$ in moving from P to P along the field will not be 'zero' because $\bar{A}$ is always in the direction of $d\bar{L}$. These types of fields whose contour line integral of $\bar{A}.d\bar{L} \neq 0$ are called non conservative or rotational fields. The shape of the field need not be circular but it can be of any shape as shown in Fig. 1.31(b).

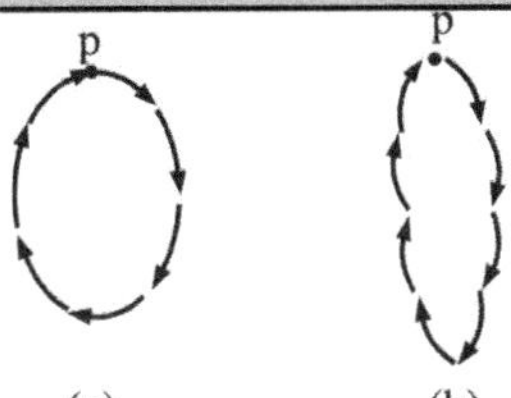

Fig. 1.31 Evaluation of non-conservative field

1.10.3 Concept of Curl

We know that in non-conservative fields as shown in Fig. 1.30 the contour line integral of $\bar{A}.d\bar{L}$ gives some finite value. This finite value is called circulation.

$\therefore$ circulation = $\oint \bar{A}.d\bar{L}$. This circulation depends upon the area chosen in the non conservative field. Let the area be ΔS. Then the ratio of $\oint \bar{A}.d\bar{L}$ to ΔS can be considered as one unit. As the field is normal to this unit we can write the above expression as $\dfrac{\oint \bar{A}.d\bar{L}}{\Delta S}\bar{a}_n$.

In general $\oint$ will be from point to point. This can be denoted by taking Limit ΔS→0 which gives curl of vector $\bar{A}$ i.e.,

$$\nabla \times \bar{A} = \lim_{\Delta S \to 0} \frac{\oint \bar{A} \cdot d\bar{L}}{\Delta S}\bar{a}_n \qquad \text{.....(1.10.1)}$$

The curl of vector $\bar{A}$ gives circulation that exists on the chosen closed surface.

As $\nabla \times \bar{A}$ or curl of a vector $\bar{A}$ is a vector. It can be represented with three components in a rectangular co-ordinate system i.e., [curl $\bar{A}$]$_1$, [curl $\bar{A}$]$_2$, [curl $\bar{A}$]$_3$ along X, Y & Z axises with $\bar{a}_x, \bar{a}_y$ & $\bar{a}_z$ as unit vectors respectively.

$$\therefore \qquad \nabla\times \bar{A} = [\text{curl } \bar{A}]_1 + [\text{curl } \bar{A}]_2 + [\text{curl } \bar{A}]_3$$

To find [curl $\bar{A}$]$_1$ consider the elemental surface Δy and Δz which is normal to x-axis as shown in Fig. 1.32.

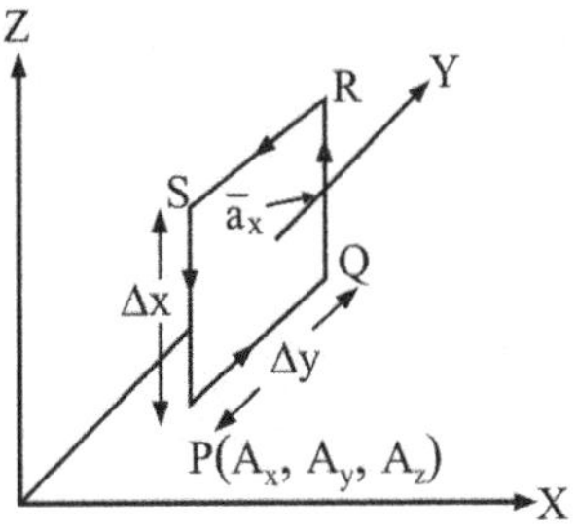

Fig. 1.32 Evaluation of curl

$$\therefore \text{ [curl } \bar{A}\text{]} = \lim_{\Delta y \Delta z \to 0} \frac{\oint \bar{A} \cdot d\bar{L}}{\Delta y \Delta z} a_x$$

Let the components of vector $\bar{A}$ at P be (A_x, A_y, A_z)

$\therefore$ The line integral from P to Q is $\int_P^Q \bar{A} \cdot d\bar{L}$.

$\because$ PQ is parallel to Y-axis, $\bar{A}$ can be taken as the Y component of $\bar{A}$ and the elemental length $d\bar{l}$ can be taken as Δy.

$\therefore$ The above integral becomes $A_y\Delta_y$. At Q we have moved a distance by Δy. To find line integral from Q to R consider the Z component at Q (because QR is parallel to Z-axis)

$\therefore$ The 'Z' component at Q is $A_z + \frac{\partial A_z}{\partial y}\Delta y$

$\therefore$ The line integral i.e., $\int_Q^R \bar{A} \cdot d\bar{L} = \left(A_z + \frac{\partial A_z}{\partial y}\Delta y\right)\Delta z$

At 'R' we have moved by a distance Δz. As RS line is parallel to Y-axis, consider the Y-component at 'R' as $A_y + \frac{\partial A_y}{\partial z}\Delta z$

$$\therefore \qquad \int_R^S \bar{A} \cdot d\bar{L} = \left(A_y + \frac{\partial A_y}{\partial z}\Delta z\right)(-\Delta y)$$

At 'S' to find $\int_S^P \bar{A} \cdot d\bar{L}$ consider the 'z' component at 'S' which is A_z

$$\therefore \qquad \int_S^P \bar{A} \cdot d\bar{L} = A_z(-\Delta z)$$

$$\therefore \qquad \oint \bar{A} \cdot d\bar{L} = \int_P^Q \bar{A} \cdot d\bar{L} + \int_Q^R \bar{A} \cdot d\bar{L} + \int_R^S \bar{A} \cdot d\bar{L} + \int_S^P \bar{A} \cdot d\bar{L}$$

$$= A_y\Delta y + A_z\Delta_z + \frac{\partial A_z}{\partial y}\Delta_y\Delta_z - A_y\Delta_y - \frac{\partial A_y}{\partial z}\Delta z\Delta y - A_z\Delta z$$

$$= \left(\frac{\partial A_z}{\partial y} - \frac{\partial A_y}{\partial z}\right)\Delta z\Delta y$$

Substitute the above equation in [Curl $\bar{A}$]$_1$ equation

$$\therefore \qquad [\text{Curl } \bar{A}]_1 = \frac{\left(\frac{\partial A_z}{\partial y} - \frac{\partial A_y}{\partial z}\right)\Delta z \Delta y}{\Delta z \Delta y}\bar{a}_x = \left(\frac{\partial A_z}{\partial y} - \frac{\partial A_y}{\partial z}\right)\bar{a}_x$$

Similarly we can also construct equations for [Curl $\bar{A}$]$_2$ by considering the elemental surface on ZX plane which is perpendicular to Y-axis

$$\therefore \qquad [\text{Curl } \bar{A}]_2 = \bar{a}_y\left(\frac{\partial A_x}{\partial z} - \frac{\partial A_z}{\partial x}\right)$$

and for [Curl $\bar{A}$]$_3$ we have to consider the elemental surface on XY plane which is perpendicular to Z-axis

$$\therefore \qquad [\text{Curl } \bar{A}]_3 = \bar{a}_z\left(\frac{\partial A_y}{\partial x} - \frac{\partial A_x}{\partial y}\right)$$

Curl $\bar{A}$ = [Curl $\bar{A}$]$_1$+ [Curl $\bar{A}$]$_2$+ [Curl $\bar{A}$]$_3$

$$\nabla \times \bar{A} = \bar{a}_x\left(\frac{\partial A_z}{\partial y} - \frac{\partial A_y}{\partial z}\right) + \bar{a}_y\left(\frac{\partial A_x}{\partial z} - \frac{\partial A_z}{\partial x}\right) + \bar{a}_z\left(\frac{\partial A_y}{\partial x} - \frac{\partial A_x}{\partial y}\right)$$

Which can be written in matrix form as

Cartesian co-ordinate system:

$$\nabla \times \bar{A} = \begin{vmatrix} \bar{a}_x & \bar{a}_y & \bar{a}_z \\ \frac{\partial}{\partial x} & \frac{\partial}{\partial y} & \frac{\partial}{\partial z} \\ A_x & A_y & A_z \end{vmatrix}$$

Cylindrical co-ordinate system:

$$\nabla \times \bar{A} = \frac{1}{\rho}\begin{vmatrix} \bar{a}_\rho & \rho\bar{a}_\phi & \bar{a}_z \\ \frac{\partial}{\partial \rho} & \frac{\partial}{\partial \phi} & \frac{\partial}{\partial z} \\ A_\rho & \rho A_\phi & A_z \end{vmatrix}$$

Spherical co-ordinate system:

$$\nabla \times \bar{A} = \frac{1}{r^2 \sin\theta}\begin{vmatrix} \bar{a}_r & r\bar{a}_\theta & r\sin\theta\,\bar{a}_\phi \\ \frac{\partial}{\partial r} & \frac{\partial}{\partial \theta} & \frac{\partial}{\partial \phi} \\ A_r & rA_\theta & r\sin\theta\, A_\phi \end{vmatrix}$$

Stoke's Theorem

Stoke's theorem gives the relation between the line integral and surface integral as

$$\int_S \nabla \times \bar{A} \cdot d\bar{s} = \oint_L \bar{A} \cdot d\bar{L} \qquad .....(1.10.2)$$

where $\bar{A}$ is the field vector. According to above equation finding curl of a vector at every point in a chosen surface and adding all those values will be equal to the contour line integral of the boundary of the chosen surface.

Proof:

Let us consider a rotational field and choose a surface on it as shown in Fig.1.33.

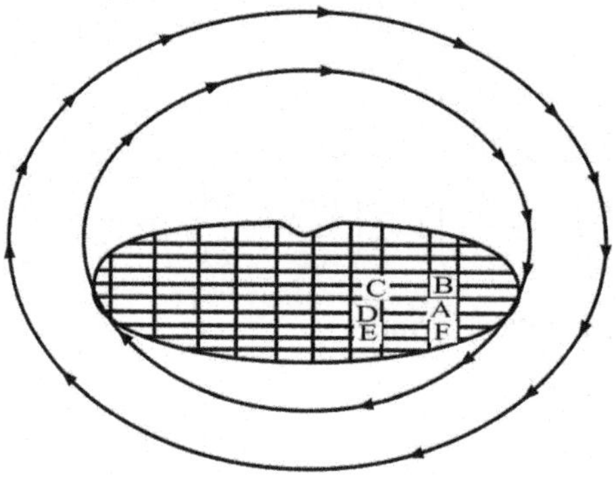

Fig. 1.33 Rotational field to explain Stoke's Theorem

We know that

$$\nabla \times \bar{A} = \lim_{\Delta s \to 0} \frac{\oint \bar{A} \cdot d\bar{L}}{\Delta S} \bar{a}_n$$

The above equation can be written as

$$\int_S \nabla \times \bar{A} \cdot d\bar{s} = \oint_L \bar{A} \cdot d\bar{L}$$

Which can be proved as

Choose a sub-surface Δs_1 (ABCDA). Then above equation becomes

$$\int_S \nabla \times \bar{A} \cdot d\bar{s}_1 = \int_A^B \bar{A} \cdot d\bar{L} + \int_B^C \bar{A} \cdot d\bar{L} + \int_C^D \bar{A} \cdot d\bar{L} + \int_D^A \bar{A} \cdot d\bar{L}$$

choose one more sub-surface Δs_2 adjacent to Δs_1 which is (ADEFA)

$$\int_S \nabla \times \bar{A} \cdot d\bar{s}_2 = \int_A^D \bar{A} \cdot d\bar{L} + \int_D^E \bar{A} \cdot d\bar{L} + \int_E^F \bar{A} \cdot d\bar{L} + \int_F^A \bar{A} \cdot d\bar{L}$$

Let $\Delta s = \Delta s_1 + \Delta s_2$

$$\int_S \nabla \times \bar{A} \cdot d\bar{s} = \int_S \nabla \times \bar{A} \cdot d\bar{s}_1 + \int_S \nabla \times \bar{A} \cdot d\bar{s}_2$$

$$= \int_A^B \bar{A} \cdot d\bar{L} + \int_B^C \bar{A} \cdot d\bar{L} + \int_C^D \bar{A} \cdot d\bar{L} + \int_D^A \bar{A} \cdot d\bar{L} - \int_D^A \bar{A} \cdot d\bar{L} + \int_D^E \bar{A} \cdot d\bar{L} + \int_E^F \bar{A} \cdot d\bar{L} + \int_F^A \bar{A} \cdot d\bar{L}$$

$$= \int_A^B \bar{A} \cdot d\bar{L} + \int_B^C \bar{A} \cdot d\bar{L} + \int_C^D \bar{A} \cdot d\bar{L} + \int_D^E \bar{A} \cdot d\bar{L} + \int_E^F \bar{A} \cdot d\bar{L} + \int_F^A \bar{A} \cdot d\bar{L}$$

From the above equation by finding curl of a vector $\bar{A}$ at all the points in a chosen surface and adding up all the values will be equal to the contour line integral of the chosen boundary surface. Adding up all the curls is nothing but integrating the curl of a vector w.r.t. chosen surface.

$$\therefore \int_S \nabla \times \bar{A} \cdot d\bar{s} = \oint_L \bar{A} \cdot d\bar{L}$$

1.11 Relation Between $\bar{E}$ and V

We know that the potential difference between points A and B is $V_{AB} = -\int_A^B \bar{E} \cdot d\bar{L}$.

Similarly the potential difference from B to A is $V_{BA} = \int_A^B \bar{E} \cdot d\bar{L}$

$\therefore$ The total potential from moving A to B and back to A is

$$V_{AB} + V_{BA} = -\int \bar{E} \cdot d\bar{L} + \int_A^B \bar{E} \cdot d\bar{L} = 0 = \oint_L \bar{E} \cdot d\bar{L} \quad .....(1.11.1)$$

The total work done in moving a point charge from A to B and back to A is ‘0’.

From equation (1.11.1) we can say that the electrostatic fields are conservative fields or irrotational fields.

According to Stoke’s theorem $\int_S \nabla \times \bar{E} \cdot d\bar{s} = \oint_L \bar{E} \cdot d\bar{L}$

$$\therefore \quad \int_S \nabla \times \bar{E} \cdot \overline{ds} = 0 \text{ or } \nabla \times \bar{E} = 0 \qquad \text{.....(1.11.2)}$$

Equation (1.11.1) is a Maxwell's second equation which is in integral form. Equation (1.11.2) is also a Maxwell's second equation which is in differential form

We know the potential difference $V = -\int \bar{E} \cdot d\bar{L}$

$$dv = -\bar{E} \cdot d\bar{L}$$

As $\bar{E}$ and $d\bar{L}$ are vectors they can be represented in rectangular co-ordinate system as

$$\bar{E} = E_x \bar{a}_x + E_y \bar{a}_y + E_z \bar{a}_z$$

$$d\bar{L} = dx\bar{a}_x + dy\bar{a}_y + dz\bar{a}_z$$

$$\therefore \quad dv = -\left(E_x dx + E_y dy + E_z dz\right) \qquad \text{.....(1.11.3)}$$

In calculus dv can be represented as

$$dv = \frac{\partial v}{\partial x} dx + \frac{\partial v}{\partial y} dy + \frac{\partial v}{\partial z} dz \qquad \text{.....(1.11.4)}$$

from (1.11.3) & (1.11.4)

$$E_x = \frac{-\partial v}{\partial x}, \quad E_y = \frac{-\partial v}{\partial y} \text{ and } E_z = \frac{-\partial v}{\partial z}$$

$$\bar{E} = \frac{-\partial v}{\partial x}\bar{a}_x - \frac{\partial v}{\partial y}\bar{a}_y - \frac{\partial v}{\partial z}\bar{a}_z$$

$$\bar{E} = \frac{-\partial v}{\partial x}\bar{a}_x - \frac{\partial v}{\partial y}\bar{a}_y - \frac{\partial v}{\partial z}\bar{a}_z$$

$$\bar{E} = -\nabla V$$

Which is the relation between $\bar{E}$ and V.

Problem 1.27

Given the potential $V = \frac{10}{r^2} \sin\theta \cos\phi$

(a) Find the electric flux density $\bar{D}$ at $(2, \pi/2, 0)$

(b) Calculate the work done in moving a 10 mC charge from point A(1, 30°, 120°) to B(4, 90°, 60°)

Solution

(a) We have

$$\overline{E} = -\nabla V$$

Since V is given in spherical co-ordinate system, consider ∇V in spherical co-ordinate system

$$\therefore \; \overline{E} = -\frac{\partial v}{\partial r}\overline{a}_r + \frac{1}{r}\frac{\partial v}{\partial \theta}\overline{a}_\theta + \frac{1}{r\sin\theta}\frac{\partial v}{\partial \phi}\overline{a}_\phi$$

$$= -\left(10\left(-2r^{-3}\right)\sin\theta\cos\phi\,\overline{a}_r + \frac{1}{r}\frac{10\cos\theta\cos\phi}{r^2}\overline{a}_\theta + \frac{1}{r\sin\theta}\frac{10\sin\theta(-\sin\phi)}{r^2}\overline{a}_\phi\right)$$

$$= -\left(10\left(-2r^{-3}\right)\sin\theta\cos\phi\,\overline{a}_r + \frac{1}{r}\frac{10\cos\theta\cos\phi}{r^2}\overline{a}_\theta + \frac{1}{r\sin\theta}\frac{10\sin\theta(-\sin\phi)}{r^2}\overline{a}_\phi\right)$$

$$= \left(\frac{20\sin\theta\cos\phi}{r^3}\overline{a}_r + \frac{-10\cos\theta\cos\phi}{r^3}\overline{a}_\theta + \frac{10\sin\phi}{r^3}\overline{a}_\phi\right)$$

$$= \frac{10}{r^3}\left(2\sin\theta\cos\phi\,\overline{a}_r - \cos\theta\cos\phi\,\overline{a}_\theta + \sin\phi\,\overline{a}_\phi\right)$$

$$\overline{D} = \overline{E}\in_0$$

$$= \frac{8.825\times10^{-11}}{r^3}\left[2\sin\theta\cos\phi\,\overline{a}_r - \cos\theta\cos\phi\,\overline{a}_\theta + \sin\phi\,\overline{a}_\phi\right]$$

$$= \frac{8.825\times10^{-11}}{r^3}\left[2.1.1\overline{a}_r - 0 + 0\right]$$

$$\overline{D}(2, \frac{\pi}{2}, 0) = 22.1\,\overline{a}_r \text{ pC/m}^2$$

(b) Work done $= -Q\int_A^B \overline{E}\cdot d\overline{L} = -Q\left(-V_{AB}\right)$

$$= Q\,(V_B - V_A)$$

$$V_B = \frac{10}{16}1.\frac{1}{2} = 0.3125\ V$$

$$V_A = \frac{10}{1}\frac{1}{2}\left(-0.5\right) = -5\times0.5 = -2.5\text{ V}$$

$$V_B - V_A = 2.8125\text{ V}$$

$$\text{W} = 10^{-3}\times10\times(V_B - V_A) = 28.125\text{ mJ}$$

Problem 1.28

Given that $\bar{E} = \left(3x^2 + y\right)\bar{a}_x + x\bar{a}_y$ kV/m. Find the work done in moving a –2 μC charge from (0, 5, 0) to (2, –1, 0) by taking the path

(a) (0, 5, 0) →(2, 5, 0) → (2, –1, 0)

(b) $y = 5 - 3x$

Solution

(a) Line equation for (0, 5, 0) to (2, 5, 0) is

$$\frac{x - x_1}{x_1 - x_2} = \frac{y - y_1}{y_1 - y_2} = \frac{z - z_1}{z_1 - z_2}$$

$$\frac{x - 0}{0 - 2} = \frac{y - 5}{5 - 5} = \frac{z - 0}{0 - 0}$$

$y = 5$ $\quad z = 0$

$dy = 0$ $\quad dz = 0$

$$W_1 = -QK \int_{(0,5,0)}^{(2,5,0)} \left(\left(3x^2 + y\right)\bar{a}_x + x\bar{a}_y\right)\left(dx\,\bar{a}_x + dy\bar{a}_y + dz\bar{a}_z\right)$$

$$= -QK \int_{(0,5,0)}^{(2,5,0)} \left(3x^2 + y\right)dx + xdy$$

$$= 2\times10^{-3} \int_{(0)}^{(2)} \left(3x^2 + 5\right)dx + 0$$

$$= 2\times10^{-3}\left(3\left(\frac{x^3}{3}\right)_0^2 + 5(2)\right)$$

$$= 36 \text{ mJ}$$

Line equation for (2, 5, 0) to (2, –1, 0)

$Z = 0$ $\quad dz = 0$

$$\frac{x - 2}{2 - 2} = \frac{y - 5}{5 + 1} = \frac{z - 0}{0 - 0}$$

$x = 2$ $\quad \mathrm{d}x = 0$

$$W_2 = -QK \int_{(2,\,5,\,0)}^{(2,\,-1,\,0)} \left(3x^2 + y\right)dx + xdy$$

$$W_2 = -QK \int_{5}^{-1} 2dy = -2QK\left(-1-5\right) = -24 \text{ mJ}$$

$W = W_1 + W_2 = 12$ mJ

(b) Line equation for (0, 5, 0) to (2, 5, 0) is $y = 5 - 3x$

$dy = -3dx$

$$W = -QK \int_{(0,\,5,\,0)}^{(2,\,-1,\,0)} \left(3x^2 + y\right)dx + xdy$$

$$W = 2\times10^{-3} \int_{0}^{2} \left(3x^2 + 5 - 3x\right)dx - 3x\,dx = 12 \text{ mJ}$$

1.12 Electric Dipole and Flux Lines

Electric dipole is formed by separating two point charges of equal magnitude but opposite in sign by a small distance.

Consider an electric dipole along Z-axis separated by a small distance 'd' as shown in Fig. 1.34.

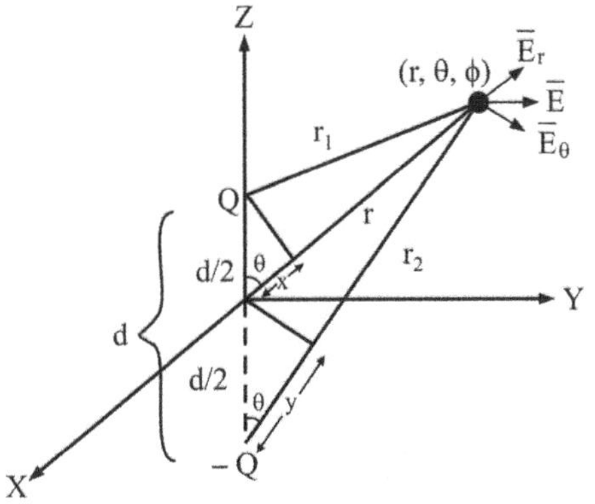

Fig. 1.34 Electric dipole to find potential

Let us find potential at $P(r,\theta,\phi)$ due to electric dipole. We know the potential at 'P' due to a point charge +Q is $V_Q = \frac{Q}{4\pi \in_0 r_1}$ and potential at 'P' due to –Q is $V_{-Q} = \frac{-Q}{4\pi \in_0 r_2}$

Potential at 'P' due to electric dipole is

$$V = \frac{Q}{4\pi \in_0 r_1} - \frac{Q}{4\pi \in_0 r_2}$$

$$= \frac{Q}{4\pi \in_0}\left(\frac{1}{r_1} - \frac{1}{r_2}\right)$$

from Fig.1.34 $\qquad \cos\theta = \frac{x}{d/2} \Rightarrow x = \frac{d}{2}\cos\theta$

$$r_1 = r - x$$

$$\therefore \quad r_1 = -\frac{d}{2}\cos\theta + r$$

$$\cos\theta = \frac{y}{d/2} \Rightarrow y = \frac{d}{2}\cos\theta$$

$$r = r_2 - y$$

$$\therefore \quad r_2 = r + \frac{d}{2}\cos\theta$$

$$\therefore \quad V = \frac{Q}{4\pi \in_0}\left(\frac{1}{r - \frac{d}{2}\cos\theta} - \frac{1}{r + \frac{d}{2}\cos\theta}\right)$$

$$= \frac{Q}{4\pi \in_0}\left(\frac{d\cos\theta}{r^2 - \left(\frac{d}{2}\cos\theta\right)^2}\right)$$

if $\qquad$ r >>d

$$V = \frac{Q}{4\pi \in_0}\left(\frac{d\cos\theta}{r^2}\right) \qquad .....(1.12.1)$$

$$V \propto \frac{1}{r^2} \text{ due to dipole}$$

$\bar{d}.\bar{a}_r$ = d cosθ and here define electric dipole moment $\bar{p} = Q\bar{d}$ whose unit is C-m.

$$\therefore \qquad V = \frac{Q(\bar{d}\cdot\bar{a}_r)}{4\pi\epsilon_0 r^2} = \frac{(\bar{p}\cdot\bar{a}_r)}{4\pi\epsilon_0 r^2} = \frac{1}{4\pi\epsilon_0 r^2}\bar{p}.\frac{\bar{r}}{|\bar{r}|} \qquad .....(1.12.2)$$

If the electric dipole center is other than origin, let it be at r' then the above equation can be generalized as

$$V = \frac{1}{4\pi\epsilon_0 |r-r'|^2}\bar{p}.\frac{r-r'}{|r-r'|} = \frac{\bar{p}.(r-r')}{4\pi\epsilon_0 |r-r'|^3} \qquad .....(1.12.3)$$

The electric field due to dipole with center at the origin can be obtained as

$\bar{E} = -\nabla V$

Since V in equation (1.12.1) is in terms of r and θ consider ∇V in spherical co-ordinate system, then

$$\bar{E} = -\frac{\partial v}{\partial r}\bar{a}_r - \frac{1}{r}\frac{\partial v}{\partial\theta}\bar{a}_\theta$$

$$\bar{E} = \left(-(-2)r^{-3}\cos\theta\,\bar{a}_r - (-\sin\theta)\frac{1}{r}\frac{1}{r^2}\bar{a}_\theta\right)\frac{Qd}{4\pi\epsilon_0}$$

$$= \frac{Qd}{4\pi\epsilon_0}\left(2r^{-3}\cos\theta\,\bar{a}_r + \frac{\sin\theta}{r^3}\bar{a}_\theta\right)$$

$$= \frac{Qd}{4\pi\epsilon_0 r^3}(2\cos\theta\,\bar{a}_r + \sin\theta\,\bar{a}_\theta)$$

$$= \frac{p}{4\pi\epsilon_0 r^3}(2\cos\theta\,\bar{a}_r + \sin\theta\,\bar{a}_\theta) \qquad .....(1.12.4)$$

Problem 1.29

An electric dipole located at the origin in free space has a moment $\bar{p} = 3\bar{a}_x - 2\bar{a}_y + \bar{a}_z$ nCm

(a) Find V at P_A (2, 3, 4)

(b) Find V at $r = 2.5$, $\theta = 30°$, $\phi = 40°$

Solution

(a) We have

$$V = \frac{1}{4\pi \in_0 |r - r'|^2} \bar{p} \cdot \frac{r - r'}{|r - r'|}$$

$r' = (0, 0, 0)$

$$|r - r'| = \sqrt{4 + 9 + 16} = \sqrt{29}$$

$$V = 9 \times 10^9 \frac{(3\bar{a}_x - 2\bar{a}_y + \bar{a}_z) \cdot (2\bar{a}_x + 3\bar{a}_y + 4\bar{a}_z)}{29\sqrt{29}} \times 10^{-9}$$

$$= \frac{9 \times (4)}{(29)^{3/2}} = 0.235 \text{ V}$$

(b) $r = 2.5 \quad \theta = 30^\circ \quad \phi = 40^\circ$

$$x = r \sin \phi \cos \theta = 0.958$$

$$y = r \sin \phi \sin \theta = 0.8035$$

$$z = r \cos \theta = 2.165$$

upon simplifying we get

$$V = 1.97 \text{ V}$$

Electric Flux Line

Electric flux line is an imaginary path or line drawn such that it's direction at any point is the direction of electric field intensity.

Equipotential Surface

Any surface which has same potential at all points is called as an equipotential surface.

Equipotential Line

The intersection line of equipotential surface with the plane is called as equipotential line. The work done to move a point charge from one point to other point along equipotential line is '0'.

The example for equipotential surface for a point charge is shown in Fig.1.35

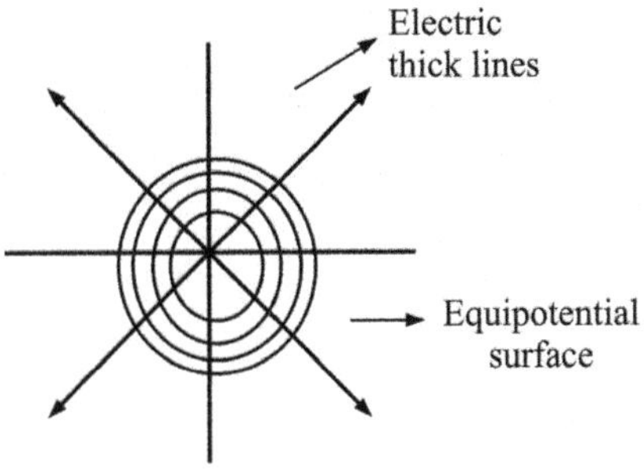

Fig. 1.35 Equipotential surface

Energy Density of Electrostatic Field

To find energy in the assembly of charges. Let us find the work required to assemble the charges. Consider a free surface and three point charges Q_1, Q_2 and Q_3 which are at infinity. The work required to move Q_4 from infinity to P_1 is $W_1 = 0$.

($\because$ initially the surface has no charge i.e., $\bar{\boldsymbol{E}} = 0 \therefore W_1 = -Q_1 \int \bar{E} \cdot Q\bar{L} = 0$)

The work required to move Q_2 from ∞ to P_2 which is shown in Fig.1.36 is $W_2 = Q_2 V_{21}$ where V_{21} is potential at P_2 due to Q_1. The work required to move Q_3 from ∞ to P_3 is $W_3 = Q_3(V_{32} + V_{31})$. Where V_{32} is potential at P_3 due to Q_2, V_{31} is potential at P_3 due to Q_1.

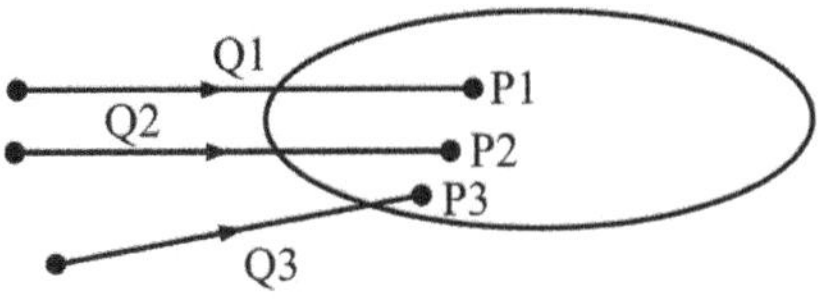

Fig. 1.36 Assembling of charges

$$\therefore W_E = W_1 + W_2 + W_3$$

$$= Q_2 V_{21} + Q_3 (V_{32} + V_{31}) \quad \text{.....(1.12.5)}$$

Suppose if we move initially Q_3 from ∞ to a free surface at P_3. The work required is $W_3 = 0$. Then work required to move Q_2 from ∞ to P_2 is $W_2 = Q_2 V_{23}$. Work required to move Q_1 from ∞ to P_1 is $W_1 = Q_1(V_{12} + V_{13})$

$$\therefore \text{Total work done } W_E = W_1 + W_2 + W_3$$

$$= 0 + Q_2 V_{23} + Q_1(V_{12} + V_{13}) \quad \text{.....(1.12.6)}$$

Add (1.12.5) and (1.12.6)

$$2W_E = Q_1 (V_{12} + V_{13}) + Q_2 (V_{21} + V_{23}) + Q_3 (V_{32} + V_{31})$$

$$= Q_1 V_1 + Q_2 V_2 + Q_3 V_3$$

$$W_E = \frac{1}{2} (Q_1 V_1 + Q_2 V_2 + Q_3 V_3) \quad \text{.....(1.12.7)}$$

Where V_1 is potential at P_1 due to Q_2 & Q_3, V_2 is potential at P_2 due to Q_1 & Q_3 and V_3 is potential at P_3 due to Q_1 & Q_2.

If we have 'n' number of charges the work required to bring them from ∞ to a surface which has initially zero charge is

$$W_E = \frac{1}{2}\sum_{k=1}^{n} Q_k V_k \quad \text{.....(1.12.8)}$$

If the surface is having continuous charge distribution then the above equation becomes

$$W_E = \frac{1}{2}\int_L \rho_L V\, dL \text{ for line charge distribution} \quad \text{.....(1.12.9)}$$

$$W_E = \frac{1}{2}\int_S \rho_s V\, dS \text{ for surface charge distribution} \quad \text{.....(1.12.10)}$$

$$W_E = \frac{1}{2}\int_v \rho_V V\, dv \text{ for volume charge distribution} \quad \text{.....(1.12.11)}$$

According to Maxwell's first equation $\rho_v = \nabla \cdot \bar{D}$

$$\therefore \quad W_E = \frac{1}{2}\int_v \left(\nabla \cdot \bar{D}\right) V\, dv \quad \text{.....(1.12.12)}$$

We know $\nabla \cdot \bar{A}V = \bar{A}\cdot\nabla V + V\left(\nabla\cdot\bar{A}\right)$ where $\bar{A}$ a general vector and V is a scalar

$$\therefore \quad \left(\nabla\cdot\bar{A}\right)V = \nabla\cdot\bar{A}V - \bar{A}\cdot\nabla V$$

i.e., $$\left(\nabla\cdot\bar{D}\right)V = \nabla\cdot\bar{D}V - \bar{D}\cdot\nabla V$$

from (1.12.12) $$W_E = \frac{1}{2}\int_v \left(\nabla\cdot\bar{D}V - \bar{D}\cdot\nabla V\right) dv$$

$$W_E = \frac{1}{2}\int_v \nabla\cdot\bar{D}V\, dv - \frac{1}{2}\int_v \bar{D}\cdot\nabla V\, dv$$

According to divergence theorem, first integral can be written as

$$W_E = \frac{1}{2}\int_S \bar{D}V.d\bar{S} - \frac{1}{2}\int_v \bar{D}\cdot\nabla V\, dv$$

For point charges the potential $V \propto \dfrac{1}{r}$, $\bar{E} \propto \dfrac{1}{r^2}$

For dipoles the potential $V \propto \frac{1}{r^2}$, $\bar{E} \propto \frac{1}{r^3}$

Surface ds $\propto$ r^2

If we consider the point charges the product of V and $\bar{E} \propto \frac{1}{r^3}$ and product of $\bar{D}V$ and $d\bar{S} \propto \frac{1}{r}$. For very large surface the first integral will become zero.

$$\therefore \qquad W_E = -\frac{1}{2}\int_v \bar{D}\cdot\nabla V\, dv$$

$$= -\frac{1}{2}\int_v \bar{D}.(-\bar{E})dv = \frac{1}{2}\int_v \bar{D}.\bar{E}\, dv$$

$$\because \bar{D} = \epsilon_0 \bar{E}$$

$$\text{Energy} = W_E = \frac{1}{2}\int_v \epsilon_0 \bar{E}\cdot\bar{E}\, dv \text{ Joules} \qquad \text{.....(1.12.13)}$$

The energy density J/m^3 is $\frac{dW_E}{dv} = \frac{1}{2}\epsilon_0 E^2 = w_E$ J/m^3 .....(1.12.14)

Problem 1.30

Three point charges –1 nC, 4 nC and 3 nC are located at (0, 0, 0), (0, 0, 1) and (1, 0, 0) respectively. Find the energy in the system.

Solution

$$W_E = W_1 + W_2 + W_3$$

$$= 0 + Q_2V_{21} + Q_3(V_{31} + V_{32})$$

$$= Q_2 \cdot \frac{Q_1}{4\pi\epsilon_0 |r_2 - r_1|} + \frac{Q_3}{4\pi\epsilon_0}\left[\frac{Q_1}{|r_3 - r_1|} + \frac{Q_2}{|r_3 - r_2|}\right]$$

$$= \frac{1}{4\pi\epsilon_0}\left(Q_1Q_2 + Q_1Q_3 + \frac{Q_2Q_3}{\sqrt{2}}\right)$$

$$= \frac{1}{4\pi . \frac{10^{-9}}{36\pi}}\left(-4 - 3 + \frac{12}{\sqrt{2}}\right).10^{-18}$$

$$= 9\left(\frac{12}{\sqrt{2}} - 7\right) \text{nJ} = 13.37\,\text{nJ}$$

Problem 1.31

Point charges $Q_1 = 1$ nC, $Q_2 = -2$ nC, $Q_3 = 3$ nC and $Q_4 = -4$ nC are positioned one at a time and in that order at (0, 0, 0), (1, 0, 0), (0, 0, –1) and (0, 0, 1) respectively. Calculate the energy in the system after each charge is positioned.

Solution

Energy after Q_1 is positioned is $W_1 = 0$

$$W_2 = Q_2V_{21} = Q_2.\frac{Q_1}{4\pi \in_0 |r_2 - r_1|} = \frac{-2\times1\times10^{-18}}{4\pi.\frac{10^{-9}}{36\pi}|(1,0,0)-(0,0,0)|} = -18 \text{ nJ}$$

Energy after Q_2 is positioned $W_2' = W_1 + W_2 = -18$ nJ

Energy after Q_3 is positioned

$$W_3' = W_2' + Q_3(V_{32} + V_{31})$$

$$= -18nJ + \frac{3\times10^{-9}}{4\pi.\frac{10^{-9}}{36\pi}}\left[\frac{-2\times10^{-9}}{|(0,0,-1)-(1,0,0)|} + \frac{1\times10^{-9}}{|(0,0,-1)-(0,0,0)|}\right]$$

$$= -29.18 \text{ nJ}$$

Energy after Q_4 is positioned

$$W_4' = W_3' + Q_4(V_{43} + V_{42} + V_{41}) = -68.27 \text{ nJ.}$$

1.13 Convection and Conduction Currents

We know that materials are classified into conductors and non conductors based on conductivity σ (siemens/m or S/m). If $\sigma > 1$, the materials are called conductors and if $\sigma < 1$, the materials are called non conductors. The materials whose conductivity lies between these two materials are called semiconductors. Technically conductors and non conductors are called metals and insulators respectively. The basic difference between conductors and dielectrics (insulators) is: Conductors posses more number of free electrons to flow current through it, Where as dielectrics contain less number of free electrons to flow current through it.

If $\sigma >> 1$, the conductors are called super conductors.

Current 'i' can be defined as charge flowing through a surface per unit time

$$\therefore \qquad i = \frac{dQ}{dt}$$

Current Density

The current Δi flowing through a surface Δs is denoted as $J_n = \Delta i/\Delta S$ A/m^2.

$$\Delta i = J_n \, \Delta S$$

If current density Jn is perpendicular to the surface ΔS

$$\Delta i = J_n \, \Delta S$$

If J_n is not perpendicular to ΔS, then $\Delta i = \bar{J} \cdot \Delta \bar{S}$

The total current flowing through surface is $I = \int_S \bar{J} \cdot d\bar{s}$.

Based on how the current I is produced, the current densities are classified in to (i) convection current density (ii) conduction current density and (iii) displacement current density.

Convection Current Density

Conductors are not involved for flowing current in case of convection current. Hence it will not satisfy ohm's law. The current flowing through an insulating material like liquid or vacuum is convection current. A beam of electrons through a vacuum tube is an example of convection current.

Consider a filament which is having volume charge density ρ_v as shown in Fig.1.37

Consider an elemental volume $\Delta V = \Delta S \Delta L$ and assume that the current is flowing in y-direction with velocity U_y.

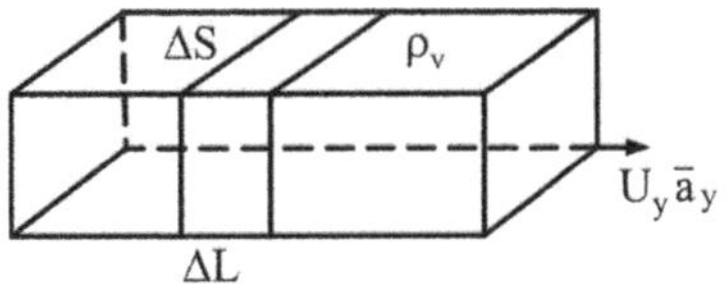

Fig.1.37 Current in a filament

We know that

$$\Delta Q = \rho_v \, \Delta V = \rho_v \, \Delta s \, \Delta l$$

Dividing with Δt

$$\frac{\Delta Q}{\Delta t} = \rho_v \Delta S \frac{\Delta l}{\Delta t}$$

$$\Delta I = \rho_v \Delta S U_y \qquad \because \frac{\Delta Q}{\Delta t} = \Delta I \quad and \quad \frac{\Delta l}{\Delta t} = U_y$$

the current density $\quad J = \frac{\Delta I}{\Delta S} = \rho_v U_y$

In general current density $\bar{J} = \rho_v \bar{U}$ .....(1.13.1)

which is convection current density and I is convection current.

Conduction Current Density

Conductors are involved in case of conduction current density. If we apply on electric field $\bar{E}$ to a conductor the force applied on electron which is having charge '– e' is

$$\bar{F} = -e\bar{E} \quad \text{.....(1.13.2)}$$

If an electron having mass 'm' is moving with a drift velocity $\bar{U}$, according to Newton's law the average change in the momentum of electron is equal to the force applied on it.

Average change in momentum is $= \dfrac{m\bar{U}}{\tau}$ .....(1.13.3)

Equations (1.13.2) = (1.13.3)

i.e.,
$$\frac{m\bar{U}}{\tau} = -e\bar{E}$$

$$\bar{U} = \frac{-e\bar{E}\tau}{m}$$

Where τ = average time interval

m = mass of electron

If we have 'n' number of electrons in the considered conductor the volume charge density

$$\rho_v = -ne$$

We know that current density

$$\bar{J} = \rho_v \bar{U}$$

∴ Conduction current density $\bar{J} = -ne\dfrac{-e\bar{E}\tau}{m}$

$$\bar{J} = ne^2\bar{E}\frac{\tau}{m} \quad \text{.....(1.13.4a)}$$

$$\bar{J} = \sigma\bar{E}$$

where

σ = conductivity of the conductor $= ne^2\dfrac{\tau}{m}$ .....(1.13.4b)

Problem 1.32

If $\bar{J} = \frac{1}{r^3}\left(2\cos\theta\,\bar{a}_r + \sin\theta\,\bar{a}_\theta\right)$ A/m^2. Calculate the current passing through

(a) Hemispherical shell of radius 20 cm.

(b) A spherical shell of radius 20 cm.

Solution

$$I = \int \bar{J}.\overline{ds}$$

Since it is sphere $\overline{ds} = r^2 \sin\theta\, d\theta\, d\phi\, \bar{a}_r$

(a) $\phi = 0$ to 2π, $\theta = 0$ to $\pi/2$ and $r = 0.2$ m for hemispherical shell

$$I = \int_{\phi=0}^{2\pi}\int_{\theta=0}^{\pi/2} \frac{1}{r^3}\left(2\cos\theta\,\bar{a}_r + \sin\theta\,\bar{a}_\theta\right)\cdot r^2 \sin\theta\, d\theta\, d\phi\, \bar{a}_r a$$

$$= \frac{1}{r}\int_{\phi=0}^{2\pi}\int_{\theta=0}^{\pi/2} 2\cos\theta \sin\theta\, d\theta\, d\phi$$

$$= \frac{1}{r}\int_{\theta=0}^{2\pi}\int_{\theta=0}^{\pi/2} \sin 2\theta\, d\theta\, d\phi$$

$$= \frac{1}{r}\int_{\phi=0}^{2\pi}\left[\frac{-\cos 2\theta}{2}\right]_0^{\pi/2} d\phi$$

$$= -\frac{1}{2r}(-1-1)(2\pi) = \frac{2\pi}{0.2} = 10\pi = 31.4A$$

(b) $\phi = 0$ to 2π, $\theta = 0$ to π and $r = 0.2$ m for spherical shell

$$I = \frac{1}{r}\int_{\phi=0}^{2\pi}\int_{\theta=0}^{\pi} sin 2\theta\, d\theta\, d\phi$$

$$= \frac{1}{r}\int_{\phi=0}^{2\pi}\left[\frac{-cos 2\theta}{2}\right]_0^{\pi} d\phi$$

$$= -\frac{1}{2r}\int_{\phi=0}^{2\pi}[1-1]\, d\phi = 0A$$

Problem 1.33

For the current density $\overline{J} = 10z\sin^2\phi\,\overline{a}_\rho$ A/m^2. Find the current through the cylindrical surface $\rho = 2$, $1 \le z \le 5\ m$.

Solution

Since it is cylinder $\overline{ds} = \rho\, d\phi\, dz\, \overline{a}_\rho$

We have

$$I = \int \overline{J} \cdot \overline{ds}$$

$$= \int_{z=1}^{5} \int_{\phi=0}^{2\pi} 10z\sin^2\phi\rho\, d\phi\, dz$$

$$= 10\rho \int_{z=1}^{5} z(1 - \cos\phi)$$

$$= 754 \text{ A}$$

*Problem 1.34

In a cylindrical conductor of radius 2 mm, the current density varies with distance from the axis according to $J = 10^3 e^{-400r}$ A/m^2. Find the total current I.

Solution

Since it is cylinder $\overline{ds} = \rho\, d\phi\, dz\, \overline{a}_\rho$

Here r = ρ = 0.02 m,

$$\therefore \qquad \overline{J} = 10^3 e^{-400\rho}\ \overline{a}_\rho\ A/m^2$$

We know the total current $I = \int_s \overline{J}.\overline{ds}$

$$\therefore \qquad I = \int_{\phi=0}^{2\pi} \int_{z=0}^{z} 10^3 e^{-400\rho} \rho d\phi\, dz$$

$$I = 2\pi z 10^3 e^{-400\rho} \rho$$

$$I = 4\pi z e^{-0.8} = z5.65 \text{ A}$$

Problem 1.35

If the current density $\bar{J} = \frac{1}{r^2}(\cos\theta \bar{a}_r + \sin\theta \bar{a}_\theta)\,\text{A/m}^2$, find the current passing through a sphere of radius 1.0 m.

Solution

We know the total current $I = \int_s \bar{J}.\bar{ds}$

Since it is spherical symmetry $\bar{ds} = r^2 \sin\theta d\theta d\phi\ \bar{a}_r$

$$\bar{J}.\bar{dS} = \frac{r^2}{r^2}\cos\theta \sin\theta\, d\phi\, d\theta$$

$$I = \int_{\theta=0}^{\pi}\int_{\phi=0}^{2\pi} \cos\theta \sin\theta\, d\phi\, d\theta$$

$$I = \pi\int_0^{\pi} \sin 2\theta\, d\theta$$

$$= \pi\left(\frac{-\cos 2\theta}{2}\right)_0^{\pi} = 0\ \text{A}$$

1.14 Polarization in Dielectrics

The basic difference between dielectrics and conductors is that dielectrics have less number of free electrons compared with the conducting material.

Consider a dielectric molecule with +Ve charge +Q (Nucleolus) and –Ve charge –Q (electron cloud) as shown in Fig. 1.38

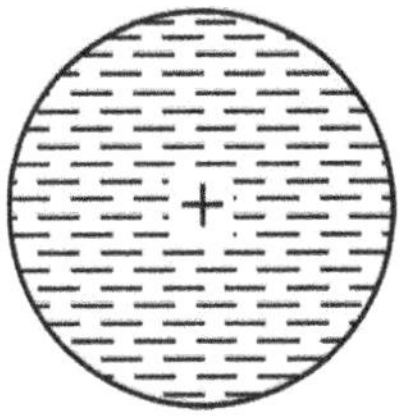

Fig. 1.38 Electron cloud

To see the effect of electric field on dielectric materials consider the dielectric molecule as shown in Fig. 1.38. If we apply electric field $\bar{E}$ on to dielectric material, the force on positive charge is $\bar{F}_+ = Q\bar{E}$ which is along the direction of electric field $\bar{E}$ and the force on negitive charges is $\bar{E} = -Q\bar{E}$ which is in opposite direction to $\bar{E}$.

After applying electric field $\bar{E}$, charge is displaced as shown in Fig.1.39. The charge displacement is equal to sum of the original charge distribution and a dipole with dipole moment $(\bar{p} = Q\bar{d})$ as shown in Fig.1.39.

After applying electric field, basically we get dipoles and hence the dielectric element is said to be polarized such dielectric material is said to be nonpolar. Examples are hydrogen, oxygen, nitrogen and the rare gases. Other types of molecules such as water, sulfur dioxide and hydrochloric acid have built-in permanent dipoles that are randomly oriented.

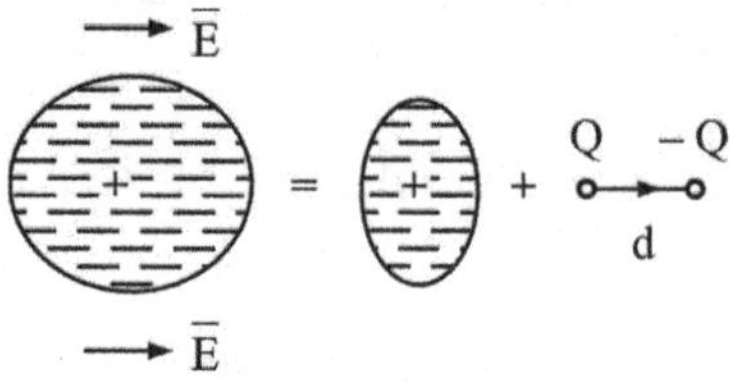

Fig. 1.39 Charge displacement after applying $\bar{E}$

Polarization

Creation of dipoles by applying electric field to the dielectric material is called polarization. Suppose 'N' numbers of dipoles are formed within 'ΔV' volume then the total number of dipole moments can be written as

$$= Q_1\bar{d}_1 + Q_2\bar{d}_2 + \ldots + Q_n\bar{d}_n = \sum_{k=1}^{N} Q_k\bar{d}_k$$

Polarization is defined as dipole moment/unit volume of the dielectric whose unit is (C/m^2)

$$\therefore \text{ Polarization} \qquad \bar{P} = \lim_{\Delta V \to 0} \frac{\sum_{k=1}^{N} Q_k\bar{d}_k}{\Delta V} \text{ C/m}^2 \qquad \ldots..(1.14.1)$$

Polarized(bounded) surface charge density $\rho_{ps} = \bar{P}\cdot\bar{a}_n$ and polarized (bounded) volume charge density $\rho_{pv} = -\nabla\cdot\bar{P}$

Consider a volume which has dielectric material with volume charge density ρ_v. Then the total volume charge density $\rho_T = \rho_v + \rho_{pv} = \nabla\cdot\bar{D}$

$$\rho_v + \rho_{pv} = \nabla\cdot \in_0 \bar{E}$$

$$\Rightarrow \qquad \rho_v = \nabla\cdot \in_0 \bar{E} - \rho_{pv}$$

$$\rho_v = \nabla\cdot \in_0 \bar{E} + \nabla\cdot\bar{P} \qquad \because \rho_{pv} = -\nabla\cdot\bar{P}$$

$$\rho_v = \nabla\cdot\left(\in_0 \bar{E} + \bar{P}\right)$$

$$\rho_v = \nabla\cdot\bar{D}$$

where $\qquad \bar{D} = \in_0 \bar{E} + \bar{P} \qquad \ldots..(1.14.2)$

The electric flux density $\bar{D}$ in free space is $\in_0 \bar{E}$ i.e., $\bar{P} = 0$ in free space.

From the above equation we can say that $\bar{D}$ is getting increased by $\bar{P}$ in dielectric materials.

From the discussion on polarization $\bar{P}$ is directly related with electric field $\bar{E}$

$$\therefore \quad \bar{P} = X_E \in_0 \bar{E} \qquad .....(1.14.3)$$

Where X_E is the electric susceptibility. The value of parameter X_E gives how susceptible the given dielectric material to the applied electric field.

Dielectric constant and strength:

Substitute equation (1.14.2) in equation (1.14.1)

$$\bar{D} = \in_0 \bar{E} + X_E \in_0 \bar{E}$$

$$= \in_0 (1 + X_E)\bar{E}$$

$$= \in_0 \in_r \bar{E}$$

$$= \in \bar{E}$$

where $\in = \in_0 . \in_r \quad \in_r = \dfrac{\in}{\in_0} = 1 + X_E$

Where $\in$ is the permittivity of dielectric material and $\in_0$ is the permittivity of free space and $\in_r$ is the dielectric constant or relative permittivity. The dielectric constant $\in_r$ can be defined as the ratio of $\in$ to $\in_0$.

If electric field strength is more such that it pulls the electrons from the outer shells of dielectric molecules, then the dielectric material becomes conducting material and we can say dielectric material has been broken.

∴ Dielectric strength can be defined as the maximum electric field with which dielectric material can tolerate or withstand.

1.15 Linear, Isotropic and Homogeneous Dielectrics

Dielectric materials can be classified into

(i) linear dielectrics

(ii) homogeneous dielectrics

(iii) isotropic dielectrics.

Linear Dielectrics: If $\in$ does not change with electric field then we can say the dielectric as linear dielectric.

Homogeneous Dielectrics: If $\in$ does not change from point to point then we can say the dielectric as homogeneous dielectric.

Isotropic dielectrics: If $\in$ does not change with the direction then we can say the dielectric as isotropic dielectric.

Similarly conducting materials are classified as

If 'σ' is independent of $\bar{E}$ then the conducting material is linear conducting material.

If 'σ' is independent of direction then the conducting material is isotropic conductor.

If 'σ' does not change from point to point then the conducting material is homogeneous conductor.

1.16 Continuity Equation and Relaxation Time

1.16.1 Continuity Equation

According to conservation of energy the rate of decrease of charge within a volume is equal to the net outward current flowing through a closed surface

$$\therefore \qquad I_{out} = \oint_S \bar{J} \cdot d\bar{s} = -\frac{dQ}{dt}$$

According to divergence theorem $\oint_S \bar{J} \cdot d\bar{s} = \int_v \nabla \cdot \bar{J}\, dv$ .....(1.16.1a)

$-\dfrac{dQ}{dt}$ can be written as $-\dfrac{dQ}{dt} = \dfrac{-d}{dt}\left[\int \rho_v dv\right]$

$$= -\int_V \left(\frac{\partial}{\partial t}\rho_v\right) dv \qquad .....(1.16.1b)$$

equations(1.16.1a) = (1.16.1b)

$$\int_v \nabla \cdot \bar{J} dv = -\int_v \left(\frac{\partial}{\partial t}\rho_v\right) dv$$

$$\therefore \qquad \nabla \cdot \bar{J} = -\frac{\partial \rho_v}{\partial t} \qquad .....(1.16.1c)$$

which is the continuity current equation.

The left side of the equation is the divergence of the Electric Current Density $(\bar{J})$. This is a measure of whether current is flowing into a volume (i.e., the divergence of $\bar{J}$ is positive if more current leaves the volume than enters).

Recall that current is the flow of electric charge. So if the divergence of $\bar{J}$ is positive, then more charge is exiting than entering the specified volume. If charge is exiting, then the amount of charge within the volume must be decreasing. This is exactly what the right side is a measure of how much electric charge is accumulating or leaving in a volume. Hence, the continuity equation is about continuity - if there is a net electric current is flowing out of a region, then the charge in that region must be decreasing. If there is more electric current flowing into a given volume than exiting, then the amount of electric charge must be increasing.

1.16.2 Relaxation Time

To derive the equation for relaxation time,

consider Maxwell's first equation i.e.,

$$\nabla \cdot \bar{D} = \rho_v$$

$$\nabla \cdot \in \bar{E} = \rho_v$$

$$\nabla \cdot \bar{E} = \frac{\rho_v}{\in} \quad \text{.....(1.16.2)}$$

Consider the conduction current equation (point form of ohm's law)

$$\bar{J} = \sigma \bar{E} \quad \text{.....(1.16.3)}$$

From (1.16.2) $\nabla \cdot \sigma \bar{E} = \sigma \frac{\rho_v}{\in}$

$$\nabla \cdot \bar{J} = \sigma \frac{\rho_v}{\in} \quad \text{from (1.16.3)}$$

$$\frac{-\partial \rho_v}{\partial t} = \sigma \frac{\rho_v}{\in} \quad \text{from continuity equation}$$

$$\frac{\partial \rho_v}{\rho_v} = -\frac{\sigma}{\in} \partial t$$

on integrating

$$\ln \rho_v = -\frac{\sigma}{\in} t + \ln \rho_{v0}$$

$$\frac{\rho_v}{\rho_{v0}} = e^{\frac{-\sigma}{\in} t} = e^{-t/(\in/\sigma)} \quad \text{.....(1.16.4)}$$

ρ_{v0} = initial volume change density

$$\frac{\in}{\sigma} = T_r$$

Which is relaxation time or rearrangement time.

Let us consider the effect of inserting the charge in the interior point of the material (Material can be conductor or dielectric).

Due to the insertion of charge in the interior point of the material, the volume charge density decreases exponentially.

Relaxation time can be defined as the time it takes a charge placed within an interior point of material to drop to $e^{-1} = 36.8\%$ of its initial value.

Relaxation time is very short for good conductors and high for good dielectrics. When we place a charge within a conductor within a short period charge disappears and it appears on the surface of conductor. Similarly when we place a charge within a dielectric material the charge remains there for a longer time.

1.17 Poisson's and Laplace's Equations

We can find $\bar{E}$ or $\bar{D}$ by using Coloumb's law or Gauss's law, (if the distribution is symmetry) if the charge distribution is known. We can also find out $\bar{E}$ or $\bar{D}$, if the potential difference is known. But in practical situation charge distribution and potential difference may not be given, in such cases either charge or potential is known only at boundary. Such type of situations or problems can be tackled either by using Poisson's equation or Laplace's equation.

We know Maxwell's first equation $\nabla \cdot \bar{D} = \rho_V$

Substitute $\bar{D} = \in \bar{E}$ in the above equation

$$\nabla \cdot \in \bar{E} = \rho_V$$

we know

$$\bar{E} = -\nabla V$$

$$\nabla \cdot (-\in \nabla V) = \rho_V \quad \text{.....(1.17.1a)}$$

which is the Poisson's equation for in-homogeneous medium.

For charge free medium $\rho_v = 0$

$$\nabla \cdot (-\in \nabla V) = 0 \quad \text{.....(1.17.1b)}$$

which is the Laplace's equation for in-homogeneous charge free medium.

For homogeneous medium since $\in$ is constant

$$\nabla^2 V = \frac{-\rho_v}{\in} \quad \text{.....(1.17.2)}$$

which is the Poisson's equation for homogeneous medium.

For charge free region $\rho_v = 0$

$$\nabla^2 V = 0 \qquad .....(1.17.3)$$

which is the Laplace's equation for homogeneous charge free medium.

We know

$$\nabla V = \frac{\partial V}{\partial x}\bar{a}_x + \frac{\partial V}{\partial y}\bar{a}_y + \frac{\partial V}{\partial z}\bar{a}_z$$

$$\nabla \cdot \nabla V = \nabla^2 V = \frac{\partial^2 V}{\partial x^2} + \frac{\partial^2 V}{\partial y^2} + \frac{\partial^2 V}{\partial z^2} = 0 \qquad .....(1.17.4)$$

Which is Laplace's equation in rectangular co-ordinate system.

where ∇^2 is Laplacian operator

In cylindrical co-ordinate system is

$$\nabla^2 V = \frac{1}{\rho}\frac{\partial}{\partial \rho}\left(\rho \frac{\partial V}{\partial \rho}\right) + \frac{1}{\rho^2}\frac{\partial^2 V}{\partial \phi^2} + \frac{\partial^2 V}{\partial z^2} = 0 \qquad .....(1.17.5)$$

In spherical co-ordinate system

$$\nabla^2 V = \frac{1}{r^2}\frac{\partial}{\partial r}\left(r^2 \frac{\partial V}{\partial r}\right) + \frac{1}{r^2 \sin\theta}\frac{\partial}{\partial \theta}\left(\sin\theta \frac{\partial V}{\partial \theta}\right) + \frac{1}{r^2 \sin^2\theta}\frac{\partial^2 V}{\partial \phi^2} = 0 \qquad .....(1.17.6)$$

Problem 1.36

Write Laplace's equation in rectangular co-ordinates for two parallel planes of infinite extent in the X and Y directions and separated by a distance 'd' in the Z-direction. Determine the potential distribution and electric field strength in the region between the planes.

Solution

$$\nabla^2 V = 0$$

$$\frac{\partial^2 V}{\partial x^2} + \frac{\partial^2 V}{\partial y^2} + \frac{\partial^2 V}{\partial z^2} = 0$$

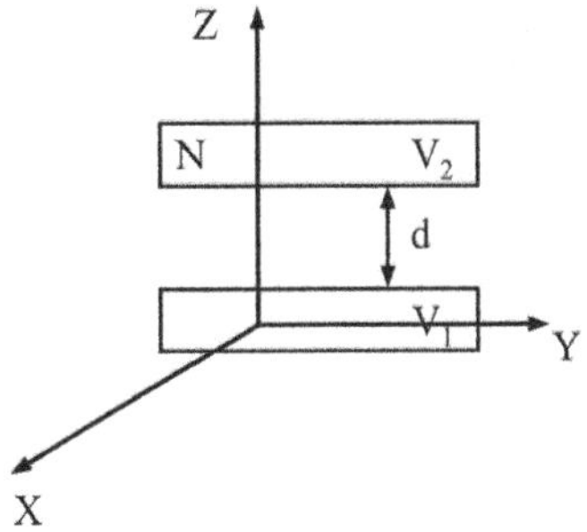

Fig. 1.40

since the potential is constant in X and Y directions

$$\frac{\partial V}{\partial x} = \frac{\partial V}{\partial y} = \frac{\partial^2 V}{\partial x^2} = \frac{\partial^2 V}{\partial y^2} = 0$$

$$\frac{\partial^2 V}{\partial z^2} = 0$$

$$\frac{\partial V}{\partial z} = A$$

$$V = Az + B$$

At $Z = 0$ $V = V_1$

$$V_1 = 0 + B$$

At $Z = d$ $V = V_2$

$$V_2 = Ad + B$$

$$V_2 = Ad + V_1$$

$$A = \frac{V_2 - V_1}{d}$$

The Potential distribution is $V = \frac{V_2 - V_1}{d} z + V_1$

The Electric field strength is

$$\bar{E} = -\nabla V = -\frac{\partial V}{\partial z}\bar{a}_z = -\frac{V_2 - V_1}{d}\bar{a}_z = \frac{V_1 - V_2}{d}\bar{a}_z$$

1.18 Parallel Plate Capacitor, Coaxial Capacitor, Spherical Capacitor

Capacitor may be obtained by separarting two conductors in some medium, which are having charges equal in magnitude but opposite in sign, such that the flux leaving from one surface of the conductor, terminates at the other conductor. Medium can be either free space or dielectric . Generally these conductors are called plates.

Let us consider two conductors with + Q and – Q charges and are connected to a voltage or potential difference 'V' as shown in the Fig.1.41.

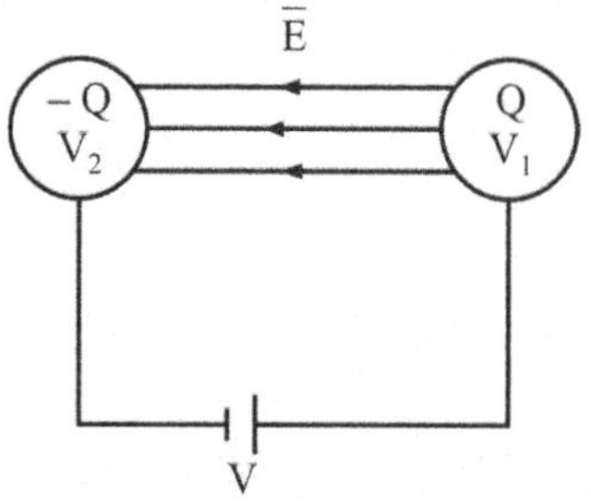

Fig. 1.41 Two conductors connected to V

The potential difference 'V' can be written in terms of $\bar{E}$ as potential difference $V = V_1 - V_2 = -\int_1^2 \bar{E} \cdot d\bar{L}$

The parameter of the capacitor i.e., 'capacitance' is defined as the ratio of charge on one of the conductors to the potential difference between two conductors.

$$C = \frac{Q}{V} = \frac{\oint_S \bar{D}\cdot d\bar{s}}{\int \bar{E}\cdot d\bar{L}} = \frac{\oint_S \in \bar{E}\cdot d\bar{s}}{\int \bar{E}\cdot d\bar{L}} \qquad .....(1.18.1)$$

1.18.1 Parallel Plate Capacitor

Consider two conductors whose area as 'A' and are separated by a distance 'd' as shown in Fig.1.42.

We know the electric field intensity $\bar{E}$ between parallel plate capacitors in free space as $\bar{E} = \frac{\rho_s}{\in_0}\bar{a}_n$

But from the Fig. 1.41 $\bar{E} = \frac{\rho_s}{\in}(-\bar{a}_x)$

$\because \bar{E}$ will be in opposite direction of x-axis

$$Q = \rho_s . A \Rightarrow \rho_s = \frac{Q}{A}$$

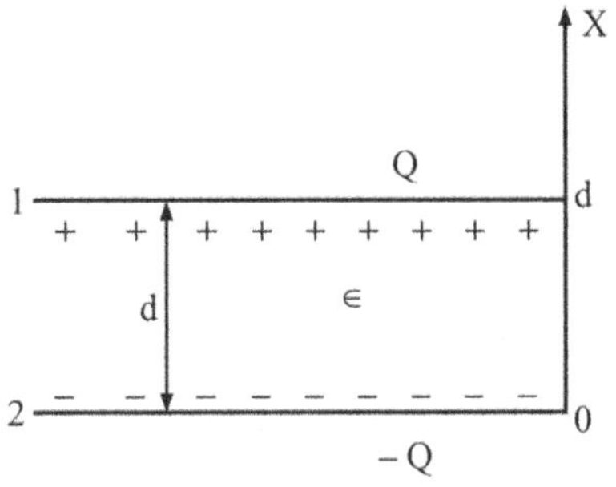

Fig. 1.42 Parallel plate capacitor

Where A = area of conductor.

$\therefore$
$$\bar{E} = -\frac{Q}{A\in}\bar{a}_x$$

We know the potential difference between two conductors which are separated by a distance 'd' as

$$V = -\int_0^d \bar{E}.d\bar{L}$$

where
$$d\bar{L} = dx\,\bar{a}_x$$

$$V = -\int_0^d \frac{-Q}{A\in}\bar{a}_x .dx\,\bar{a}_x$$

$$V = \int_0^d \frac{Q}{A\in}dx = \frac{Qd}{A\in}$$

$\therefore$
$$C = \frac{Q}{V} = \frac{A\in}{d} \qquad .....(1.18.2)$$

Energy stored in the parallel plate capacitor is

$$W_E = \frac{1}{2}\int_v \in \bar{E} \cdot \bar{E}\, dv$$

$$W_E = \frac{1}{2}\int_v \in \frac{\rho_s}{\in}\bar{a}_x \cdot \frac{\rho_s}{\in}\bar{a}_x dv$$

$$W_E = \frac{1}{2}\int_v \in \frac{\rho_s^2}{\in^2} dv$$

$$W_E = \frac{1}{2}\frac{\rho_s^2}{\in}\int_V dv = \frac{1}{2}\frac{\rho_s^2}{\in}(A \times d) = \frac{\rho_s^2 Ad}{2\in}$$

Replace ρ_s by $\frac{Q}{A}$

$$W_E = \frac{1}{2}\frac{Q^2}{A^2}\frac{Ad}{\in} = \frac{1}{2}\frac{Q^2}{C} = \frac{1}{2}VQ \qquad .....(1.18.3)$$

*Problem 1.37

Calculate the capacitance of a parallel plate capacitor with a dielectric, mica filled between plates. $\in_r$ of mica is 6. The plates of the capacitor are square in shape with 0.254 cm side. Separation between the two plates is 0.254 cm.

Solution

We have $C = \frac{\in A}{d}$

Here $\in = \in_0 \in_r = 8.854 \times 10^{-12} \times 6$

$$C = \frac{8.854 \times 10^{-12} \times 6 \times 0.254 \times 0.254 \times 10^{-4}}{0.254 \times 10^{-2}} = 0.1349 \text{ pF}$$

*Problem 1.38

A parallel plate capacitance has 500 mm side plates of square shape separated by 10 mm distance. A sulphur slab of 6 mm thickness with $\in_r = 4$ is kept on the lower plate find the capacitance of the set-up. If a voltage of 100 volts is applied across the capacitor, calculate the voltages at both the regions of the capacitor between the plates.

Solution

Given

Area of parallel plates, A = 500 mm × 500 mm = $500 \times 500 \times 10^{-6}$ m^2.

Distance of separation d = 10 mm = 10 × 10^{-3} m.

Thickness of sulphur slab d_2 = 6 mm = 6 × 10^{-3} m.

Relative permittivity of sulphur slab $\in_r = 4$.

Voltage applied across the capacitor V = 100 V.

Here the capacitor has two dielectric media,

One medium is the sulphur slab of thickness (d_2)6 mm,

since the distance between the plates(d) is 10 mm

The remaining distance is air $d_1 = d - d_2 = 4$ mm.

∴ The other dielectric medium is air with thickness (d_1) 4 mm.

The capacitance of the parallel plate capacitor with two dielectric media is

$$C = \frac{\in_0 A}{\left(\frac{d_1}{\in_{r1}} + \frac{d_2}{\in_{r2}}\right)} \text{ F}$$

Here $\in_{r1}$ (air) = 1, $\in_{r2} = \in_r = 4$

$$C = \frac{8.854 \times 10^{-12} \times 500 \times 500 \times 10^{-6}}{\left(\frac{4 \times 10^{-3}}{1} + \frac{6 \times 10^{-3}}{4}\right)} = 0.402 \text{ nF}$$

The charge Q = CV = 0.402 × 10^{-9} × 100 = 4.02 × 10^{-8} C

The value of capacitance (C_1) in delectric-1 i.e., air is

$$C_1 = \frac{\in_0 A}{d_1} = \frac{8.854 \times 10^{-12} \times 500 \times 500 \times 10^{-6}}{4 \times 10^{-3}} = 0.55 \text{ nF}$$

Similarly, The value of capacitance (C_2) in delectric-2 i.e., sulphur is

$$C_2 = \frac{\in A}{d_2} = \frac{4 \times 8.854 \times 10^{-12} \times 500 \times 500 \times 10^{-6}}{6 \times 10^{-3}} = 1.48 \text{ nF}$$

We have V = $V_1 + V_2$

Where V_1 is the voltage at the region of the capacitor plate near dielectric-1 i.e., air.

and V_2 is the voltage at the region of the capacitor plate near dielectric-2 i.e., sulphur.

$$V_1 = \frac{Q_1}{C_1} = \frac{Q}{C_1} = \frac{4.02 \times 10^{-8}}{0.55 \times 10^{-9}} = 73.1 \text{ V}$$

∴ V_2 = 100 – 73.1 = 26.9 V

1.18.2 Co-axial Capacitor

Consider two co-axial cables or co-axial cylinders of length 'L' where inner cylinder radius is 'a' and outer cylinder radius is 'b' as shown in Fig.1.43. The space between two cylinders is filled up with a homogeneous dielectric material with permittivity $\in$. Assume the charge on inner cylinder as Q and on the outer cylinder as –Q.

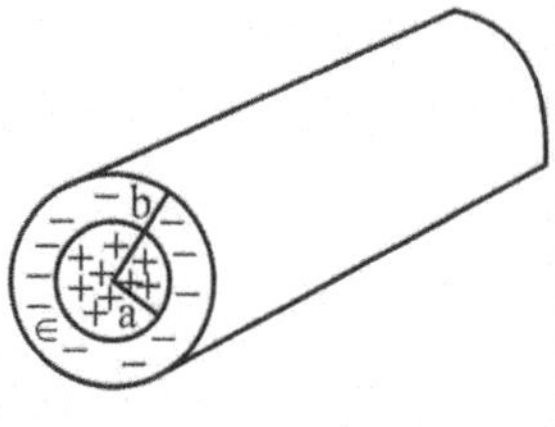

Fig. 1.43 Co-axial capacitor

we have charge enclosed by the cylinder as

$$Q = \oint \bar{D} \cdot \bar{ds} \quad \text{where } \bar{D} = D_\rho \, \bar{a}_\rho \text{ and } \bar{ds} = \rho \, d\phi \, dz \, \bar{a}_\rho$$

$$\therefore \quad Q = D_\rho \rho \int_{\phi=0}^{2\pi} d\phi \int_{z=0}^{L} dz = 2\pi D_\rho \rho L = 2\pi \in E_\rho \rho L$$

$$\text{i.e.,} \quad E_\rho = \frac{Q}{\in 2\pi \rho L} \Rightarrow \bar{E} = \frac{Q}{2\pi \in \rho L} \bar{a}_\rho$$

To find the capacitance of co-axial capacitor. We need to find the potential difference between the two cylinders.

$$\therefore \quad V = -\int_b^a \bar{E}.\bar{dl} \quad \text{where } \bar{dl} = d\rho \bar{a}_\rho$$

$$V = -\int_b^a \bar{E}.d\rho \bar{a}_\rho$$

$$= -\int_b^a \frac{Q}{2\pi \in \rho L} d\rho$$

$$V = \frac{Q}{2\pi \in L} \ln\left(\frac{b}{a}\right)$$

$$C = \frac{Q}{V} = \frac{2\pi \in L}{\ln\left(b/a\right)} \qquad \text{.....(1.18.4)}$$

Which is the expression for Coaxial capacitance.

1.18.3 Spherical Capacitor

Consider two spheres i.e., inner sphere of radius 'a' and outer sphere of radius 'b' which are separated by a dielectric medium with permittivity $\in$ as shown in Fig.1.44. The charge on the inner sphere is +Q and on the outer sphere is –Q.

We have charge enclosed by the sphere as

$$Q = \oint_S \bar{D}.d\bar{s}$$

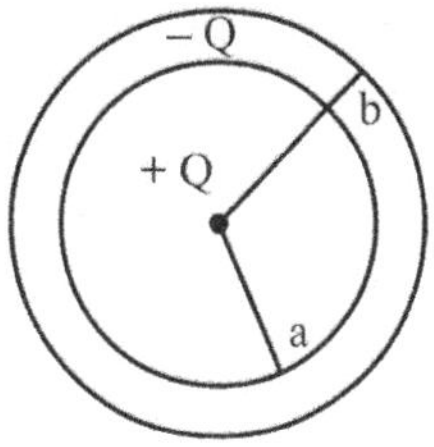

Fig. 1.44 Spherical capacitor

where $\bar{D} = D_r \bar{a}_r$;

$$d\bar{s} = r^2 \sin\theta \, d\theta \, d\phi \, \bar{a}_r$$

$$Q = \int_{\phi=0}^{2\pi} d\phi \int_{\theta=0}^{\pi} r^2 \sin\theta \, D_r \, d\theta$$

$$D_r = \frac{Q}{4\pi r^2}$$

$$E_r = \frac{Q}{4\pi \in r^2}$$

$$\bar{E} = \frac{Q}{4\pi \in r^2} \bar{a}_r$$

To find the capacitance of spherical capacitor. We need to find the potential difference between the two spheres.

$\therefore$ $$V = -\int_b^a \bar{E} \cdot d\bar{l}$$

where $$d\bar{l} = dr \, \bar{a}_r$$

$$V = \frac{-Q}{4\pi \in} \int_b^a \frac{1}{r^2} dr$$

$$= \frac{Q}{4\pi \in} \left[\frac{1}{a} - \frac{1}{b}\right]$$

$$C = \frac{Q}{V} = \frac{4\pi \in}{\left(\frac{1}{a} - \frac{1}{b}\right)} \quad \text{.....(1.18.5)}$$

Which is the expression for Spherical capacitance.

Review Questions and Answers

1. State stokes theorem.

Ans. The line integral of a vector around a closed path is equal to the surface integral of the normal component of its curl over any surface bounded by the path.

$$\int_S \nabla \times \bar{A} \cdot d\bar{s} = \oint_L \bar{A} \cdot d\bar{L}$$

2. State coulombs law.

Ans. Coulombs law states that the force between any two point charges is directly proportional to the product of their magnitudes and inversely proportional to the square of the distance between them. It is directed along the line joining the two charges.

$$\bar{F}_{12} = \frac{Q_1 Q_2}{4\pi \in_0 R^2} \bar{a}_{R12}$$

3. State Gauss law for eelectric fields.

Ans. The total electric flux passing through any closed surface is equal to the total charge enclosed by that surface.

4. Define electric flux.

Ans. The lines of electric force is electric flux.

5. Define electric flux density.

Ans. Electric flux density is defined as electric flux per unit area.

6. Define electric field intensity.

Ans. Electric field intensity is defined as the electric force per unit positive charge.

7. Name few applications of Gauss law in electrostatics.

Ans. Gauss law is applied to find the electric field intensity from a closed surface, i.e., Electric field can be determined for shell, two concentric shell or cylinders etc.

8. What is a point charge?

Ans. Point charge is one whose maximum dimension is very small in comparison with any other length.

9. Define linear charge density.

Ans. It is the charge per unit length.

10. Write poisson's and laplace's equations.

Ans. Poisson's eqn:

$$\nabla^2 V = \frac{-\rho_v}{\in}$$

Laplace's eqn:

$$\nabla^2 V = 0$$

11. Define potential difference.

Ans. Potential difference is defined as the work done in moving a unit positive charge from one point to another point in an electric field.

12. Define potential.

Ans. Potential at any point is defined as the work done in moving a unit positive charge from infinity to that point in an electric field.

13. Give the relation between electric field intensity and electric flux density.

Ans. $$\bar{D} = \in \bar{E} \text{ C/m}^2$$

14. Give the relationship between potential gradiant and electric field.

Ans. $$\bar{E} = -\nabla V$$

15. What is the physical significance of div D ?

Ans. $$\nabla . \bar{D} = -\rho_v$$

The divergence of a vector flux density is electric flux per unit volume leaving a small volume. This is equal to the volume charge density.

16. Define current density

Ans. Current density is defined as the current per unit area.

$$J = \frac{I}{A} \text{ Amp/m}^2$$

17. Write the point form of continuity equation and explain its significance.

Ans. $$\therefore \quad \nabla \cdot \bar{J} = -\frac{\partial \rho_v}{\partial t}$$

which is the continuity current equation and it's significance is:

The left side of the equation is the divergence of the Electric Current Density $(\bar{J})$. This is a measure of whether current is flowing into a volume (i.e., the divergence of $\bar{J}$ is positive if more current leaves the volume than enters).

Recall that current is the flow of electric charge. So if the divergence of $\bar{J}$ is positive, then more charge is exiting than entering the specified volume. If charge is exiting, then the amount of charge within the volume must be decreasing. This is exactly what the right side is a measure of - how much electric charge is accumulating or leaving in a volume. Hence, the continuity equation is about continuity - if there is a net electric current is flowing out of a region, then the charge in that region must be decreasing. If there is more electric current flowing into a given volume than exiting, then the amount of electric charge must be increasing.

18. Write the expression for energy density in electrostatic field.

Ans.

$$w_E = \frac{1}{2} \in E^2$$

19. Write down the expression for capacitance between two parallel plates.

Ans.

$$C = \frac{\in A}{d}$$

20. What is meant by displacement current?

Ans. Displacement current is the current flowing through the capacitor.

Multiple Choice Questions

1. Q_1 and Q_2 are two point charges, which are at a distance 8 cm apart. The force acting on Q_2 is given by $\bar{F}_{21} = \bar{a}_y 9\times10^{-12}$ N. Now we replace Q_2 with a charge of the same magnitude but opposite polarity, $Q_3 = -Q_2$, and we place Q_3 at a distance 24 cm away from Q_1. What is the vector F_{31} of the force acting on Q_3?

   (a) $\bar{F}_{31} = 3\times10^{-12}\bar{a}_y$ N (b) $\bar{F}_{31} = -3\times10^{-12}\bar{a}_y$ N

   (c) $\bar{F}_{31} = -1\times10^{-12}\bar{a}_y$ N (d) $\bar{F}_{31} = 1\times10^{-12}\bar{a}_y$ N

2. The intensity of the field due to a point charge Q_1 at a distance R_1 = 1 cm away from it is E_1 = 1 V/m. What is the intensity E_2 of the field of a charge $Q_2 = 4Q_1$ at a distance R_2 = 2 cm from it?

   (a) E_2 = 1 V/m (b) E_2 = 4 V/m

   (c) E_2 = 2 V/m (d) E_2 = ½ V/m

3. The intensity of the field due to a line charge p_{L1} at a distance $r_1 = 1$ cm away from it is $E_1 = 1$ V/m. What is the intensity E_2 of the field of the line charge $p_{L2} = 4$ at a distance $r_2 = 2$ cm from it?

   (a) $E_2 = 1$ V/m (b) $E_2 = 4$ V/m

   (c) $E_2 = 2$ V/m (d) $E_2 = ½$ V/m

4. Charge Q is uniformly distributed in a sphere of radius a_1. How is the charge density going to change if this same charge is now occupying a sphere of radius $a_2 = a_1/4$?

   (a) It will increase 4 times (b) It will increase 64 times

   (c) It will increase 16 times (d) It will increase 2 times

5. A line charge $p_L = 5 \times 10^{-3}$ C/m is located at $(x, y) = (0, 0)$, and is along the z-axis. Calculate the surface charge density p_s ($p_s > 0$) and the location x_p ($x_p > 0$) of an infinite planar charge distributed on the plane at $x = x_p$, so that the total field at the point P $(0.5 \times 10^{-3}, 0)$ m, is zero.

   (a) $\rho_s = 1/(2\pi)$ C/m^2, $x_p = 5 \times 10^{-3}$ m (b) $\rho_s = 1/(2\pi)$ C/m^2, $\forall x_p$

   (c) $\rho_s = 1/\pi$ C/m^2, $x_p = 10 \times 10^{-3}$ m (d) $\rho_s = 1/\pi$ C/m^2, $\forall x_p$

6. The volume charge density associated with the electric displacement vector in spherical coordinates $\left(\sin\theta \sin\phi \, a_r + \cos\theta \sin\phi \, a_\phi + \cos\ \phi \, a_\phi\right)$ is

   (a) 0 (b) 1

   (c) Not compatible (d) $\sin\theta$

7. The divergence theorem

   (a) Relates a line integral to a surface integral

   (b) Holds for specific vector fields only

   (c) Works only for open surfaces

   (d) Relates a surface integral to a volume integral

8. The flux of a vector quantity crossing a closed surface

   (a) is always zero

   (b) is related to the quantity's component normal to the surface

   (c) is related to the quantity's component tangential to the surface

   (d) is not related in any way to the divergence of that vector quantity

9. The flux produced by a given set of fixed charges enclosed in a given closed region is
   (a) Dependent on the surface shape of the region, but not the volume
   (b) Dependent on the total volume of the region, but not the surface shape
   (c) Dependent on the ratio of volume to surface area of the region
   (d) Not dependent on any of these as long as the charges are inside the region

10. Consider charges placed inside a closed hemisphere. Consider the flux due to these charges through the curved regions (Flux A) and through the flat region (Flux B)
    (a) Flux A = Flux B
    (b) Flux A > Flux B
    (c) Flux A < Flux B
    (d) Not enough information to decide the relation between Flux A and Flux B

11. An electron ($q_e = 1.602 \times 10^{-19}$ C) leaves the cathode of a cathode ray tube (CRT) and travels in a uniform electrostatic field toward the anode, which is at a potential V_a = 500 V with respect to the cathode. What is the work W done by the electrostatic field involved in moving the electron from the cathode to the anode?
    (a) $W = 5\text{kJ}$ (b) $W = 8\times10^{-19}\,\text{J}$
    (c) $W = 8\times10^{-17}\,\text{J}$ (d) $W = 5\,\text{J}$

12. In the previous question, what is the electric field strength $E = |E|$ if the distance between the cathode and the anode is 10 cm?
    (a) $E = 5$ V/m (b) $E = 500$ V/m
    (c) $E = 50$ V/m (d) $E = 5$ kV/m

13. The electrostatic potential due to a point charge Q_1 at a distance r_1 = 1 cm away from it is V_1 = 1 V. What is the potential V_2 of a charge $Q_2 = 4Q_1$ at a distance r_2 = 2 cm from it?
    (a) $V_2 = 0.5$ V (b) $V_2 = 1$ V
    (c) $V_2 = 4$ V (d) $V_2 = 2$ V

14. The electrostatic potential due to a dipole $\mathbf{p}_1 = p_1\mathbf{a}_z$ at a distance r_1 = 1 cm away from it along the z-axis, is V_1 = 1 V. What is the potential V_2 of a dipole $\mathbf{p}_2 = 4p_1\mathbf{a}_z$ at a distance r_2 = 2 cm from it along the z-axis?
    (a) $V_2 = 0.5$ V (b) $V_2 = 1$ V
    (c) $V_2 = 4$ V (d) $V_2 = 2$ V

15. The electrostatic potential $V = \frac{2\times10^{-3}}{\sqrt{\epsilon_0}} x$ V. Where x is measured in meters and ϵ_0 is the permittivity of vacuum, exists in a region of space (vacuum) in the shape of a parallelogram of size 10 × 10 × 1 cm. What is the electrostatic energy W_E stored in this region?

(a) $W_E = 2\times10^{-10}$ J
(b) $W_E = 1\times10^{-9}$ J
(c) $W_E = 4\times10^{-10}$ J
(d) $W_E = 3\times10^{-9}$ J

16. Which statement is not true?

(a) The static electric field in a conductor is zero
(b) The conductor surface is equipotential
(c) Zero tangential electric field on the surface of a conductor leads to zero potential difference between points on the surface
(d) The normally directed electrical field on the surface of a conductor is zero

17. The "skin" effect results in

(a) Current flowing in the entire volume as frequency increases
(b) Current flowing only near the surface as frequency increases
(c) Current flowing only near the surface as frequency decreases
(d) Current flowing near the surface at any frequency

18. As frequency increases, skin effect results in

(a) Decreased resistance
(b) Increased resistance
(c) No change in resistance
(d) Increase or decrease depending on material properties.

19. In a parallel-plate capacitor, the charge on the plates is C. What is the electric flux density magnitude D, if the area of each plate is $A = 10^{-4}$ m^2. Assume uniform field distribution.

(a) $D = 10^{-5}$ C/m^2
(b) $D = 10^{-5}/\epsilon_0$ C/m^2
(c) $D = 10^{-5}\,\epsilon_0$ C/m^2
(d) $D = 10^{-13}$ C/m^2

20. For the capacitor in Previous question, find the voltage between its plates, provided its capacitance is $C = 10$ pF.

(a) $V \simeq 885$ V
(b) $V = 0$ V
(c) $V = 100$ V
(d) $V = 10^{-5}$ V

21. The capacitor in above Q no. 19 and 20 uses dielectric of permittivity $\epsilon = \epsilon_0$. The maximum allowable field intensity (*dielectric strength*) of this dielectric is E_{ds} = 3 MV/m. (If $E > E_{ds}$, the material breaks down.) What is the maximum voltage V_{max}, up to which the capacitor can operate safely (*its breakdown voltage*)?

(a) $V_{max} = 885\,\text{V}$ (b) $V_{max} = 1000\,\text{V}$

(c) $V_{max} = 3\times10^6\,\text{V}$ (d) $V_{max} = 265\,\text{V}$

22. A coaxial capacitor whose cross-section is shown in the figure below has a central conductor of radius r_1 and an outer conductor of radius r_3. The region between the two conductors consists of two regions: (i) the region $r_1 < p < r_2$ has a relative permittivity of $\varepsilon_{r1} = 2$ and (ii) the region $r_2 < p < r_3$ has a relative permittivity of $\varepsilon_{r2} = 1$. The radius r_2 is such that $r_2/r_1 = e^2$ and $r_3/r_2 = e$ where $e \approx 2.71$.

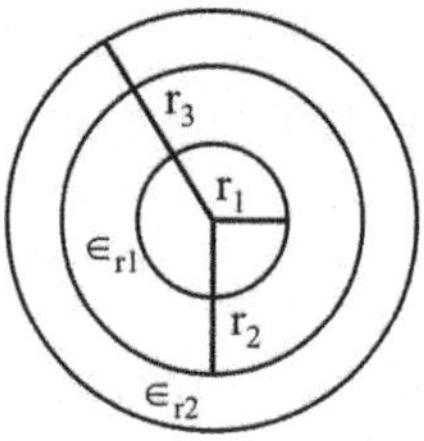

What is the capacitance per unit length?

(a) $C_1 = 4\pi \epsilon_0$ (b) $C_1 = \pi \epsilon_0$

(c) $C_1 = 2\pi \epsilon_0$ (d) $C_1 = \pi \epsilon_0/2$

23. Poisson's and Laplace's equations are different in terms of

(a) Definition of potential (b) Presence of non-zero charge

(c) Boundary conditions on potential (d) No difference

Answers

1.	(c)	9.	(d)	17.	(b)
2.	(a)	10.	(a)	18.	(b)
3.	(c)	11.	(c)	19.	(a)
4.	(b)	12.	(d)	20.	(c)
5.	(d)	13.	(d)	21.	(d)
6.	(a)	14.	(b)	22.	(b)
7.	(d)	15.	(a)	23.	(b)
8.	(b)	16.	(d)		

Exercise Questions

1. State the Coulomb's law in SI units and indicate the parameters used in the equations with the aid of a diagram.
2. State Gauss's law. Using divergence theorem and Gauss's law, relate the density D to the volume charge density ρ_v.
3. Explain the following terms:
   (a) Homogeneous and isotropic medium and
   (b) Line, surface and volume charge distributions.
4. State and Prove Gauss's law. List the limitations of Gauss's law.
5. Express Gauss's law in both integral and differential forms. Discuss the salient features of Gauss's law.
6. Derive Poisson's and Laplace's equations starting from Gauss's law.
7. Using Gauss's law derive expressions for electric field intensity and electric flux density due to an infinite sheet of conductor of charge density ρ C/m.
8. Find the force on a charge of –100 mC located at P(2, 0, 5) in free space due to another charge 300 μC located at Q(1, 2, 3).
9. Find the force on a 100 μC charge at(0, 0, 3) m, if four like charges of 20 μC are located on X and Y axes at ± 4 m.
10. Derive an expression for the electric field intensity due to a finite length line charge along the Z-axis at an arbitrary point Q(x, y, z).
11. A point charge of 15 nC is situated at the origin and another point charge of –12 nC is located at the point (3, 3, 3) m. Find $\bar{E}$ and V at the point(0, –3, –3).
12. Obtain the expressions for the field and the potential due to a small Electric dipole oriented along Z-axis.
13. Define conductivity of a material. Explain the equation of continuity for time varying fields.
14. As an example of the solution of Laplace's equation, derive an expression for capacitance of a parallel plate capacitors.
15. In a certain region $\bar{J} = 3r^2 \cos\theta \bar{a}_r - r^2 \sin\theta \bar{a}_\theta$ A/m, find the current crossing the surface defined by $\theta = 30^o, 0 < \phi < 2\pi, 0 < r < 2\,\text{m}$.

Chapter

2

Static Magnetic Fields

2.1 Introduction

The electro static field is characterized by electric field intensity $\overline{E}$ and electric flux density $\overline{D}$ and they are related as $\overline{D}=\in\overline{E}$. Similarly static magnetic fields are characterized by magnetic field intensity $\overline{H}$ and magnetic flux density $\overline{B}$ and they are related as $\overline{B}=\mu\overline{H}$. Static electric fields are obtained by static charges. Static magnetic fields are obtained when the static charges are moving with constant velocity or static magnetic fields are generated by constant current flow.

Electrostatic fields are described by Coulomb's law and Gauss's law. Similarly magneto static fields are described by Biot-Savart's law and Ampere's circuit law. In general case we use Biot-Savart's law to find either $\overline{H}$ or $\overline{B}$. If the distribution is symmetry we use Ampere's circuit law to find either $\overline{B}$ or $\overline{H}$.

2.2 Biot-Savart's Law

It states that the elemental magnetic field intensity $d\overline{H}$ produced at point 'p' as shown in Fig.2.1, by a differential current element *Idl* is proportional to the product *Idl* and sine of the angle 'α' between the Idl and line joining the element to point 'p' and is inversely proportional to square of the distance between element and point 'p'.

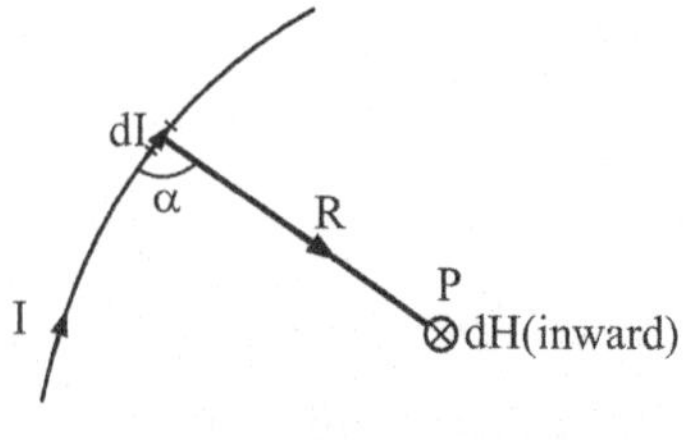

Fig. 2.1 Magnetic field due to current element

$$d\bar{H} \propto \frac{Idl\sin\alpha}{R^2}$$

$$d\bar{H} = \frac{K\,Idl\sin\alpha}{R^2}$$

$$K = \text{proportional constant} = \frac{1}{4\pi}$$

$$\therefore \qquad d\bar{H} = \frac{Idl\sin\alpha}{4\pi R^2}$$

The above equation can be written as

$$d\bar{H} = \frac{Id\bar{l} \times \bar{a}_R}{4\pi R^2} \qquad \text{where} \qquad \bar{a}_R = \frac{\bar{R}}{|\bar{R}|}$$

$$d\bar{H} = \frac{I\,d\bar{l} \times \bar{R}}{4\pi R^3} \qquad .....(2.2.1)$$

To indicate the direction of magnetic field intensity we use right hand thumb rule. In this thumb indicates direction of current and the fingers encircling the wire indicates the direction of magnetic field intensity. We can also use right hand screw rule to find the direction of magnetic field intensity. The direction of current is indicated along the screw and the rotation of screw in such a way that it advances indicates the direction of magnetic field intensity.

In Fig.2.1 at 'p' there is a small circle with cross mark which indicates the direction of magnetic field intensity and is into the page. To indicate the direction of magnetic field intensity out of the page the symbol is small circle with dot mark.

Similar to the line charge density, surface charge density and volume charge density in electrostatic fields, here we have line current density, surface current density and volume current density.

Surface current density is denoted with K and volume current density is denoted with J.

For surface current density $d\bar{H} = \dfrac{\bar{K}ds \times \bar{R}}{4\pi R^3}$ .....(2.2.2)

For volume current density $d\bar{H} = \dfrac{\bar{J}dv \times \bar{R}}{4\pi R^3}$ .....(2.2.3)

2.2.1 Infinite Line Conductor

Let us find out the magnetic field intensity at 'p' due to a conductor AB in which current I is flowing and is placed along 'z' axis as shown in Fig.2.2.

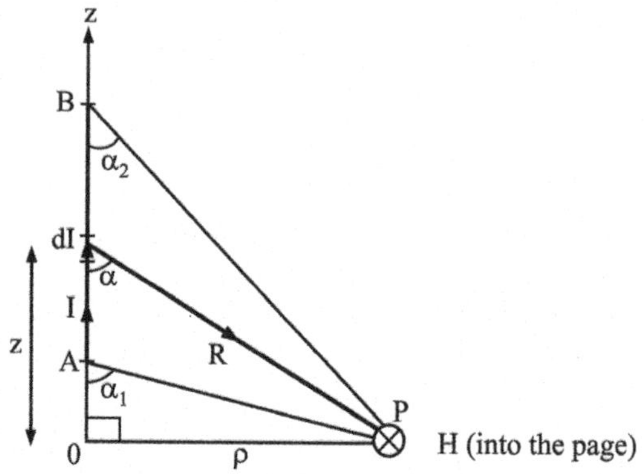

Fig. 2.2 Field at P due to a filamentary conductor

Since it is a conductor, the problem has cylindrical symmetry, hence we can use cylindrical coordinate system to solve the problem.

From Fig. 2.2,

$$\bar{R} = \rho\bar{a}_\rho - z\bar{a}_z$$

and $d\bar{l} = dz\,\bar{a}_z$

$$d\bar{H} = \frac{I\,dz\,\bar{a}_z \times \left(\rho\bar{a}_\rho - z\bar{a}_z\right)}{4\pi\left(\rho^2 + z^2\right)^{3/2}}$$

$$d\bar{H} = \frac{I\,dz\,\rho\,\bar{a}_\phi - 0}{4\pi\left(\rho^2 + z^2\right)^{3/2}}$$

From Fig.2.2,

$$\tan\alpha = \frac{\rho}{z}$$

$$z = \rho \cot \alpha$$

$$dz = -\rho \operatorname{cosec}^2 \alpha \, d\alpha$$

$$d\overline{H} = \frac{I \rho \overline{a}_\phi \left(-\rho \operatorname{cosec}^2 \alpha \, d\alpha\right)}{4\pi \left(\rho^2 + \rho^2 \cot^2 \alpha\right)^{3/2}}$$

$$= \frac{-I \rho^2 \operatorname{cosec}^2 \alpha \, \overline{a}_\phi \, d\alpha}{4\pi \rho^3 \operatorname{cosec}^3 \alpha}$$

$$= \frac{-I \, \overline{a}_\phi \, d\alpha}{4\pi \rho \operatorname{cosec} \alpha}$$

$$\overline{H} = \frac{-I \overline{a}_\phi}{4\pi \rho} \int_{\alpha_1}^{\alpha_2} \sin \alpha \, d\alpha$$

$$= \frac{-I \overline{a}_\phi}{4\pi \rho} \left[-\cos \alpha\right]_{\alpha_1}^{\alpha_2}$$

$$= \frac{-I \overline{a}_\phi}{4\pi \rho} \left(-\cos \alpha_2 + \cos \alpha_1\right)$$

$$\overline{H} = \frac{I \overline{a}_\phi}{4\pi \rho} \left(\cos \alpha_2 - \cos \alpha_1\right) \qquad .....(2.2.4)$$

For a semi infinite conductor the conductor lies from '0' to '∞'. Then $\alpha_1 = 90^\circ$ and $\alpha_2 = 0^\circ$

$$\therefore \qquad \overline{H} = \frac{I}{4\pi\rho} \overline{a}_\phi \left(\cos 0^\circ - \cos 90^\circ\right)$$

$$= \frac{I}{4\pi\rho} \overline{a}_\phi \qquad .....(2.2.5)$$

For an infinite conductor, conductor lies from $-\infty$ to ∞ i.e.,

$\alpha_1 = 180^\circ$ and $\alpha_2 = 0^\circ$

$$\therefore \qquad \overline{H} = \frac{I}{4\pi\rho} \overline{a}_\phi \left(\cos 0^\circ - \cos 180^\circ\right)$$

$$= \frac{I}{2\pi\rho} \overline{a}_\phi \qquad .....(2.2.6)$$

Problem 2.1

Find $\bar{H}$ at (0, 0, 5) due to side OA and side OB of the triangular current loop shown in Fig.2.3.

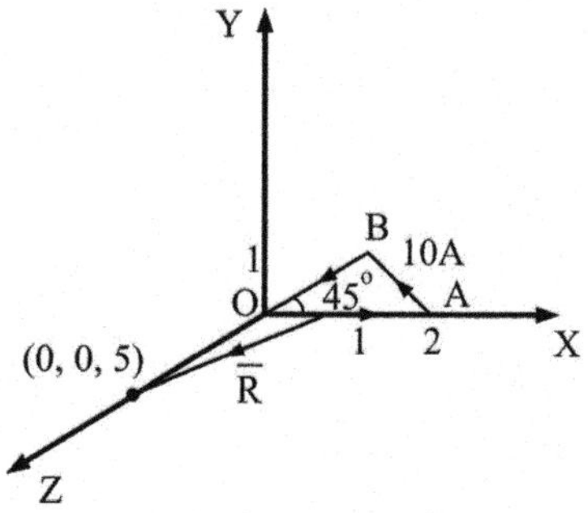

Fig. 2.3

Solution

Magnetic field intensity due to OA

$$d\bar{H}_{OA} = \frac{Id\bar{l} \times \bar{R}}{4\pi R^3} \quad \text{where } \bar{R} = (0, 0, 5) - (x, 0, 0)$$

$$= -x\bar{a}_x + 5\bar{a}_z$$

$d\bar{l} = dx\,\bar{a}_x$

$$d\bar{H}_{OA} = \frac{10dx\,\bar{a}_x \times \left(-x\bar{a}_x + 5\bar{a}_z\right)}{4\pi\left(x^2 + 5^2\right)^{3/2}}$$

$$d\bar{H}_{OA} = \frac{10\left(-5dx\bar{a}_y\right)}{4\pi\left(x^2 + 5^2\right)^{3/2}}$$

$$\bar{H}_{OA} = -\int_0^2 \frac{50\bar{a}_y}{4\pi\left(x^2 + 25\right)^{3/2}}\,dx$$

$$= \frac{-50}{4\pi}\bar{a}_y\left[\frac{x/25}{\left(x^2 + 25\right)^{1/2}}\right]_0^2$$

$$= \frac{-50}{4\pi}\frac{1}{25}\bar{a}_y \frac{2}{(29)^{1/2}}$$

$$= -59.1\bar{a}_y \text{ mA/m}$$

Magnetic field intensity due to OB

$$\bar{R} = (0, 0, 5) - (1, 1, z) = -\bar{a}_x - \bar{a}_y + (5-z)\bar{a}_z$$

$$d\bar{l} = dz\,\bar{a}_z$$

$$d\bar{H}_{OB} = \frac{(10)\,dz\,\bar{a}_z \times \left(-\bar{a}_x - \bar{a}_y + (5-z)\bar{a}_z\right)}{4\pi\left[2+(5-z)^2\right]^{3/2}}$$

$$= \frac{(10)\left(-\bar{a}_y dz + \bar{a}_x dz\right)}{4\pi\left[2+(5-z)^2\right]^{3/2}}$$

$$\bar{H}_{OB} = \frac{10}{4\pi}\int_{\sqrt{2}}^{0} \frac{\left(-\bar{a}_y dz + \bar{a}_x dz\right)}{\left[2+(5-z)^2\right]^{3/2}}$$

$$= \frac{10}{4\pi}\left(-\bar{a}_y + \bar{a}_x\right)\int_{\sqrt{2}}^{0} \frac{dz}{\left(2+(5-z)^2\right)^{3/2}}$$

$$= \frac{10}{4\pi}\left(\bar{a}_y - \bar{a}_x\right)\left[\frac{(5-z)/2}{\left(2+(5-z)^2\right)^{1/2}}\right]_{\sqrt{2}}^{0}$$

$$= \frac{5}{4\pi}\left(\bar{a}_y - \bar{a}_x\right)\left[\frac{5}{\sqrt{27}} - \frac{\left(5-\sqrt{2}\right)}{\left(2+\left(5-\sqrt{2}\right)^2\right)^{1/2}}\right]$$

$$= \frac{5}{4\pi}\left(\bar{a}_y - \bar{a}_x\right)(0.9623 - 0.9303)$$

$$= -12.73\bar{a}_x + 12.73\bar{a}_y \text{ mA / m}$$

***Problem 2.2**

Show that the magnetic field due to a finite current element along Z-axis at point 'P', 'r' distance away along Y-axis is given by $\bar{H} = (I/4\pi r)(\sin\alpha_1 - \sin\alpha_2)\bar{a}_\phi$ where I is the current through the conductor, α_1 and α_2 are the angles made by the tips of the conductor at 'P'.

Solution

Consider Fig.2.4,

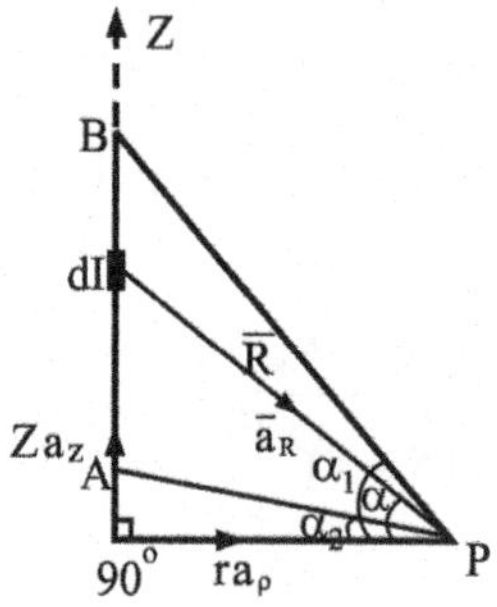

Fig. 2.4

From Fig.2.4, $\bar{R} = r\bar{a}_\rho - z\bar{a}_z$

and $d\bar{l} = dz\,\bar{a}_z$

$$d\bar{H} = \frac{I\,dz\,\bar{a}_z \times \left(r\bar{a}_\rho - z\bar{a}_z\right)}{4\pi\left(r^2 + z^2\right)^{3/2}}$$

$$d\bar{H} = \frac{I\,dz\,r\,\bar{a}_\phi - 0}{4\pi\left(r^2 + z^2\right)^{3/2}}$$

From Fig.2.4, $tan\alpha = \frac{z}{r}$

$\therefore \quad z = r\ tan\alpha \quad dz = r\sec^2\alpha\ d\alpha$

$$d\bar{H} = \frac{I\,r\,\bar{a}_\phi\left(r\sec^2\alpha\ d\alpha\right)}{4\pi\left(r^2 + r^2\tan^2\alpha\right)^{3/2}}$$

$$= \frac{I\, r^2 \sec^2\alpha\; \bar{a}_\phi\, d\alpha}{4\pi\, r^3 \sec^3\alpha}$$

$$= \frac{I\; \bar{a}_\phi\, d\alpha}{4\pi\, r \sec\alpha}$$

$$\bar{H} = \frac{I\,\bar{a}_\phi}{4\pi\, r}\int_{\alpha_2}^{\alpha_1} \cos\alpha\, d\alpha$$

$$= \frac{I\,\bar{a}_\phi}{4\pi\, r}\left[\sin\alpha\right]_{\alpha_2}^{\alpha_1}$$

$$= \frac{I\,\bar{a}_\phi}{4\pi\, r}\left(\sin\alpha_1 - \sin\alpha_2\right)$$

$$\bar{H} = \frac{I\,\bar{a}_\phi}{4\pi\, r}\left(\sin\alpha_1 - \sin\alpha_2\right)$$

***Problem 2.3**

Derive an expression for magnetic field strength H, due to a current carrying conductor of finite length placed along the Y-axis, at a point P in X–Z plane and 'r' distance from the origin. Hence deduce expressions for H due to semi-infinite length of the conductor.

Solution

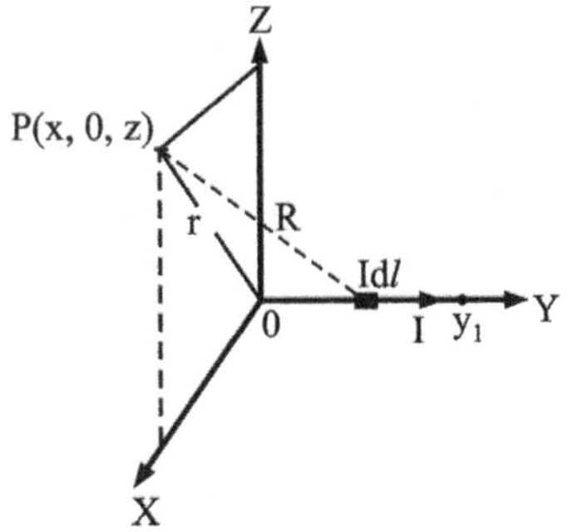

Fig. 2.5

The geometry of the given problem is shown in Fig. 2.5 with the finite length (y_1 m) current carrying conductor lying along Y-axis.

Since the point P lies in the XZ plane, for all values of X and Z the line (OP = r) makes 90° with Y-axis.

Where $\bar{r} = x\bar{a}_x + z\bar{a}_z$

The above figure can be modified as shown in Fig.2.6

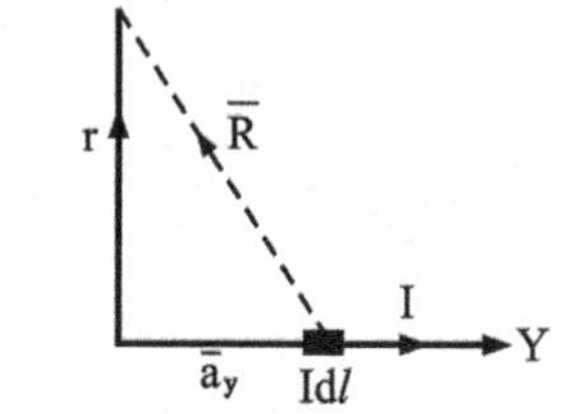

Fig. 2.6

From Fig.2.6, $\bar{R} = (x,0,z) - (0,y,0)$

$$\bar{R} = x\bar{a}_x - y\bar{a}_y + z\bar{a}_z$$

$$r^2 = x^2 + z^2$$

Consider a small differential current element *Idl* along Y-axis

According to Biot-Savart's law $d\bar{H} = \dfrac{Id\bar{l} \times \bar{R}}{4\pi R^3}$

Here $d\bar{l} = dy\bar{a}_y$

$$d\bar{H} = \frac{Idy\ \bar{a}_y \times (x\bar{a}_x + z\bar{a}_z - y\bar{a}_y)}{4\pi\left(x^2 + z^2 + y^2\right)^{3/2}}$$

$$d\bar{H} = \frac{Idy}{4\pi\left(r^2 + y^2\right)^{3/2}}\left(z\bar{a}_x - x\bar{a}_z\right)$$

Integrating w.r.t. y from y = 0 to y_1, we get total magnetic field strength

$$\bar{H} = \int_{y=0}^{y_1} \frac{Idy}{4\pi\left(r^2 + y^2\right)^{3/2}}\left(z\bar{a}_x - x\bar{a}_z\right)$$

∴

$$= \frac{I\left(z\bar{a}_x - x\bar{a}_z\right)}{4\pi}\int_{y=0}^{y_1} \frac{dy}{\left(r^2 + y^2\right)^{3/2}}$$

$$= \frac{I\left(z\bar{a}_x - x\bar{a}_z\right)}{4\pi}\left[\frac{y/r^2}{\sqrt{r^2+y^2}}\right]_0^{y_1} \qquad \because \int \frac{dx}{\left(a^2+x^2\right)^{3/2}} = \frac{x/a^2}{\sqrt{a^2+x^2}}$$

$$\bar{H} = \frac{I}{4\pi r^2\sqrt{\frac{r^2}{y_1^2}+1}}\left(z\bar{a}_x - x\bar{a}_z\right)$$

For a semi-infinite length conductor, $y_1 = \infty$

$$\therefore \qquad \bar{H} = \frac{I}{4\pi r^2\sqrt{0+1}}\left(z\bar{a}_x - x\bar{a}_z\right) = \frac{I}{4\pi r^2}\left(z\bar{a}_x - x\bar{a}_z\right)$$

Problem 2.4

Find $\bar{H}$ at (–3, 4, 0) due to the current filament shown in Fig.2.7

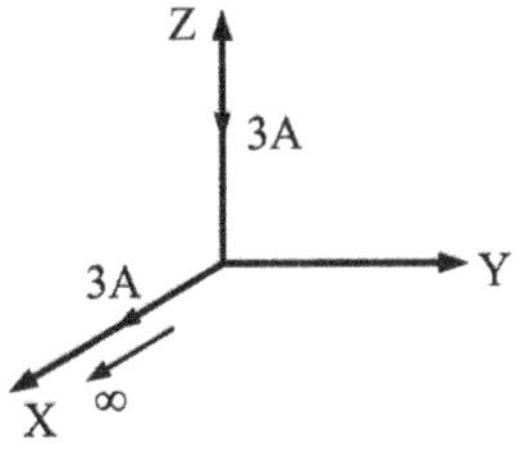

Fig. 2.7

Solution

For the element along X-axis is

$\bar{R} = (-3, 4, 0) - (x, 0, 0)$

$d\bar{l} = dx\,\bar{a}_x$

$$\bar{H}_x = \int_0^{\infty} \frac{3\,dx\,\bar{a}_x \times \left[(-3-x)\bar{a}_x + 4\bar{a}_y\right]}{4\pi\left[(-3-x)^2+16\right]^{3/2}}$$

$$= \frac{3}{4\pi}4\bar{a}_z \int_0^{\infty} \frac{dx}{\left[16+(-3-x)^2\right]^{3/2}}$$

$$= \frac{3}{\pi}\bar{a}_z \int_0^\infty \frac{dx}{\left[16+(3+x)^2\right]^{3/2}}$$

$$= \frac{3\bar{a}_z}{\pi}\left[\frac{(3+x)/16}{\left[16+(3+x)^2\right]^{1/2}}\right]_0^\infty$$

$$= \frac{3\bar{a}_z}{16\pi}\left[1-\left(\frac{3}{5}\right)\right] = 23.88\bar{a}_z \text{ m A/m}$$

For the element along Z-axis is

$\bar{R} = (-3, 4, 0) - (0, 0, z)$

$d\bar{l} = dz\,\bar{a}_z$

$$\bar{H}_z = \int_\infty^0 \frac{3\,dz\,\bar{a}_z \times \left[\left(-3\bar{a}_x + 4\bar{a}_y - Z\,\bar{a}_z\right)\right]}{4\pi\left[9+16+z^2\right]^{3/2}}$$

$$= \frac{3}{4\pi}\int_0^\infty \frac{\left(3\bar{a}_y + 4\bar{a}_x\right)dz}{\left(25+z^2\right)^{3/2}}$$

$$= \frac{3\left(3\bar{a}_y + 4\bar{a}_x\right)}{4\pi}\int_0^\infty \frac{dz}{\left(25+z^2\right)^{3/2}}$$

$$= \frac{3\left(3\bar{a}_y + 4\bar{a}_x\right)}{4\pi}\left[\frac{z/25}{\left(25+z^2\right)^{1/2}}\right]_0^\infty$$

$$= \frac{3\left(3\bar{a}_y + 4\bar{a}_x\right)}{100\pi}(1-0) = 38.2\bar{a}_x + 28.65\bar{a}_y \text{ mA/m}$$

$\bar{H} = \bar{H}_x + \bar{H}_z = 38.2\bar{a}_x + 28.65\bar{a}_y + 23.88\bar{a}_z$ mA/m

Problem 2.5

The +Ve Y-axis (semi infinite line w.r.t origin) carries a filamentary current of 2A in the $-\bar{a}_y$ direction. Find $\bar{H}$ at (a) A(2, 3, 0) (b) B(3, 12, –4).

Solution

(a) $\bar{R} = (2, 3, 0) - (0, y, 0)$

$d\bar{l} = dy\,\bar{a}_y$

Since 2A is along $-\bar{a}_y$, limits are ∞ to 0.

$$\bar{H}_A = \int_{\infty}^{0} \frac{2dy\,\bar{a}_y \times \left[\left(2\bar{a}_x + (3-y)\bar{a}_y\right)\right]}{4\pi\left[4+(3-y)^2\right]^{3/2}}$$

$$\bar{H}_A = \frac{\bar{a}_z}{\pi}\int_0^{\infty} \frac{dy}{\left[4+(3-y)^2\right]^{3/2}}$$

$$= -\frac{\bar{a}_z}{\pi}\left[\frac{(3-y)/4}{\left[4+(3-y)^2\right]^{1/2}}\right]_0^{\infty}$$

$$= -\frac{\bar{a}_z}{4\pi}\left(-1-\frac{3}{\sqrt{13}}\right) = 145.8\bar{a}_z \text{ mA/m}$$

(b) $\bar{R} = (3, 12, -4) - (0, y, 0)$

$$\therefore \quad \bar{H}_B = \int_{\infty}^{0} \frac{2dy\,\bar{a}_y \times \left[3\bar{a}_x + (12-y)\bar{a}_y - 4\bar{a}_z\right]}{4\pi\left[25+(12-y)^2\right]^{3/2}}$$

$$\bar{H}_B = \frac{6\bar{a}_z + 8\bar{a}_x}{4\pi}\int_0^{\infty} \frac{dy}{\left[25+(12-y)^2\right]^{3/2}}$$

$$= -\frac{6\bar{a}_z + 8\bar{a}_x}{4\pi}\left[\frac{(12-y)/25}{\left[25+(12-y)^2\right]^{1/2}}\right]_0^{\infty}$$

$$= -\frac{6\bar{a}_z + 8\bar{a}_x}{100\pi}\left(-1-\frac{12}{13}\right) = 48.97\bar{a}_x + 36.73\bar{a}_z \text{ mA/m}$$

Problem 2.6

Show that the magnetic field intensity $\bar{H}$ at (0, 0, h) due to a circle which lies on XY-plane with radius 'ρ' carries a current I as

$$\bar{H} = \frac{I\rho^2\,\bar{a}_z}{2\left[\rho^2 + h^2\right]^{3/2}}$$

Solution

Consider the circular loop shown in Fig.2.8

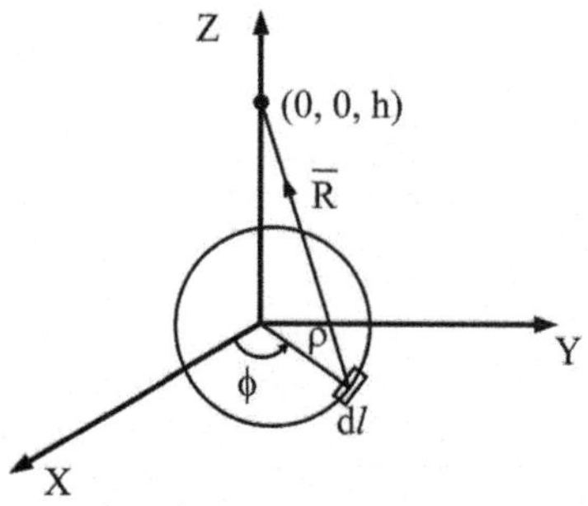

Fig. 2.8

From Fig.2.8 $\bar{R} = -\rho\bar{a}_\rho + h\bar{a}_z$

and $d\bar{l} = \rho d\phi \bar{a}_\phi$

We have $$d\bar{H} = \frac{I d\bar{l} \times \bar{R}}{4\pi R^3}$$

$$d\bar{H} = \frac{I\rho d\phi \bar{a}_\phi \times (-\rho\bar{a}_\rho + h\bar{a}_z)}{4\pi(\rho^2 + h^2)^{3/2}}$$

$$d\bar{H} = \frac{I\rho d\phi(\rho\bar{a}_z + h\bar{a}_\rho)}{4\pi(\rho^2 + h^2)^{3/2}}$$

due to symmetry of circle the components in ρ direction will get cancelled.

$$d\bar{H} = \frac{I\rho^2 d\phi \bar{a}_z}{4\pi(\rho^2 + h^2)^{3/2}}$$

Integrating

$$\bar{H} = \frac{I}{4\pi}\int_0^{2\pi}\left[\frac{\rho^2\bar{a}_z d\phi}{\left(\rho^2 + h^2\right)^{3/2}}\right]$$

$$\bar{H} = \frac{I\rho^2\bar{a}_z}{2\left[\rho^2 + h^2\right]^{3/2}}$$

Problem 2.7

A circular loop located on $x^2 + y^2 = 9$, $z = 0$ carries a direct current of 10 A along $\bar{a}_\phi$ direction. Determine $\bar{H}$ at (0, 0, 4) and (0, 0, –4).

Solution

Here $\rho = 3$, h = 4 and I = 10 A

$$\therefore \qquad \bar{H} = \frac{10}{4\pi}\frac{3^2}{\left(3^2+4^2\right)^{3/2}}\bar{a}_z\int_0^{2\pi} d\phi$$

$$= 0.36\bar{a}_z \text{ A/m}$$

Similarly $\bar{H}_{(0,0,-4)} = \bar{H}_{(0,0,4)} = 0.36\bar{a}_z$ A/m

*Problem 2.8

Find the field at the centre of a circular loop of radius 'a', carrying current I along ϕ direction in Z = 0 plane.

Solution

We have
$$\bar{H} = \frac{I\rho^2\bar{a}_z}{2\left[\rho^2+h^2\right]^{3/2}}$$

Here $\rho = a$, and h = 0

$$\bar{H} = \int_{\phi=0}^{2\pi}\frac{I a^2 d\phi\bar{a}_z}{4\pi\left[a^2+0\right]^{3/2}} = \frac{I\bar{a}_z}{2a}\text{ A/m}$$

Problem 2.9

A thin ring of radius 5 cm is placed on plane z = 1 cm so that its center is at (0, 0, 1) cm. If the ring carries 50 mA along $\bar{a}_\phi$. Find $\bar{H}$ at

(a) (0, 0, –1) cm

(b) (0, 0, 10) cm

Solution

(a) Here ρ = 5 cm, h = 2 cm and I = 50 mA

We have $\bar{H} = \dfrac{I\rho^2 \bar{a}_z}{2\left[\rho^2 + h^2\right]^{3/2}}$

$$\bar{H} = \frac{50\times10^{-3}(5\times10^{-2})^2 \bar{a}_z}{2\left[(5\times10^{-2})^2 + (2\times10^{-2})^2\right]^{3/2}}$$

$$\bar{H} = \frac{125\,\bar{a}_z}{2[29]^{3/2}}$$

$\bar{H} = 400\bar{a}_z$ mA/m

(b) Here ρ = 5 cm, h = 9 cm and I = 50 mA

We have $\bar{H} = \dfrac{I\rho^2 \bar{a}_z}{2\left[\rho^2 + h^2\right]^{3/2}}$

$$\bar{H} = \frac{50\times10^{-3}(5\times10^{-2})^2 \bar{a}_z}{2\left[(5\times10^{-2})^2 + (9\times10^{-2})^2\right]^{3/2}}$$

$$\bar{H} = \frac{125\,\bar{a}_z}{2[106]^{3/2}}$$

$\bar{H} = 57.3\bar{a}_z$ mA/m

Problem 2.10

A square conducting loop 2a m on each side carries a current of I amp. Calculate the magnetic field intensity at the center of the loop.

Solution

Consider a square loop with each side 2a m as shown in the Fig.2.9.

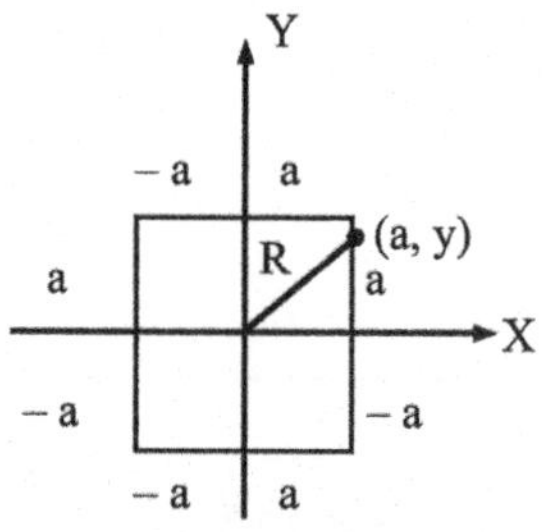

Fig. 2.9 A square loop

We need to find magnetic field intensity at (0, 0) due to elemental current flowing in one side of square loop at (a, y)

According to Biot Savart's law, we have

$$d\bar{H} = \frac{Id\bar{l} \times \bar{R}}{4\pi R^3}$$

From Fig.2.9, $d\bar{l} = dy\bar{a}_y$

$$\bar{R} = (0,0) - (a,y) = -a\bar{a}_x - y\bar{a}_y$$

$$\therefore \quad d\bar{H}_{oneside} = \frac{Idy\bar{a}_y \times (-a\bar{a}_x - y\bar{a}_y)}{4\pi(a^2 + y^2)^{3/2}}$$

$$\therefore \quad d\bar{H}_{oneside} = \frac{Iady\bar{a}_z}{4\pi(a^2 + y^2)^{3/2}}$$

$$\bar{H}_{oneside} = \int_{-a}^{a} \frac{Iady\bar{a}_z}{4\pi(a^2 + y^2)^{3/2}}$$

$$\bar{H}_{oneside} = \frac{Ia}{4\pi}\bar{a}_z \int_{-a}^{a} \frac{dy}{(a^2 + y^2)^{3/2}}$$

$$\bar{H}_{oneside} = \frac{Ia}{4\pi}\bar{a}_z \left[\frac{y/a^2}{(a^2 + y^2)^{1/2}}\right]_{-a}^{a}$$

$$\bar{H}_{oneside} = \frac{Ia}{4\pi a^2}\bar{a}_z \left[\frac{a}{(a^2 + a^2)^{1/2}} + \frac{a}{(a^2 + a^2)^{1/2}}\right]$$

$$\bar{H}_{oneside} = \frac{I}{4\pi a}\bar{a}_z \left[\frac{1}{\sqrt{2}} + \frac{1}{\sqrt{2}}\right]$$

$$\bar{H}_{oneside} = \frac{I\sqrt{2}}{4\pi a}\bar{a}_z$$

Magnetic field intensity at (0, 0) due to four sides is

$$\bar{H} = 4\bar{H}_{oneside} = \frac{4I\sqrt{2}}{4\pi a}\bar{a}_z$$

$$\bar{H} = \frac{I\sqrt{2}}{\pi a}\bar{a}_z$$

Problem 2.11

A square conducting loop 3 cm on each side carries a current of 10 A. Calculate the magnetic field intensity at the center of the loop.

Solution

We have $$\bar{H} = \frac{I\sqrt{2}}{\pi a}\bar{a}_z$$

Here a = 1.5 × 10^{-2} m and I = 10 A

$$\therefore \quad \bar{H} = \frac{10\sqrt{2}}{\pi \times 1.5 \times 10^{-2}}\bar{a}_z = 300.105\bar{a}_z \text{ A/m}$$

2.3 Ampere's Circuit Law or Ampere's Work Law

Ampere's circuit law states that the line integral of $\bar{H}$ around a closed path is the same as the net current enclosed (I_{enc}) by the path.

i.e., the circulation of $\bar{H}$ equals I_{enc}

$$\therefore \quad \oint_L \bar{H} \cdot d\bar{l} = I_{enc} \qquad \text{.....(2.3.1a)}$$

Where $\bar{H}$ = magnetic field intensity

$d\bar{l}$ = elemental length in a closed path and

I_{enc} = current enclosed by closed path.

according to Stoke's theorem

$$\oint_L \bar{H} \cdot d\bar{l} = \int_S \nabla \times \bar{H} \cdot d\bar{s} \qquad \text{.....(2.3.1b)}$$

If we know current density $\bar{J}$ of conductor. We can have current

$$I = \int_S \bar{J} \cdot d\bar{s} \qquad \text{......(2.3.2)}$$

Equating Equations (2.31b) and (2.3.2) we get

$$\nabla \times \bar{H} = \bar{J}$$

which is Maxwell's 3rd equation.

2.3.1 Applications of Ampere's Circuit Law- Infinite Line Conductor

Let us consider an infinite line conductor that carries current I and which lies along Z-axis as shown in Fig. 2.10. Since conductor has cylindrical symmetry. We use cylindrical coordinate system to find $\bar{H}$ at any point 'p'.

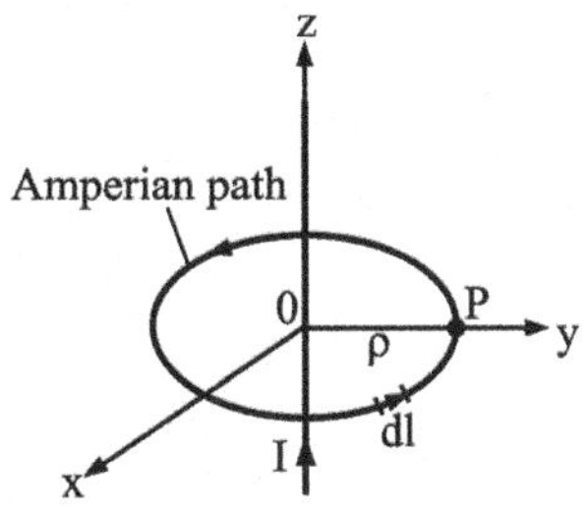

Fig. 2.10 Infinite line conductor

∴ choose a closed path which encloses the conductor.

∵ current carrying conductor has symmetry, we use Ampere's circuit law to find $\bar{H}$.

According to Ampere's circuit law $\oint_L \bar{H} \cdot d\bar{l} = I_{enc}$.

Here $I_{enc} = I$

By applying Right hand thumb rule, we can say that the $\bar{H}$ will have component only along φ axis.

∴ $\bar{H} = H_\phi \bar{a}_\phi$,

$d\bar{l}$ can be written as $d\bar{l} = \rho\, d\phi\ \bar{a}_\phi$

and φ ranges from 0 to 2π.

∴ The above integral is $I = \int_0^{2\pi} H_\phi \bar{a}_\phi . \rho\, d\phi\, \bar{a}_\phi$

$$= H_\phi\, \rho\, 2\pi$$

$$H_\phi = \frac{I}{2\pi\rho}$$

$$\bar{H} = \frac{I}{2\pi\rho}\bar{a}_\phi \qquad .....(2.3.3)$$

2.3.2 Applications of Ampere's Circuit Law- Infinite Sheet

Consider an infinite sheet which lies on XY plane carrying current density $\bar{K}$ along Y-direction as shown in Fig. 2.11.

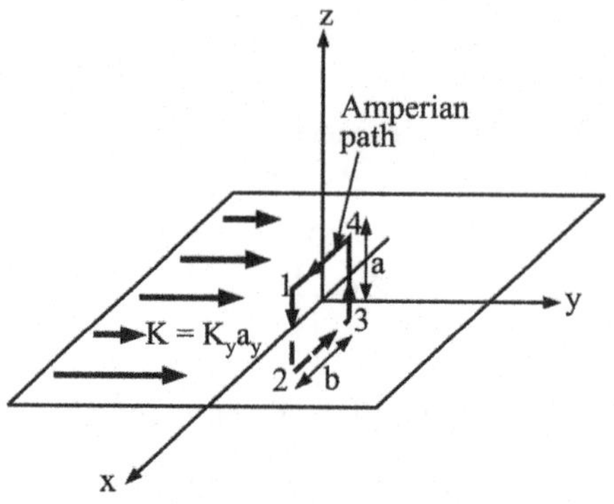

Fig. 2.11 Infinite sheet

To find magnetic field intensity at any point above or below an infinite sheet, consider a small rectangular path which lies above and below the sheet with width 'b' and height 'a'.

Current density $\bar{K}$ is defined as current carried by sheet per a meter width.

$\therefore$ If a sheet of width 'b' is carrying uniform current density $\bar{K}$, then the current enclosed by Amperian path is $I_{enc} = \bar{K}b$

As $\bar{K}$ is in Y-direction $\bar{K} = K_y \bar{a}_y$.

$\therefore \; I_{enc} = K_y b$

According to Ampere's circuit law the line integral $\bar{H}$ around a chosen path 1-2-3-4-1 is equal to the current enclosed by that path.

$$\therefore \qquad \oint_L \bar{H} \cdot d\bar{l} = I_{enc} = K_y b$$

By applying Right hand thumb rule, we can say that

$\bar{H}$ will have it's component only in the X-direction. It will not have components either along Y-axis or Z-axis.

$\therefore \; \bar{H} = H_x \bar{a}_x \quad z > 0$ (above the sheet)

and

$\bar{H} = -H_x \bar{a}_x \quad z < 0$ (below the sheet)

Now $$\oint_L \bar{H}\cdot d\bar{l} = \int_1^2 \bar{H}\cdot d\bar{l} + \int_2^3 \bar{H}\cdot d\bar{l} + \int_3^4 \bar{H}\cdot d\bar{l} + \int_4^1 \bar{H}\cdot d\bar{l}$$

$$= 0(-a) + \int_0^b (-H_x\bar{a}_x)\cdot(-dx\,\bar{a}_x) + 0(a) + \int_0^b H_x\bar{a}_x\cdot dx\,\bar{a}_x$$

$$= H_x b + H_x b$$

$$\oint_L \bar{H}\cdot d\bar{l} = 2H_x b$$

$$\therefore\ 2H_x b = K_y b$$

$$\Rightarrow \qquad H_x = \frac{1}{2}K_y$$

$$\therefore \qquad \bar{H} = \frac{1}{2}K_y\bar{a}_x \quad z>0 \qquad \because \frac{1}{2}K_y\bar{a}_y\times\bar{a}_z = \frac{1}{2}K_y\bar{a}_x$$

$$= -\frac{1}{2}K_y\bar{a}_x \quad z<0$$

$\bar{H}$ can be generalized as

i.e., $$\bar{H} = \frac{1}{2}\bar{K}\times\bar{a}_n \qquad .....(2.3.4)$$

where $\bar{a}_n$ is the unit vector normal to the surface.

2.3.3 Applications of Ampere's Circuit Law- Infinitely Long Co-axial Cable

Consider an infinitely long co-axial transmission line which is along Z-axis with an internal conductor of radius 'a' and external conductor of internal radius 'b' and width 't' and assume internal conductor is carrying current I which is along +ve Z-axis and external conductor is carrying current in -ve Z-axis i.e., –I. The cross sectional view of coaxial transmission line is as shown in Fig.2.12. We can find $\bar{H}$ by considering 4 cases.

(i) $0\le\rho\le a$

(ii) $a\le\rho\le b$

(iii) $b\le\rho\le b+t$

(iv) $\rho\ge b+t$

(i) $0 \le \rho \le a$

choose a closed path L_1 as shown in Fig.2.12.

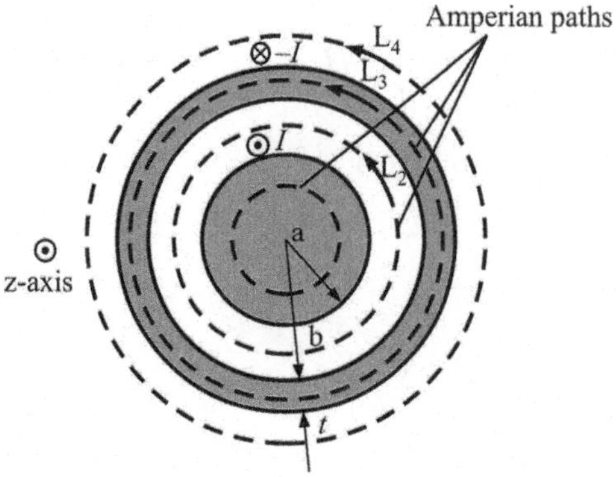

Fig. 2.12 Cross section of transmission line

According to Ampere's circuit law

$$\oint_L \bar{H}.d\bar{l} = I_{enc} = \int_S \bar{J} \cdot d\bar{s}$$

$$\oint_L \bar{H} \cdot d\bar{l} = \int_0^{2\pi} H_\phi \bar{a}_\phi \cdot \rho d\phi \bar{a}_\phi = H_\phi 2\pi\rho$$

$\bar{J}$ of the internal conductor is $\frac{I}{\pi a^2}\bar{a}_z$ and $d\bar{s} = \rho d\phi d\rho \bar{a}_z$

$$\therefore \qquad I_{enc} = \int_s \bar{J} \cdot d\bar{s}$$

$$= \iint \frac{I}{\pi a^2}\bar{a}_z.\rho d\phi d\rho \bar{a}_z$$

$$= \int_0^{2\pi}\int_0^{\rho} \frac{I}{\pi a^2}\rho d\phi d\rho$$

$$= \frac{I}{\pi a^2}\int_0^{\rho} \rho(2\pi) d\rho = \frac{2\pi I}{\pi a^2}\left(\frac{\rho^2}{2}\right) = \frac{I\rho^2}{a^2}$$

$$\therefore \qquad 2\pi H_\phi \rho = \frac{I\rho^2}{a^2}$$

$$H_\phi = \frac{I\rho}{a^2 2\pi} \Rightarrow \bar{H} = \frac{I\rho}{2\pi a^2}\bar{a}_\phi \qquad .....(2.3.5)$$

(ii) $a \le \rho \le b$

Choose closed path L_2 as shown in Fig.2.12.

Then according to Ampere's circuit law

$$\oint_L \bar{H} \cdot d\bar{l} = I_{enc}$$

Here $I_{enc} = I$

$$\therefore \quad \int_0^{2\pi} H_\phi \bar{a}_\phi . \rho d\phi \bar{a}_\phi = I$$

$$\Rightarrow \quad 2\pi H_\phi \rho = I$$

$$H_\phi = \frac{I}{2\pi\rho}$$

$$\bar{H} = \frac{I}{2\pi \rho} \bar{a}_\phi \qquad \text{.....(2.3.6)}$$

(iii) $b \le \rho \le b + t$

Choose closed path L_3 as shown in Fig.2.12.

According to Ampere's circuit law

$$\oint_L \bar{H} \cdot d\bar{l} = I_{enc}$$

$$\oint_L \bar{H} \cdot d\bar{l} = \int_0^{2\pi} H_\phi \bar{a}_\phi \cdot \rho d\phi \bar{a}_\phi = H_\phi 2\pi \rho$$

$$I_{enc} = I + \int_s \bar{J} . d\bar{s}$$

Here current density is

$$\bar{J} = \frac{-I}{\pi\left[(b+t)^2 - b^2\right]} \bar{a}_z \qquad \text{where} \quad d\bar{s} = \rho d\phi d\rho \bar{a}_z$$

$$\therefore \quad \int_s \bar{J} \cdot d\bar{s} = \iint \frac{-I}{\pi\left[(b+t)^2 - b^2\right]} \bar{a}_z \cdot \rho d\phi d\rho \bar{a}_z$$

$$= \frac{-I}{\pi\left[(b+t)^2 - b^2\right]} \int_0^{2\pi} \int_b^{\rho} \rho\, d\phi\, d\rho$$

$$= \frac{-I}{\pi\left[(b+t)^2 - b^2\right]} 2\pi \left[\frac{\rho^2}{2}\right]_b^{\rho}$$

$$= \frac{-2I}{2\left[(b+t)^2 - b^2\right]}\left[\rho^2 - b^2\right] = \frac{-I\left[\rho^2 - b^2\right]}{\left[(b+t)^2 - b^2\right]}$$

$$= \frac{-I\left[\rho^2 - b^2\right]}{t^2 + 2bt}$$

$$I_{enc} = I - \frac{I\left[\rho^2 - b^2\right]}{t^2 + 2bt}$$

$$\therefore \quad H_\phi 2\pi\rho = I\left[1 - \frac{\left[\rho^2 - b^2\right]}{t^2 + 2bt}\right]$$

$$H_\phi = \frac{I}{2\pi\rho}\left[1 - \frac{\left[\rho^2 - b^2\right]}{t^2 + 2bt}\right]$$

$$\bar{H} = \frac{I}{2\pi\rho}\left[1 - \frac{\left[\rho^2 - b^2\right]}{t^2 + 2bt}\right]\bar{a}_\phi \qquad \text{.....(2.3.7)}$$

(iv) $\rho \geq b + t$

Choose closed path L_4 as shown in Fig.2.12

According to Ampere's circuit law

$$\oint_L \bar{H} \cdot d\bar{l} = I_{enc}$$

Here $I_{enc} = I - I = 0$

$$\therefore \quad H_\phi 2\pi\rho = 0$$

$$H_\phi = 0$$

Problem 2.12

Planes z = 0 and z = 4 carry current $\overline{K} = -10\overline{a}_x$ A/m and $\overline{K} = 10\overline{a}_x$ A/m respectively. Determine $\overline{H}$ at

(a) (1, 1, 1)

(b) (0, –3, 10)

Solution

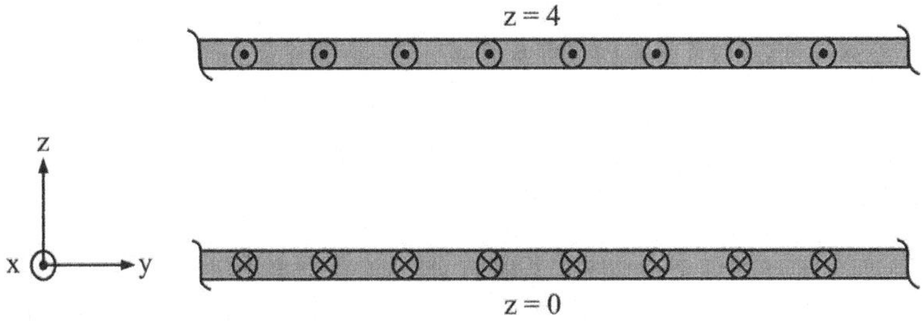

Fig. 2.13

(a) We have $\overline{H} = \frac{1}{2}\overline{K} \times \overline{a}_n$

here $\overline{K} = -10\overline{a}_x$ A/m on Z = 0 plane and Since point (1, 1, 1) is lying above the plane Z = 0, $\overline{a}_n = \overline{a}_z$

$$\therefore \quad H_0 = \frac{1}{2}(-10)\overline{a}_x \times \overline{a}_z$$

$$= -5(-\overline{a}_y)$$

$$= 5\overline{a}_y$$

here $\overline{K} = 10\overline{a}_x A/m$ on Z = 4 plane and Since point (1, 1, 1) is lying below the plane Z = 4, $\overline{a}_n = -\overline{a}_z$

$$\therefore \quad H_4 = \frac{1}{2}10\overline{a}_x \times (-\overline{a}_z)$$

$$= -5(-\overline{a}_y)$$

$$= 5\overline{a}_y$$

$$\overline{H} = \overline{H}_0 + \overline{H}_4 = 10\overline{a}_y \text{ A/m}$$

(b) We have $\bar{H} = \frac{1}{2}\bar{K} \times \bar{a}_n$

here $\bar{K} = -10\bar{a}_x$ A/m on Z = 0 plane and Since point (0, –3, 10) is lying above the plane Z = 0, $\bar{a}_n = \bar{a}_z$

$$\therefore \quad \bar{H}_0 = \frac{1}{2}(-10\bar{a}_x) \times \bar{a}_z$$

$$= -5(-\bar{a}_y)$$

$$= 5\bar{a}_y$$

here $\bar{K} = 10\bar{a}_x$ A/m on Z = 4 plane and Since point (0, –3, 10) is lying above the plane Z = 4, $\bar{a}_n = \bar{a}_z$

$$\therefore \quad \bar{H}_4 = \frac{1}{2}10\bar{a}_x \times \bar{a}_z$$

$$= 5(-\bar{a}_y)$$

$$= -5\bar{a}_y$$

$$\bar{H} = \bar{H}_0 + \bar{H}_4 = 5\bar{a}_y - 5\bar{a}_y = 0$$

Problem 2.13

Plane Y = 1 carries current $\bar{K} = 50\bar{a}_z$ mA/m. Find $\bar{H}$ at (a) (0, 0, 0) (b) (1, 5, –3)

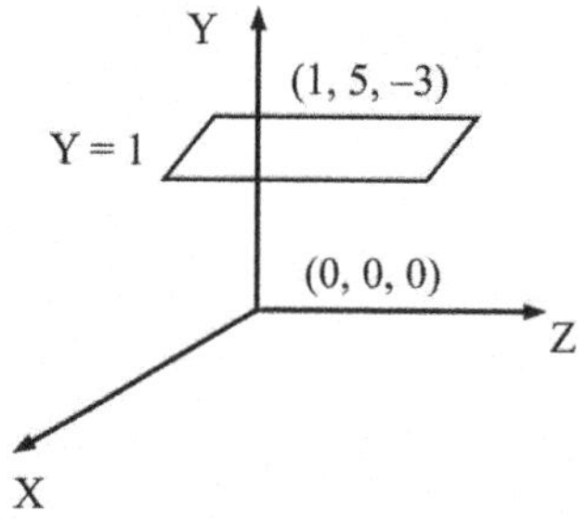

Fig. 2.14

Solution

(a) Here $\bar{K} = 50\bar{a}_z$ mA/m on Y = 1 plane and Since point (0, 0, 0) is lying below the plane Y = 1, $\bar{a}_n = -\bar{a}_y$

$$\therefore \qquad \bar{H} = \frac{1}{2} 50\bar{a}_z \times (-\bar{a}_y)$$

$$= -25(-\bar{a}_x)$$

$$= 25\bar{a}_x \text{ mA/m}$$

(b) here $\bar{K} = 50\bar{a}_z$ mA/m on Y = 1 plane and Since point (1, 5, –3) is lying above the plane Y = 1, $\bar{a}_n = \bar{a}_y$

$$H = \frac{1}{2} 50\bar{a}_z \times \bar{a}_y$$

$$= 25(-\bar{a}_x)$$

$$= -25\bar{a}_x \text{ mA/m}$$

***Problem 2.14**

A long coaxial cable has an inner conductor carrying a current of 1 mA along +ve Z direction, its axis coinciding with Z-axis. Its inner conductor diameter is 6 mm. If its outer conductor has an inside diameter of 12 mm and thickness of 2 mm, determine $\bar{H}$ at (0, 0, 0), (0, 4.5 mm, 0) and (0, 1 cm, 0). (No derivations).

Solution

As per the derivations in the section "Application's of Ampere's Circuit Law-Infinitely long co-axial transmission line"

Given I = 1 mA, a = 3 mm, b = 6 mm and t = 2 mm.

Let the given points be $P_1(0, 0, 0)$, $P_2(0, 4.5 \text{ mm}, 0)$ and $P_3(0, 1 \text{ cm}, 0)$.

Given points are in rectangular coordinate system

$$\therefore \qquad \rho = \sqrt{x^2 + y^2}$$

For P_1, $\rho = 0$, i.e., $\rho < a$, Hence case(i) formula from the section **"Infinitely long co-axial transmission line"**

$$\bar{H} = \frac{I\rho}{2\pi a^2}\bar{a}_\phi = 0 \text{ A/m}$$

For P_2, $\rho = 4.5$ mm, i.e., $a < \rho < b$, Hence case(ii) formula from the section **"Infinitely long co-axial transmission line"**

$$\bar{H} = \frac{I}{2\pi\rho}\bar{a}_\phi = \frac{1\times10^{-3}}{2\pi\times4.5\times10^{-3}}\bar{a}_\phi = 0.03537\bar{a}_\phi \text{ A/m}$$

For P_3, $\rho = 1$ cm = 10 mm, i.e., $\rho > b + t$, Hence case (iv) formula from the section **"Infinitely long co-axial transmission line"**

$$\bar{H} = 0 \text{ A/m}$$

Problem 2.15

A solenoid of length 'l' and radius 'a' consists of 'N' turns of wire carrying current 'I'. Show that at point p along it's axis $\bar{H} = \frac{nI}{2}(\cos\theta_2 - \cos\theta_1)\bar{a}_z$ where $n = \frac{N}{l}$, θ_1, θ_2 are the angles subtended at 'p' by the end turns as illustrated in Fig:2.15. Also Show that if $l >> a$ at the center of the solenoid $\bar{H} = nI\,\bar{a}_z$

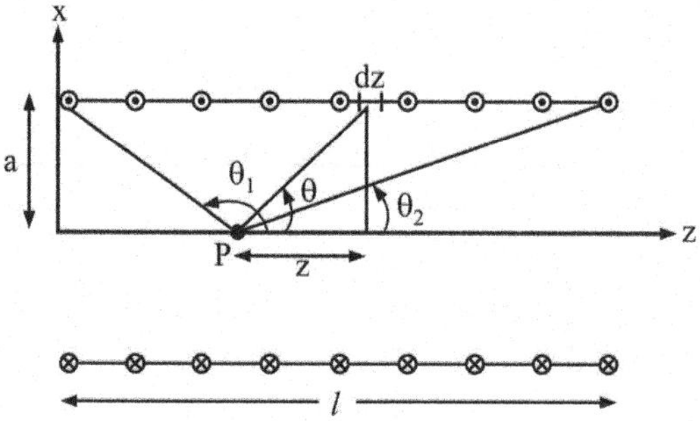

Fig. 2.15 Cross section of a solenoid

Solution

We know the elemental magnetic field intensity dH_z due to one turn (circle) at point p is

$$dH_z = \frac{I\,a^2 d\phi}{4\pi\left(a^2 + z^2\right)^{3/2}}\bar{a}_z$$

$H_z = \frac{I\,a^2}{2\left(a^2 + z^2\right)^{3/2}}\bar{a}_z$ which is the magnetic field intensity at point 'p' due to one turn.

As the solenoid contains 'N' number of turns by considering the elemental length dl, the elemental magnetic field intensity due to a solenoid of length 'L' and having 'N' number of turns at point 'P' is

$$dH_z = \frac{I\,a^2 dl\,\bar{a}_z}{2\left(a^2 + z^2\right)^{3/2}}$$

where dl = ndz = $\frac{N}{l}dz$

from Fig.2.15 $\tan\theta = \frac{a}{z}$

$z = a\cot\theta$

$dz = -a\,\mathrm{cosec}^2\theta\, d\theta$

$$dH_z = \frac{I\,a^2 n\left(-a\,\mathrm{cosec}^2\theta\, d\theta\right)}{2\left(a^2 + a^2\cot^2\theta\right)^{3/2}}\bar{a}_z$$

$$= \frac{I\,a^2 n\left(-a\,\mathrm{cosec}^2\theta\, d\theta\right)}{2\,a^3\,\mathrm{cosec}^3\theta}\bar{a}_z$$

$$= -\frac{nI}{2}\sin\theta\, d\theta\,\bar{a}_z$$

$$H_z = -\frac{nI}{2}\int_{\theta_2}^{\theta_1}\sin\theta\, d\theta\,\bar{a}_z$$

$$= -\frac{nI}{2}(\cos\theta_1 - \cos\theta_2)\bar{a}_z$$

$$= \frac{nI}{2}(\cos\theta_2 - \cos\theta_1)\bar{a}_z$$

At the center of the solenoid we can write

$$\cos\theta_2 = \frac{l/2}{\sqrt{a^2 + (l/2)^2}} = -\cos\theta_1$$

as $l >> a$

$$H_z = \frac{nI}{2}2\cos\theta_2\,\bar{a}_z$$

$$= nI\frac{l/2}{\sqrt{a^2 + (l/2)^2}}\bar{a}_z$$

$$= nI\frac{l/2}{l/2}\bar{a}_z = nI\,\bar{a}_z$$

Problem 2.16

A Toroid whose dimensions are shown in Fig.2.16 has 'N' turns and carries current I. Determine $\overline{H}$ inside and outside the Toroid.

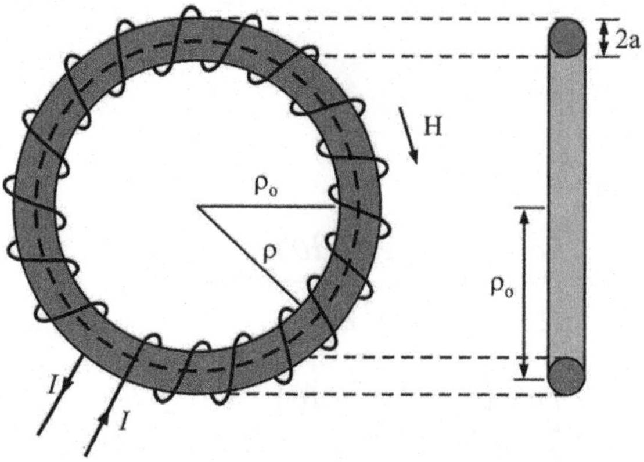

Fig. 2.16 A toroid with a circular cross section

Solution

Inside the toroid consider the closed path (i.e., dotted circle), According to Ampere's circuit law

$$\oint_L \bar{H} \cdot d\bar{l} = I_{enc}$$

$$\Rightarrow H_\phi 2\pi\rho = nI$$

$$H_\phi = \frac{nI}{2\pi\rho}$$

$$\bar{H} = \frac{nI}{2\pi\rho}\bar{a}_\phi$$

outside the Toroid $\oint_L \bar{H} \cdot d\bar{l} = I_{enc}$

$$= \text{nI} - \text{nI}$$

$$\Rightarrow \quad H_\phi = 0 \quad \Rightarrow \quad \bar{H} = 0;$$

***Note*:** By bending a solenoid in to a form of circle we get a toroid.

***Problem 2.17**

A long straight conductor with radius 'a' has a magnetic field strength $\bar{H} = \frac{Ir}{2\pi a^2}\bar{a}_\phi$ within the conductor (r < a) and $\bar{H} = \frac{I}{2\pi r}\bar{a}_\phi$ outside the conductor (r > a). Find the current density $\bar{J}$ in both the regions (r < a and r > a).

Solution

we have $\bar{J} = \nabla \times \bar{H}$

Given $\bar{H} = \dfrac{Ir}{2\pi a^2}\bar{a}_\phi$ within the conductor (r < a)

which has cylindrical symmetry, here ρ = r,

$$\bar{J} = \nabla \times \bar{H} = \frac{1}{r}\begin{vmatrix} \bar{a}_r & r\bar{a}_\phi & \bar{a}_z \\ \dfrac{\partial}{\partial r} & \dfrac{\partial}{\partial \phi} & \dfrac{\partial}{\partial z} \\ H_r & rH_\phi & H_z \end{vmatrix}$$

$$= \frac{1}{r}\begin{vmatrix} \bar{a}_r & r\bar{a}_\phi & \bar{a}_z \\ \dfrac{\partial}{\partial r} & \dfrac{\partial}{\partial \phi} & \dfrac{\partial}{\partial z} \\ 0 & r\dfrac{Ir}{2\pi a^2} & 0 \end{vmatrix} = \frac{I}{\pi a^2}\bar{a}_z \ \text{A/m}^2$$

And also given $\bar{H} = \dfrac{I}{2\pi r}\bar{a}_\phi$ outside the conductor (r > a)

$$\bar{J} = \frac{1}{r}\begin{vmatrix} \bar{a}_r & r\bar{a}_\phi & \bar{a}_z \\ \dfrac{\partial}{\partial r} & \dfrac{\partial}{\partial \phi} & \dfrac{\partial}{\partial z} \\ 0 & r\dfrac{I}{2\pi r} & 0 \end{vmatrix} = 0 \ \text{A/m}^2$$

2.4 Magnetic Flux Density

Similar to electric flux density $\bar{D}$ and electric field intensity $\bar{E}$ in electrostatic fields, magnetic flux density $\bar{B}$ and magnetic field intensity $\bar{H}$ in magneto static fields are related as $\bar{B} = \mu_0\bar{H}$ where

μ_0 = permeability of free space

= $4\pi \times 10^{-7}$ Henry/meter

Magnetic flux density is defined as flux flowing per unit area, whose unit is $\mathrm{Wb/m^2}$ *or Tesla*.

2.4.1 Magnetic Flux Line

It is a line indicated by the needle of magnetic compass by itself when it is placed in magnetic field. In magnetic fields we cannot have isolated poles (charges), because if we have a magnetic bar which is having north and south poles as shown in Fig. 2.17. If we make the bar in to pieces each individual bar will have 'N' and 'S' poles as shown in Fig. 2.17.

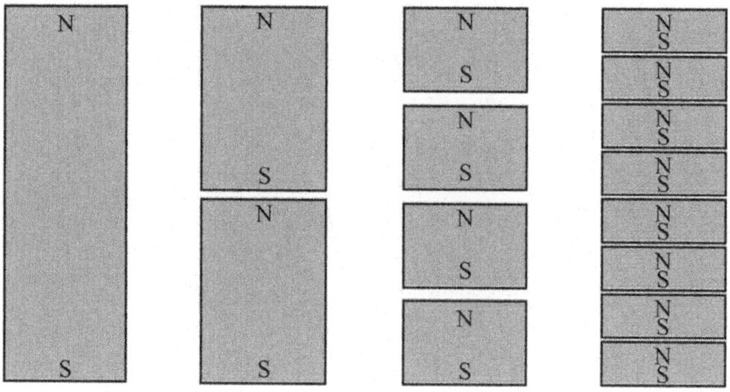

Fig. 2.17 Division of a bar magnet results in pieces with north and south poles

The flux (ψ) enclosed by a closed surface can be written as

$$\psi = \oint_S \bar{B} \cdot \overline{ds} \quad \text{....(2.4.1)}$$

As the magnetic fields are rotational fields, the total flux enclosed by closed surface = 0, i.e.,

$$\oint_S \bar{B} \cdot \overline{ds} = 0 \quad \text{.....(2.4.2)}$$

Applying divergence theorem, we get

$$\int_V \nabla \cdot \bar{B}\, dV = 0 \text{ or } \nabla \cdot \bar{B} = 0 \quad \text{.....(2.4.3)}$$

which is Maxwell's 4th equation.

Note: Maxwell's equations for static electromagnetic fields are

1. $\nabla \cdot \bar{D} = \rho_V$
2. $\nabla \times \bar{E} = 0$

3. $\nabla \times \bar{H} = \bar{J}$

4. $\nabla \cdot \bar{B} = 0$

***Problem 2.18**

A conducting plane at y = 1 carries a surface current of $10\,\bar{a}_z$ mA/m. Find H and B at (a) (0, 0, 0) and (b) (2, 2, 2).

Solution

(a) We have $\bar{H} = \frac{1}{2}\bar{K} \times \bar{a}_n$

here $\bar{K} = 10\bar{a}_z$ mA/m .

Since the point (0, 0, 0) is lying below the plane Y = 1, $\bar{a}_n = -\bar{a}_y$

$$\therefore \quad \bar{H} = \frac{1}{2}10\bar{a}_z \times (-\bar{a}_y)$$

$$= -5(-\bar{a}_x)$$

$$= 5\,\bar{a}_x \text{ mA/m}$$

$$\bar{B} = \mu_0\bar{H} = 4\pi \times 10^{-7} \times 5\bar{a}_x = 62.83 \times 10^{-10}\,\bar{a}_x \; T$$

(b) We have $\bar{H} = \frac{1}{2}\bar{K} \times \bar{a}_n$

here $\bar{K} = 10\bar{a}_z$ mA/m

Since the point (2, 2, 2) is lying above the plane Y = 1, $\bar{a}_n = \bar{a}_y$

$$\therefore \quad \bar{H} = \frac{1}{2}10\bar{a}_z \times \bar{a}_y$$

$$= 5(-\bar{a}_x)$$

$$= -5\,\bar{a}_x \text{ mA/m}$$

$$\bar{B} = \mu_0\bar{H} = 4\pi \times 10^{-7} \times -5\bar{a}_x = -62.83 \times 10^{-10}\,\bar{a}_x \text{T}$$

***Problem 2.19**

An infinitely long straight conducting rod of radius ‘a’ carries a current of I in +ve Z-direction. Using Ampere’s Circuital Law, find $\bar{H}$ in all regions and sketch the variation of H as a function of radial distance. If I = 3 mA and a = 2 cm, find $\bar{H}$ and $\bar{B}$ at (0, 1 cm, 0) and (0, 4 cm, 0).

Solution

Consider cylindrical co-ordinate system

Case(i): inside the conductor (ρ < a)

According to Ampere's circuit law

$$\oint_L \bar{H}\cdot d\bar{l} = I_{enc} = \int_s \bar{J}\cdot d\bar{s}$$

$$\oint_L \bar{H}\cdot d\bar{l} = \int_0^{2\pi} H_\phi \bar{a}_\phi \cdot \rho d\phi \bar{a}_\phi = H_\phi 2\pi\rho$$

$\bar{J}$ of the internal conductor is $\frac{I}{\pi a^2}\bar{a}_z$ and $d\bar{s} = \rho d\phi d\rho \bar{a}_z$

$$\therefore \qquad I_{enc} = \int_s \bar{J}\cdot d\bar{s}$$

$$= \iint \frac{I}{\pi a^2}\bar{a}_z . \rho d\phi d\rho \bar{a}_z$$

$$= \int_0^{2\pi}\int_0^{\rho} \frac{I}{\pi a^2}\rho d\phi d\rho$$

$$= \frac{I}{\pi a^2}\int_0^{\rho} \rho(2\pi) d\rho = \frac{2\pi I}{\pi a^2}\left(\frac{\rho^2}{2}\right) = \frac{I\rho^2}{a^2}$$

$$\therefore \qquad 2\pi H_\phi \rho = \frac{I\rho^2}{a^2}$$

$$H_\phi = \frac{I\rho}{a^2 2\pi} \Rightarrow \bar{H} = \frac{I\rho}{2\pi a^2}\bar{a}_\phi$$

Case(ii): outside the conductor(ρ > a)

According to Ampere's circuit law

$$\oint_L \bar{H}\cdot d\bar{l} = I_{enc}$$

$$\Rightarrow \qquad \int_0^{2\pi} H_\phi \bar{a}_\phi . \rho\, d\phi\, \bar{a}_\phi = I_{enc}$$

$$\Rightarrow 2\pi H_\phi \rho = I$$

$$H_\phi = \frac{I}{2\pi\rho}$$

$$\bar{H} = \frac{I}{2\pi\,\rho}\bar{a}_\phi$$

Given points let be $P_1(0, 1 \text{ cm}, 0)$ and $P_2(0, 4 \text{ cm}, 0)$

For P_1 radial distance $\rho = \sqrt{0^2 + 1^2 + 0^2} = 1$ cm

Also given a = 2 cm and I = 3 mA i.e., $\rho < a$ (inside the conductor)

$$\therefore \qquad \bar{H} = \frac{I\rho}{2\pi\,a^2}\bar{a}_\phi = \frac{3\times10^{-3}\times1\times10^{-2}}{2\pi(2\times10^{-2})^2}\bar{a}_\phi = 0.0119\bar{a}_\phi \text{ A/m and } \bar{B} = \mu_0\bar{H}$$

$$\therefore \qquad \bar{B} = 4\pi\times10^{-7}\times0.0119\bar{a}_\phi = 0.15\times10^{-7}\,\bar{a}_\phi \text{ T}$$

For P_2 radial distance $\rho = \sqrt{0^2 + 4^2 + 0^2} = 4$ cm

Here $\rho > a$ (outside the conductor)

$$\therefore \qquad \bar{H} = \frac{I}{2\pi\,\rho}\bar{a}_\phi = \frac{3\times10^{-3}}{2\pi\times4\times10^{-2}}\bar{a}_\phi = 0.0119\bar{a}_\phi \text{ A/m and } \bar{B} = \mu_0\bar{H}$$

$$\therefore \qquad \bar{B} = 4\pi\times10^{-7}\times0.0119\bar{a}_\phi = 0.15\times10^{-7}\,\bar{a}_\phi \text{ T}$$

Sketch of H_ϕ:

We have $\qquad H_\phi = \dfrac{I\rho}{2\pi\,a^2}$ for $\rho < a$ i.e., $\quad H_\phi \propto \rho$

and $\qquad H_\phi = \dfrac{I}{2\pi\,\rho}$ for $\rho > a$

i.e., $\qquad H_\phi \propto 1/\rho$

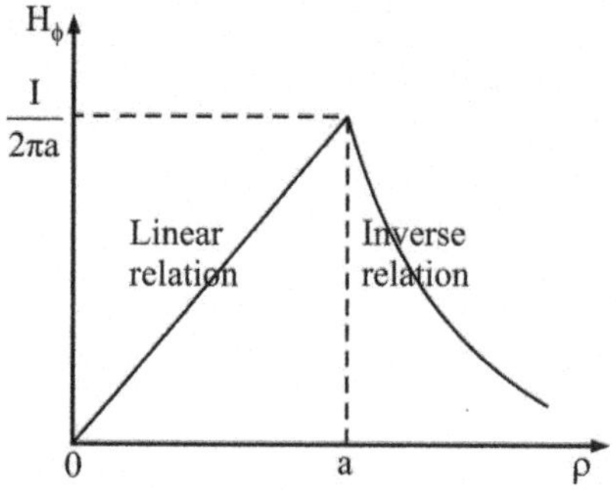

Fig. 2.18

***Problem 2.20**

Determine the magnetic flux, for the surface described by

(a) $\rho = 1\text{m.}, 0 \le \phi \le \pi/2, 0 \le z \le 2m$

(b) a sphere of radius 2 m.

If the magnetic field is of the form $\bar{H} = \frac{1}{\rho}\cos\phi \bar{a}_\rho$ A/m

---

Solution

We have magnetic flux $\psi = \int_s \bar{B}.\bar{ds} = \mu_0 \int_s \bar{H}.\bar{ds}$

(a) here $\bar{ds} = \rho d\phi dz \bar{a}_\rho$ (cylindrical symmetry)

$$\psi = \mu_0 \int_s \frac{1}{\rho}\cos\phi \bar{a}_\rho . \rho d\phi dz \bar{a}_\rho = \mu_0 \int_{z=0}^{2} dz \int_{\phi=0}^{\pi/2} \cos\phi d\phi = 2\mu_0 = 25.13\times 10^{-7} \text{ Wb}$$

(b) here $\bar{ds} = r^2 \sin\theta d\theta d\phi \bar{a}_r$ (spherical symmetry)

$$\psi = \mu_0 \int_s \frac{1}{r}\cos\phi \bar{a}_r r^2 \sin\theta d\theta d\phi \bar{a}_r = \mu_0 r \int_{\theta=0}^{\pi} \sin\theta d\theta \int_{\phi=0}^{2\pi} \cos\phi d\phi = 0 \text{ Wb}$$

***Problem 2.21**

In a conducting medium $\bar{H} = y^2 z \bar{a}_x + 2(x+1)yz\bar{a}_y - (x+1)z^2\bar{a}_z$ A/m. Find the current density at (1, 0, −3) and calculate the current passing through Y = 1 plane, $0 \le x \le 1$, $0 \le z \le 1$.

Solution

We have current density $\bar{J} = \nabla \times \bar{H}$

$$\bar{J} = \begin{vmatrix} \bar{a}_x & \bar{a}_y & \bar{a}_z \\ \frac{\partial}{\partial x} & \frac{\partial}{\partial y} & \frac{\partial}{\partial z} \\ H_x & H_y & H_z \end{vmatrix} = \begin{vmatrix} \bar{a}_x & \bar{a}_y & \bar{a}_z \\ \frac{\partial}{\partial x} & \frac{\partial}{\partial y} & \frac{\partial}{\partial z} \\ y^2 z & 2(x+1)yz & -(x+1)z^2 \end{vmatrix}$$

$$\bar{J} = -2(x+1)y\bar{a}_x + (z^2 + y^2)\bar{a}_y \text{ A/m}^2$$

$$\bar{J}_{(1,0,-3)} = 9\bar{a}_y \text{ A/m}^2$$

Current is $I = \int_s \bar{J}.d\bar{s}$, here $d\bar{s} = dxdz\bar{a}_y$

$$\therefore \quad I = \int_s \left(-2(x+1)y\bar{a}_x + (z^2 + y^2)\bar{a}_y\right) dx\, dz\, \bar{a}_y = \int_{x=0}^{1} dx \int_{z=0}^{1} (z^2 + y^2) dz = 1.33 \text{ A}$$

Problem 2.22

Find the flux density at the center of a square loop of 10 turns carrying a current of 10 A. The loop is in air and has a side of 2 m.

Solution

We have $\bar{H} = \frac{I\sqrt{2}}{\pi a}\bar{a}_z$

$\therefore$ magnetic flux density is $\bar{B} = \frac{\mu_0 I\sqrt{2}}{\pi a}\bar{a}_z$

Here no. of turns N = 10

Total current I = N × current in each turn

I = 10 × 10 = 100 A

and a = 1 m

$$\therefore \qquad \bar{B} = \frac{\mu_0 100\sqrt{2}}{\pi \times 1}\bar{a}_z$$

$$\bar{B} = \frac{4\pi \times 10^{-7} \times 100\sqrt{2}}{\pi \times 1}\bar{a}_z = 56.569\mu\bar{a}_z \text{Tesla}$$

2.5 Magnetic Scalar and Vector Potentials

In electrostatics we have only scalar potential which is related with electric field intensity as $\bar{E} = -\nabla V$, where as in magneto static fields we have both scalar and vector potentials.

2.5.1 Magnetic Scalar Potential

To define magnetic scalar potential V_m as $\bar{H} = -\nabla V_m$ the $\bar{J}$ must be 0, which is explained as follows

we know the identity $\nabla \times \nabla V = 0$ ....(2.5.1)

and Maxwell's 3rd equation is $\nabla \times \bar{H} = \bar{J}$

by substituting $\bar{H} = -\nabla V_m$ in the above equation, we get

$$\nabla \times -\nabla V_m = \bar{J} \qquad .....(2.5.2)$$

By comparing equations (2.5.1) and (2.5.2), we can say that to define $\bar{H} = -\nabla V_m$, $\bar{J}$ must be 0. The magnetic scalar potential also satisfies Laplace's equation i.e., $\nabla^2 V_m = 0$. The unit for magnetic scalar potential is **ampere**.

2.5.2 Magnetic Vector Potential

To define magnetic vector potential, consider Maxwell's 4th equation $\nabla \cdot \bar{B} = 0$ and the identity $\nabla \cdot \nabla \times \bar{A} = 0$.

From the above two equations, we can say that the magnetic vector potential $\bar{A}$ can be related with $\bar{B}$ as

$$\bar{B} = \nabla \times \bar{A} \qquad .....(2.5.3a)$$

Like we have expression for Electric potential 'V' as $V = \int \frac{dQ}{4\pi\varepsilon_0 R}$, we can have expressions for Magnetic vector potential as

$$\bar{A} = \int_L \frac{\mu_0 I\, d\bar{l}}{4\pi R} \qquad \text{for line current}$$

$$\bar{A} = \int_s \frac{\mu_0 \bar{K}\, ds}{4\pi R} \qquad \text{for surface current}$$

$$\bar{A} = \int_v \frac{\mu_0 \bar{J}\, dv}{4\pi R} \qquad \text{for volume current}$$

To derive the equation for $\bar{A}$, Consider the elemental length $d\bar{l}'$ at (x', y', z') and field at (x, y, z) as shown in the Fig. 2.19

We have

$$\bar{H} = \int_L \frac{I\, d\bar{l}' \times \bar{R}}{4\pi R^3}$$

$$\bar{B} = \int_L \mu_0 \frac{I\, d\bar{l}' \times \bar{R}}{4\pi R^3} \quad \text{.....(2.5.3b)}$$

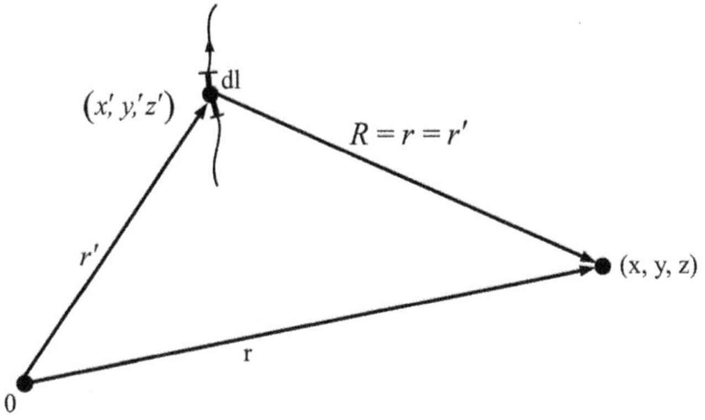

Fig. 2.19 Evaluation of magnetic vector potential

From Fig. 2.19

$$\bar{R} = (x - x')\bar{a}_x + (y - y')\bar{a}_y + (z - z')\bar{a}_z$$

$$|\bar{R}| = \sqrt{(x - x')^2 + (y - y')^2 + (z - z')^2}$$

Evaluate $\nabla\left(\frac{1}{R}\right)$ as follows

$$\nabla\left(\frac{1}{R}\right) = \nabla\left(\frac{1}{\sqrt{(x - x')^2 + (y - y')^2 + (z - z')^2}}\right)$$

$$\nabla\left(\frac{1}{R}\right) = \nabla\left(\left[(x - x')^2 + (y - y')^2 + (z - z')^2\right]^{-1/2}\right)$$

$$\Delta\left(\frac{1}{R}\right) = -\frac{1}{2}\left\{2(x-x')\bar{a}_x\left[(x-x')^2+(y-y')^2+(z-z')^2\right]^{-\frac{1}{2}-1}\right.$$
$$+2(y-y')\bar{a}_y\left[(x-x')^2+(y-y')^2+(z-z')^2\right]^{-\frac{1}{2}-1}$$
$$\left.+2(z-z')\bar{a}_z\left[(x-x')^2+(y-y')^2+(z-z')^2\right]^{-\frac{1}{2}-1}\right\}$$

$\because$
$$\nabla\left(\frac{1}{R}\right) = -\frac{(x-x')\bar{a}_x+(y-y')\bar{a}_y+(z-z')\bar{a}_z}{\left(\sqrt{(x-x')^2+(y-y')^2+(z-z')^2}\right)^3} = \frac{-\bar{R}}{R^3}$$

$\therefore$ $\frac{\bar{R}}{R^3}$ can be replaced in equation (2.5.3 (b)) with $-\nabla\left(\frac{1}{R}\right)$

$$\bar{B} = -\int_L \frac{\mu_0 I}{4\pi} d\bar{l}' \times \nabla\left(\frac{1}{R}\right) \qquad \text{.....(2.5.3c)}$$

From the identity

$\nabla\times(f\bar{F}) = f\nabla\times\bar{F}+\nabla(f)\times\bar{F}$ where f is scalar and $\bar{F}$ is vector and consider $f=\frac{1}{R}$ and $\bar{F}=d\bar{l}'$

$$\nabla\times\left(\frac{d\bar{l}'}{R}\right) = \frac{1}{R}\nabla\times d\bar{l}'+\nabla\left(\frac{1}{R}\right)\times d\bar{l}'$$

Since ∇ is in terms of x, y, z and $d\bar{l}'$ is in terms of x', y', z', the first term in the above equation becomes zero.

$\therefore$
$$d\bar{l}'\times\nabla\left(\frac{1}{R}\right) = -\nabla\times\left(\frac{d\bar{l}'}{R}\right) \qquad \text{.....(2.5.3d)}$$

Substituting in eq. (2.5.3(c)), we get

$\therefore$
$$\bar{B} = \nabla\times\int_L \frac{\mu_0 I\, d\bar{l}'}{4\pi R}$$

Comparing the eq.s (2.5.3(a)) and (2.5.3(d)), we get

$$\bar{A} = \int_L \frac{\mu_0 I\, d\bar{l}'}{4\pi R}$$

We know the magnetic flux $\psi = \oint_s \bar{B} \cdot d\bar{s}$

In terms of magnetic vector potential, $\bar{A}$ it is

$$\psi = \oint_s \left(\nabla \times \bar{A}\right) \cdot d\bar{s}$$

$$\psi = \oint_L \bar{A} \cdot d\bar{l} \qquad \text{(according to Stoke's theorem)} \qquad \text{....(2.5.4)}$$

which is magnetic flux in terms of magnetic vector potential.

The unit for magnetic vector potential is **Wb/m**

Problem 2.23

Given the magnetic vector potential $\bar{A} = \dfrac{-\rho^2}{4}\bar{a}_z$ Wb/m. Calculate the total magnetic flux crossing the surface $\phi = \dfrac{\pi}{2}$, $1 \le \rho \le 2$ m, $0 \le z \le 5$ m.

Solution

We have

$$\nabla \times \bar{A} = \frac{1}{\rho}\begin{vmatrix} \bar{a}_\rho & \rho\bar{a}_\phi & \bar{a}_Z \\ \dfrac{\partial}{\partial\rho} & \dfrac{\partial}{\partial\phi} & \dfrac{\partial}{\partial z} \\ A_\rho & \rho A_\phi & A_Z \end{vmatrix}$$

$$\nabla \times \bar{A} = \frac{1}{\rho}\begin{vmatrix} \bar{a}_\rho & \rho\bar{a}_\phi & \bar{a}_Z \\ \dfrac{\partial}{\partial\rho} & \dfrac{\partial}{\partial\phi} & \dfrac{\partial}{\partial z} \\ 0 & 0 & -\dfrac{\rho^2}{4} \end{vmatrix}$$

$$= \frac{1}{\rho}\left[\bar{a}_\rho(0) - \rho\bar{a}_\phi \frac{\partial}{\partial\rho}\left(\frac{-\rho^2}{4}\right) + 0\right]$$

$$= \frac{1}{\rho}\left[-\rho\bar{a}_\phi\left(-\frac{1}{4}2\rho\right)\right] = \frac{\rho}{2}\bar{a}_\phi$$

$\because$ ϕ is constant and ρ and z are varying, $d\bar{s} = d\rho\, dz\bar{a}_\phi$

The magnetic flux crossing the surface is $\psi = \int_S (\nabla \times \bar{A}) \cdot d\bar{s}$

$$= \int_{\rho=1}^{2} \int_{z=0}^{5} \frac{\rho}{2} d\rho dz$$

$$= 3.75 \text{ Wb}$$

Problem 2.24

A current distribution gives rise to the vector magnetic potential $\bar{A} = x^2 y\bar{a}_x + y^2 x\bar{a}_y - 4xyz\,\bar{a}_z$ Wb/m. Calculate

(a) $\bar{B}\ at(-1,2,5)$

(b) The flux through the surface defined by Z = 1, $0 \le x \le 1, -1 \le y \le 4$.

Solution

(a) $$\bar{B} = \nabla \times \bar{A} = \begin{vmatrix} \bar{a}_x & \bar{a}_y & \bar{a}_z \\ \frac{\partial}{\partial x} & \frac{\partial}{\partial y} & \frac{\partial}{\partial z} \\ x^2 y & y^2 x & -4xyz \end{vmatrix}$$

$$\bar{B} = \bar{a}_x(-4xz - 0) - \bar{a}_y(-4yz - 0) + \bar{a}_z(y^2 - x^2)$$

$$\bar{B} = -4xz\bar{a}_x + 4yz\,\bar{a}_y + (y^2 - x^2)\bar{a}_z$$

$$\bar{B}_{(-1,\,2,\,3)} = 20\,\bar{a}_x + 40\,\bar{a}_y + 3\bar{a}_z \text{ Wb/m}^2$$

(b) Here $d\bar{s} = dx\,dy\,\bar{a}_z$

$$\psi = \int_S \nabla \times \bar{A} \cdot d\bar{s}$$

$$\psi = \int_{x=0}^{1} \int_{y=-1}^{4} \left(-4xz\bar{a}_x + 4yz\bar{a}_y + (y^2 - x^2)\bar{a}_z\right) \cdot dxdy\bar{a}_z$$

$$\psi = \int_{x=0}^{1} \int_{y=-1}^{4} (y^2 - x^2) dxdy$$

$$\psi = \int_{y=-1}^{4}\left[y^2x - \frac{x^3}{3}\right]_0^1 y = \int_{y=-1}^{4}\left[y^2 - \frac{1}{3}\right]dy$$

$$\psi = \frac{1}{3}\left[y^3 - y\right]_{-1}^{4} = \frac{1}{3}\left[64 - 4 + 1 - 1\right] = 20 \text{ Wb}$$

2.6 Forces due to Magnetic Fields

There are three ways in which force due to magnetic field can be experienced. Force can be due to a moving charge particle in magnetic field $\bar{B}$, on current carrying element by an external magnetic field $\bar{B}$ and between two current carrying elements.

2.6.1 Force due to Moving Charge Particle in $\bar{B}$

The force $\bar{F}$ due to a static charge Q in electrostatic fields is $\bar{F} = Q\bar{E}$. In magnetic fields force can exert only when the charge moves. Let us assume the charge Q is moving with velocity $\bar{u}$ in magnetic field $\bar{B}$, then force due to moving charge particle in $\bar{B}$ is $\bar{F}_m = Q\bar{u} \times \bar{B}$

The direction of $\bar{F}_m$ will be perpendicular to the directions of $\bar{u}$ and $\bar{B}$.

The force due to a charge Q which is moving with velocity $\bar{u}$ in electromagnetic field is

$$\bar{F} = \bar{F}_e + \bar{F}_u = Q\left(\bar{E} + \bar{u} \times \bar{B}\right)$$

which is called Lorentz force equation.

If the charge is having mass 'm' and it is moving with velocity $\bar{u}$. The above equation can be equated to the mechanical force according to Newton's second law.

$$ma = \bar{F} = Q\left(\bar{E} + \bar{u} \times \bar{B}\right) = m\frac{du}{dt}$$

2.6.2 Force on a Current Carrying Conductor

We know $Id\bar{l} = \bar{K}ds = \bar{J}dv$

Where $\bar{k}$ = surface current density and $\bar{J}$ = Volume current density.

also $$Id\bar{l} = \frac{dQ}{dt}d\bar{l} = dQ\bar{u}$$

We know that the force due to a moving charge Q with velocity $\bar{u}$ in $\bar{B}$ is $\bar{F}_m = Q\bar{u} \times \bar{B}$.

The elemental force $d\bar{F}_m$ on the elemental charge dQ is $d\bar{F}_m = dQ\bar{u} \times \bar{B}$

$$d\bar{F}_m = I\,d\bar{l} \times \bar{B}$$

Force in a closed path or in a circuit due to a current carrying conductor is $\bar{F}_m = \oint_L I\,d\bar{l} \times \bar{B}$.

Similarly force due to surface current distribution in a closed surface is

$$\oint_S \bar{K}\,ds \times \bar{B} \qquad \text{.....(2.6.1)}$$

and due to volume current distribution is $\int_v \bar{J}\,dv \times \bar{B}$ .....(2.6.2)

2.6.3 Force between Two Current Carrying Conductors (Ampere's Force Law)

Let us consider two current elements $I_1 d\bar{l}_1$ and $I_2 d\bar{l}_2$ as shown in Fig.2.20. Let us find the force $\bar{F}_1$ on $I_1 d\bar{l}_1$ due to $\bar{B}_2$ which is created by $I_2 d\bar{l}_2$. We know the elemental force $d\bar{F}$ due to the current element $Id\bar{l}$ is $d\bar{F} = I\,d\bar{l} \times \bar{B}$. The elemental force $d\left(d\bar{F}_1\right)$ on $I_1 d\bar{l}_1$ in the elemental magnetic field $d\bar{B}_2$ is $d\left(d\bar{F}_1\right) = I_1 d\bar{l}_1 \times d\bar{B}_2$.

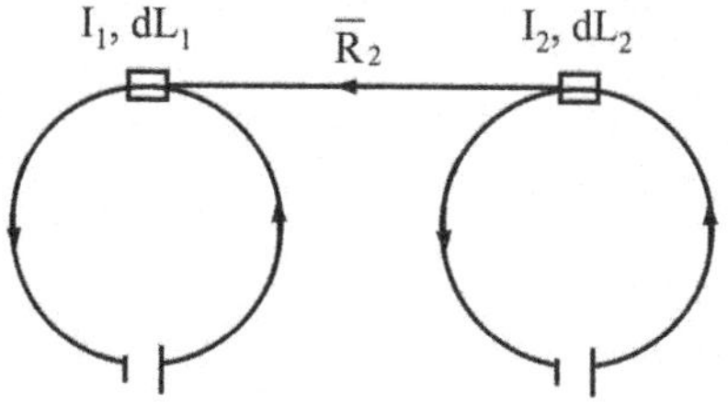

Fig. 2.20 Evaluation of force between two current carrying conductors

According to Biot-Savart's law

$$d\bar{H} = \frac{I\,d\bar{l} \times \bar{a}_R}{4\pi R^2}$$

$$d\bar{B} = \frac{\mu_0 I\,d\bar{l} \times \bar{a}_R}{4\pi R^2}$$

$$\therefore \qquad d\bar{B}_2 = \frac{\mu_0 I_2 d\bar{l}_2 \times \bar{a}_{R21}}{4\pi R_{21}^2}$$

where $\bar{a}_{R21}$ is the unit vector along $\bar{R}_{21}$

$$d\left(d\bar{F}_1\right) = \frac{I_1 d\bar{l}_1 \times \mu_0 I_2 d\bar{l}_2 \times \bar{a}_{R21}}{4\pi R_{21}^2}$$

By integrating twice we get $\bar{F}_1 = \oint\limits_{L_1}\oint\limits_{L_2} \frac{\mu_0}{4\pi R_{21}^2} I_1 d\bar{l}_1 \times \left(I_2 d\bar{l}_2 \times \bar{a}_{R21}\right)$ .....(2.6.3)

Which is Ampere's Force Law equation and that resembles the Coulomb's law force equation.

We get force on $I_2 d\bar{l}_2$ *i.e.*, $\bar{F}_2$ due to $I_1 d\bar{l}_1$ in magnetic field $\bar{B}_1$ by interchanging the suffixes in the above equation

$$\bar{F}_2 = \oint\limits_{L_1}\oint\limits_{L_2} \frac{\mu_0}{4\pi R_{12}^2} I_2 d\bar{l}_2 \times \left(I_1 d\bar{l}_1 \times \bar{a}_{R12}\right) \qquad ......(2.6.4)$$

Here i.e., in Ampere's Force law $\bar{F}_1 \neq \bar{F}_2$, where as in Coulombs law $\bar{F}_1 = -\bar{F}_2$

Problem 2.25

A rectangular loop carrying current I_2 is placed parallel to an infinitely long filamentary wire carrying current I_1 as shown in Fig.2.21. Show that the force experienced by the loop is given by

$$\bar{F} = -\frac{\mu_0 I_1 I_2 b}{2\pi}\left[\frac{1}{\rho_0} - \frac{1}{\rho_0 + a}\right]\bar{a}_\rho \text{ N}$$

Solution

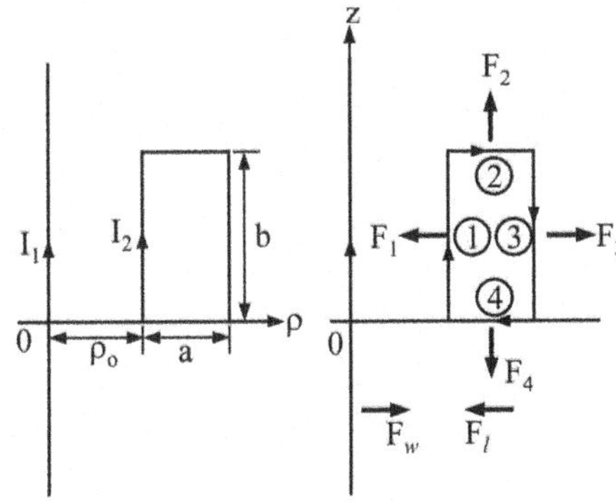

Fig. 2.21

Let $\bar{F}_\ell$ be the force on the loop

$$\bar{F}_\ell = \bar{F}_1 + \bar{F}_2 + \bar{F}_3 + \bar{F}_4$$

$$= I_2 \oint d\bar{l}_2 \times \bar{B}_1$$

Where $\bar{F}_1, \bar{F}_2, \bar{F}_3$ *and* $\bar{F}_4$ are the forces exerted on sides of the loop.

We know for infinitely long wire $\bar{B} = \dfrac{\mu_0 I}{2\pi\rho}\bar{a}_\phi$

To evaluate $\bar{F}_1$, $d\bar{l}_2 = dz\bar{a}_z$, z ranges from 0 to b, $I = I_1$ and $\rho = \rho_0$

$$\bar{F}_1 = I_2 \int d\bar{l}_2 \times \bar{B}_1 = I_2 \int_{z=0}^{b} dz\bar{a}_z \times \frac{\mu_0 I_1}{2\pi\rho_0}\bar{a}_\phi$$

$$\therefore \qquad \bar{F}_1 = -\frac{\mu_0 I_1 I_2 b}{2\pi\rho_0}\bar{a}_\rho \qquad (attractive)$$

To evaluate $\bar{F}_3$, $d\bar{l}_2 = dz\bar{a}_z$, z ranges from b to 0, $I = I_1$ and $\rho = \rho_0 + a$

$$\bar{F}_3 = I_2 \int d\bar{l}_2 \times \bar{B}_1 = I_2 \int_{z=b}^{0} dz\bar{a}_z \times \frac{\mu_0 I_1}{2\pi(\rho_0 + a)}\bar{a}_\phi$$

$$\therefore \qquad \bar{F}_3 = \frac{\mu_0 I_1 I_2 b}{2\pi(\rho_0 + a)}\bar{a}_\rho \qquad (repulsive)$$

To evaluate $\bar{F}_2$, $d\bar{l}_2 = d\rho\bar{a}_\rho$, ρ ranges from ρ_0 to $\rho_0 + a$, $I = I_1$ and $\rho = \rho$

$$\bar{F}_2 = I_2 \int_{\rho=\rho_0}^{\rho_0 + a} d\rho\bar{a}_\rho \times \frac{\mu_0 I_1}{2\pi\rho}\bar{a}_\phi$$

$$\therefore \qquad \bar{F}_2 = \frac{\mu_0 I_1 I_2}{2\pi}\ln\left(\frac{\rho_0 + a}{\rho_0}\right)\bar{a}_z \qquad (parallel)$$

To evaluate $\bar{F}_4$, $d\bar{l}_2 = d\rho\bar{a}_\rho$, ρ ranges from $\rho_0 + a$ to ρ_0, $I = I_1$ and $\rho = \rho$

$$\bar{F}_4 = I_2 \int_{\rho=\rho_0 + a}^{\rho_0} d\rho\bar{a}_\rho \times \frac{\mu_0 I_1}{2\pi\rho}\bar{a}_\phi$$

$$\therefore \qquad \bar{F}_4 = -\frac{\mu_0 I_1 I_2}{2\pi} \ln\left(\frac{\rho_0 + a}{\rho_0}\right) \bar{a}_z \qquad (parallel)$$

Then the total force $\bar{F}_\ell$ on the loop is

$$\bar{F}_\ell = \bar{F}_1 + \bar{F}_2 + \bar{F}_3 + \bar{F}_4$$

$$\bar{F}_\ell = -\frac{\mu_0 I_1 I_2 b}{2\pi}\left[\frac{1}{\rho_0} - \frac{1}{\rho_0 + a}\right] \bar{a}_\rho \text{ N}$$

and

$$\bar{F}_w = -\bar{F}_\ell$$

Problem 2.26

A charged particle of mass 2 kg and 1 C starts at origin with velocity $3\bar{a}_y$ m/s and travels in a region of uniform magnetic field $\bar{B} = 10\bar{a}_z$ Wb/m^2 at t = 4 sec. Calculate

(a) velocity and acceleration of particle

(b) the magnetic force on it.

Solution

(a) We have

$$\bar{F} = m\frac{d\bar{u}}{dt} = Q\bar{u} \times \bar{B}$$

Acceleration is

$$\bar{a} = \frac{d\bar{u}}{dt} = \frac{Q}{m}\bar{u} \times \bar{B}$$

Hence $$\frac{d}{dt}(u_x \bar{a}_x + u_y \bar{a}_y + u_z \bar{a}_z) = \frac{1}{2}\begin{vmatrix} \bar{a}_x & \bar{a}_y & \bar{a}_z \\ u_x & u_y & u_z \\ 0 & 0 & 10 \end{vmatrix} = 5(u_y \bar{a}_x - u_x \bar{a}_y)$$

By equating components, we get

$$\frac{du_x}{dt} = 5u_y, \quad \frac{du_y}{dt} = -5u_x, \quad \frac{du_z}{dt} = 0 \Rightarrow u_z = c_0$$

u_x or u_y can be eliminated in the above equations by taking second derivative of one equation and utilizing the other. Thus

$$\frac{d^2u_x}{dt^2} = 5\frac{du_y}{dt} = -25u_x \quad \text{or} \quad \frac{d^2u_x}{dt^2} + 25u_x = 0$$

Which is a linear differential equation whose solution is

$$u_x = c_1 \cos 5t + c_2 \sin 5t$$

$$\frac{du_x}{dt} = 5u_y = -5c_1 \sin 5t + 5c_2 \cos 5t \quad \text{or} \quad u_y = -c_1 \sin 5t + c_2 \cos 5t$$

Let us determine c_0, c_1 and c_2 using initial conditions. At t = 0, $\bar{u} = 3\bar{a}_y$.

Hence,

$$u_x = 0 \Rightarrow 0 = c_1.1 + c_2.0 \Rightarrow c_1 = 0$$

$$u_y = 3 \Rightarrow 3 = -c_1.0 + c_2.1 \Rightarrow c_2 = 3$$

$$u_z = 0 \Rightarrow 0 = c_0$$

Substituting the values of c_0, c_1 and c_2, gives velocity as

$$\bar{u} = (u_x, u_y, u_z) = (3\sin 5t, 3\cos 5t, 0)$$

Hence, velocity at t = 4 sec is

$$\bar{u} = (3\sin 20, 3\cos 20, 0)$$

$$= 2.739\bar{a}_x + 1.224\bar{a}_y \text{ m/s}$$

Acceleration is $\bar{a} = \dfrac{d\bar{u}}{dt} = (15\cos 5t, -15\sin 5t, 0)$

Hence, acceleration at t = 4 sec is

$$\bar{a} = 6.101\bar{a}_x - 13.703\bar{a}_y \;\; m/s^2$$

(b) $\bar{F} = m\dfrac{d\bar{u}}{dt} = m\bar{a} = 12.2\bar{a}_x - 27.4\bar{a}_y \; N$

2.7 Magnetic Dipole, Torque and Moment

2.7.1 Magnetic Dipole

Two poles of equal in magnitude but opposite in sign are separated by a small distance is known as a magnetic dipole.

Ex: a bar magnet, a small filamentary current loop.

Let us determine magnetic flux density at an arbitrary point $P(r,\theta,\phi)$ due to a small circular loop carrying current.

To find magnetic flux density at P, consider a circular loop of radius a, carrying current I and lying on XY plane as shown in Fig. 2.22.

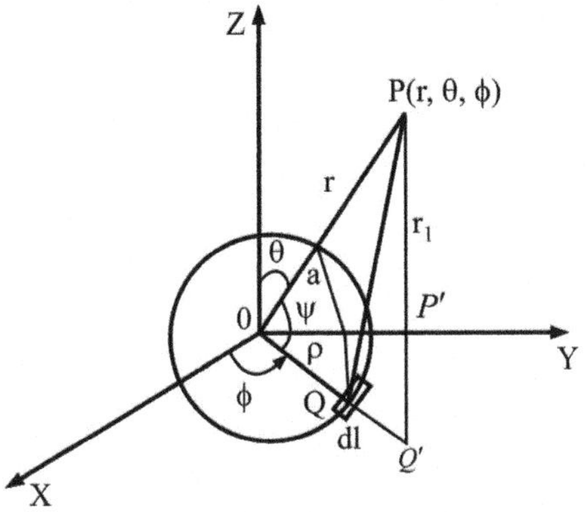

Fig. 2.22 A circular loop lying on XY plane

Before going to find magnetic flux density $\overline{B}$, let us determine $\overline{A}$

We have magnetic vector potential due to current element as

$$\overline{A} = \frac{\mu_0 I}{4\pi} \oint \frac{d\overline{l}}{r_1} \qquad \text{.....(2.7.1)}$$

From Fig.2.22

The component of r on OP' is $r\sin\theta$

i.e., $$OP' = r\sin\theta \qquad \text{.....(2.7.2)}$$

The component of r on OQ' is $r\cos\psi$

i.e., $$OQ' = r\cos\psi \qquad \text{.....(2.7.3)}$$

The component of OP' on OQ' is $OP'\cos(90-\phi)$

$$OQ' = OP'\cos(90-\phi)$$

Substituting, OP' from equation (2.7.2)

$$OQ' = r\sin\theta\sin\phi \qquad \text{.....(2.7.4)}$$

Equating eqs. (2.7.3) and (2.7.4)

$$r\cos\psi = r\sin\theta\sin\phi$$

$$\cos\psi = \sin\theta \sin\phi \qquad .....(2.7.5)$$

From the Fig.2.22

$$d\bar{l} = \rho d\phi \bar{a}_\phi$$

The component of 'a' on ρ is $a\cos\psi$

i.e., $\rho = a\cos\psi$

$$\therefore \; d\bar{l} = a\cos\psi d\phi \bar{a}_\phi$$

Substituting $\cos\psi$ from eq. (2.7.5)

$$d\bar{l} = a\sin\theta \sin\phi \; d\phi \; \bar{a}_\phi$$

Since the circle lies at the half of the sphere, co-latitude angle $\theta = \frac{\pi}{2}$

$$\therefore \; d\bar{l} = a\sin\phi d\phi \bar{a}_\phi \qquad .....(2.7.6)$$

Apply cosine rule to the triangle OPQ

$$r_1^2 = r^2 + a^2 - 2ra\cos\psi$$

Substituting $\cos\psi$ from eq. (2.7.5)

$$r_1^2 = r^2 + a^2 - 2ra\sin\theta\sin\phi$$

For circle to be small, r >> a;

$$r_1^2 = r^2 - 2ra\sin\theta\sin\phi$$

$$r_1 = r\sqrt{1 - \frac{2a\sin\theta\sin\phi}{r}}$$

$$\frac{1}{r_1} = \frac{1}{r}\left(1 - \frac{2a\sin\theta\sin\phi}{r}\right)^{\frac{-1}{2}}$$

Expanding series and neglecting higher terms, we get

$$\frac{1}{r_1} = \frac{1}{r}\left(1 + \frac{a\sin\theta\sin\phi}{r}\right) \qquad ....(2.7.7)$$

Substituting $d\bar{l}$ and $\frac{1}{r_1}$ from eqs.(2.7.6) and (2.7.7) in eq.(2.7.1), we get

$$\overline{A}=\frac{\mu_0 I}{4\pi}\int_{\phi=0}^{2\pi}\frac{1}{r}\left(1+\frac{a\sin\theta\sin\phi}{r}\right)a\sin\phi d\phi\overline{a}_\phi$$

$$\overline{A}=\frac{\mu_0 Ia\overline{a}_\phi}{4\pi r}\left\{\int_{\phi=0}^{2\pi}\sin\phi d\phi+\int_{\phi=0}^{2\pi}\frac{a\sin\theta\sin^2\phi d\phi}{r}\right\}$$

$$\overline{A}=\frac{\mu_0 Ia\overline{a}_\phi}{4\pi r}\left\{0+\int_{\phi=0}^{2\pi}\frac{a\sin\theta\sin^2\phi d\phi}{r}\right\}$$

$$\overline{A}=\frac{\mu_0 Ia^2\sin\theta\overline{a}_\phi}{4\pi r^2}\int_{\phi=0}^{2\pi}\sin^2\phi d\phi$$

$$\overline{A}=\frac{\mu_0 Ia^2\sin\theta\overline{a}_\phi}{4\pi r^2}\int_{\phi=0}^{2\pi}\frac{1-\cos 2\phi}{2}d\phi$$

$$\overline{A}=\frac{\mu_0 Ia^2\sin\theta\overline{a}_\phi}{4\pi r^2}\{\pi-0\}$$

$$\overline{A}=\frac{\mu_0 \pi Ia^2\sin\theta\overline{a}_\phi}{4\pi r^2} \qquad \text{......(2.7.8)}$$

We know $\overline{B}=\nabla\times\overline{A}$

Since $\overline{A}$ is in spherical coordinate system, write $\nabla\times\overline{A}$ in spherical coordinate system, then

$$\overline{B}=\frac{1}{r^2\sin\theta}\begin{vmatrix}\overline{a}_r & r\overline{a}_\theta & r\sin\theta\,\overline{a}_\phi\\ \frac{\partial}{\partial r} & \frac{\partial}{\partial\theta} & \frac{\partial}{\partial\phi}\\ A_r & rA_\theta & r\sin\theta\,A_\phi\end{vmatrix}$$

$$\overline{B}=\frac{1}{r^2\sin\theta}\begin{vmatrix}\overline{a}_r & r\overline{a}_\theta & r\sin\theta\,\overline{a}_\phi\\ \frac{\partial}{\partial r} & \frac{\partial}{\partial\theta} & \frac{\partial}{\partial\phi}\\ 0 & 0 & r\sin\theta\frac{\mu_0\pi Ia^2\sin\theta}{4\pi r^2}\end{vmatrix}$$

$$\bar{B} = \frac{1}{r^2 \sin\theta}\left\{\bar{a}_r\left(\frac{\mu_0 \pi I a^2 2\sin\theta\cos\theta}{4\pi r}\right) + r\bar{a}_\theta \frac{\mu_0 \pi I a^2 \sin^2\theta}{4\pi r^2}\right\}$$

$$\bar{B} = \frac{1}{r^2 \sin\theta}\frac{\mu_0 \pi I a^2 \sin\theta}{4\pi r}(2\cos\theta\bar{a}_r + \sin\theta\bar{a}_\theta)$$

$$\bar{B} = \frac{\mu_0 I \pi a^2}{4\pi r^3}(2\cos\theta\bar{a}_r + \sin\theta\bar{a}_\theta) \quad .....(2.7.9)$$

2.7.2 Magnetic Torque and Moment

Magnetic dipole moment is the product of current flowing through the loop and area of the loop. It's direction is always normal to the loop.

The magnetic moment of a circle of radius 'a' and carries current I is $\bar{m} = I\pi a^2 \bar{a}_n$ A-m^2 s.

Eq.(2.7.8) in terms of magnetic moment can be written as

$$\bar{A} = \frac{\mu_0 \bar{m} \times \bar{a}_r}{4\pi r^2} \quad .....(2.7.10)$$

$\because \ \bar{a}_z \times \bar{a}_r = \sin\theta \bar{a}_\phi$

Eq.(2.7.9) in terms of magnetic moment can be written as

$$\bar{B} = \frac{\mu_0 m}{4\pi r^3}(2\cos\theta\bar{a}_r + \sin\theta\bar{a}_\theta)$$

Magnetic Torque is the vector product of magnetic moment and magnetic flux density i.e., $\bar{T} = \bar{m} \times \bar{B}$

It's unit is N-m.

2.8 Magnetization in Materials

The magnetization $\bar{m}$, in amperes per meter, is the magnetic dipole moment per unit volume.

If there are N atoms in a given volume Δv and the kth atom has a magnetic moment $\bar{m}_k$,

$$\bar{M} = \lim_{\Delta v \to 0} \frac{\sum_{k=1}^{N} \bar{m}k}{\Delta v} \quad .....(2.8.1)$$

A medium for which $\bar{M}$ is not zero everywhere is said to be magnetized.

In free space $\overline{M}=0$ and we have $\nabla\times\overline{H}=\overline{J}_f$ or $\nabla\times\left(\dfrac{\overline{B}}{\mu_0}\right)=\overline{J}_f$

$\overline{J}_f$ = free volume current density

In a material medium $M\neq 0$

$$\nabla\times\left(\frac{\overline{B}}{\mu_0}\right)=\overline{J}_f+\overline{J}_b=\overline{J}$$

$\overline{J}_b$ = bounded volume current density

$\overline{J}_f=\nabla\times\overline{H}$ *and* $\overline{J}_b=\nabla\times\overline{M}$

$\therefore \qquad \nabla\times\left(\dfrac{\overline{B}}{\mu_0}\right)=\nabla\times\overline{H}+\nabla\times\overline{M}$

$\Rightarrow \qquad \overline{B}=\mu_0(\overline{H}+\overline{M})$

For linear materials $\overline{M}\neq x_m\overline{H}$ ....(2.8.2)

Where χ_m is called magnetic susceptibility of the medium, defined as how susceptible the material to a magnetic field.

$\therefore\ \overline{B}=\mu_0(1+\chi_m)\overline{H}=\mu\overline{H}$ .....(2.8.3)

or $\qquad \overline{B}=\mu_0\mu_r\overline{H}$

where $\qquad \mu_r=1+\chi_m=\dfrac{\mu}{\mu_0}$ .....(2.8.4)

The quantity μ (henry/meter) is permeability of the material and μ_r is the dimensionless quantity defined as ratio of the permeability of a given material to that of free space and is known as relative permeability of the material.

Problem 2.27

A flux density of $0.05\overline{a}_y$ tesla in a material having magnetic susceptibility 2.5, find magnetic field current density and magnetization.

Solution

Given $\overline{B}=0.05\overline{a}_y$ and $\chi_m=2.5$

Relative permiability $\mu_r = 1 + \chi_m = 1 + 2.5 = 3.5$

Permiability of material $\mu = \mu_0 \mu_r = 4\pi \times 10^{-7} \times 3.5 = 4.398 \times 10^{-6}$ H/m

Magnetic field intensity $\bar{H} = \dfrac{\bar{B}}{\mu} = \dfrac{0.05\bar{a}_y}{4.398 \times 10^{-6}} = 11368.8\bar{a}_y$ A/m

$$\text{Magnetic field current density } \bar{J} = \nabla \times \bar{H} = \begin{vmatrix} \bar{a}_x & \bar{a}_y & \bar{a}_z \\ \dfrac{\partial}{\partial x} & \dfrac{\partial}{\partial y} & \dfrac{\partial}{\partial z} \\ 0 & 11368.8 & 0 \end{vmatrix} = 0 \text{ A/m}^2$$

Magnetization $\bar{M} \neq x_m \bar{H} = 25 \times 11368.8\bar{a}_y = 28422\bar{a}_y$ A/m

Problem 2.28

In a certain material, $\chi_m = 4.2$ and $\bar{H} = 0.2x\bar{a}_y$ A/m. Determine:

(a) μ_r (b) μ (c) $\bar{M}$ (d) $\bar{B}$

(e) $\bar{J}$ (f) $\bar{J}_b$

Solution

(a) $\mu_r = 1 + \chi_m = 1 + 4.2 = 5.2$

(b) $\mu = \mu_0 \mu_r = 4\pi \times 10^{-7} \times 5.2 = 6.534 \times 10^{-6}$ H/m

(c) $\bar{M} = \chi_m \bar{H} = 4.2(0.2x\bar{a}_y) = 0.84x\bar{a}_y$ A/m

(d) $\bar{B} = \mu\bar{H} = 6.534 \times 10^{-6} \times 0.2x\bar{a}_y = 1.307x\bar{a}_y$ μWb/m^2

(e) $$\bar{J} = \nabla \times \bar{H} = \begin{vmatrix} \bar{a}_x & \bar{a}_y & \bar{a}_z \\ \dfrac{\partial}{\partial x} & \dfrac{\partial}{\partial y} & \dfrac{\partial}{\partial z} \\ 0 & 0.2x & 0 \end{vmatrix} = 0.2\bar{a}_z \text{ A/m}^2$$

(f) $\bar{J}_b = \chi_m \bar{J} = 4.2 \times 0.2\bar{a}_z = 0.84\bar{a}_z$ A/m^2

2.9 Inductance and Magnetic Energy

2.9.1 Inductance

To define inductance let us consider a circuit which has 'N' number of turns and current 'I' flows through it. Due to current 'I' magnetic field $\overline{B}$ is produced. Circuit is shown in Fig.2.23.

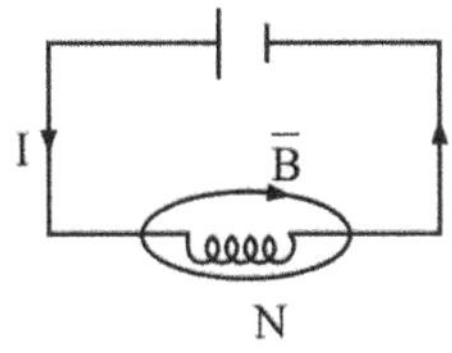

Fig. 2.23 Circuit which has 'N' number of turns and current 'I' flows through it

The flux $\psi = \int \overline{B} \cdot \overline{ds}$ is caused due to the magnetic field $\overline{B}$ which flows through each turn of the circuit. Then the flux linkage $\lambda = N.\psi$

If the surrounding medium is linear the flux linkage λ is directly proportional to the current flowing through the circuit.

$$\therefore \qquad \lambda \propto I$$

$$\lambda = LI \qquad \text{.....(2.9.1)}$$

Where L = proportionality constant which is the inductance of the given circuit.

Unit for inductance is 'Henry'

$$\therefore \qquad L = \frac{\lambda}{I}$$

When we have two circuits with the currents I_1 and I_2 and turns N_1 and N_2 respectively as shown in the Fig. 2.24.

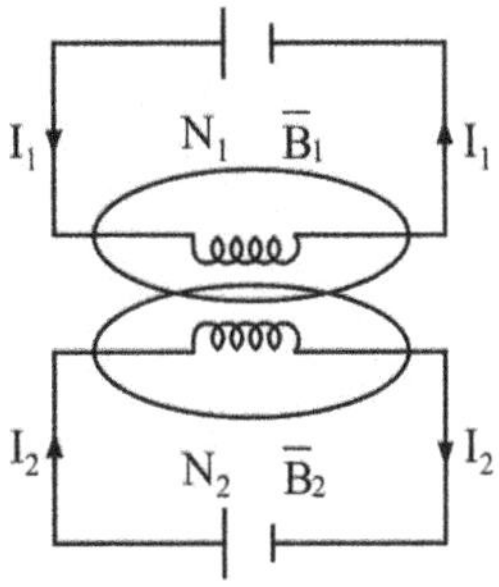

Fig. 2.24 Two circuits with the currents I_1 and I_2 and turns N_1 and N_2

When the circuits are as shown in Fig.2.24, we get the mutual inductances M_{12}, M_{21} along with the self inductances L_{11}, L_{22}. The mutual inductance on circuit 1 due to circuit 2 is

$$M_{12} = \frac{\lambda_{12}}{I_2} = \frac{N_1 \psi_{12}}{I_2} \quad .....(2.9.2)$$

The mutual inductance on circuit 2 due to circuit 1 is

$$M_{21} = \frac{\lambda_{21}}{I_1} = \frac{N_2 \psi_{21}}{I_1} \quad .....(2.9.3)$$

The self inductance in circuit 1 is L_{11} and in circuit 2 is L_{22}

$$L_{11} = \frac{\lambda_{11}}{I_1} = \frac{N_1 \psi_{11}}{I_1} \quad .....(2.9.4)$$

$$L_{22} = \frac{\lambda_{22}}{I_2} = \frac{N_2 \psi_{22}}{I_2} \quad .....(2.9.5)$$

The coupling coefficient k is a measure of the magnetic coupling between two coils

$$k = \frac{M}{\sqrt{L_1 L_2}} \quad where \quad 0 \le k \le 1 \;\; and \quad M = M_{12} = M_{21} \quad .....(2.9.6)$$

$k < 0.5$ loosely coupled;

$k > 0.5$ tightly coupled.

2.9.2 Magnetic Energy

We know the energy stored in inductor as $W_m = \frac{1}{2} L I^2$. Let us represent this in terms of either $\bar{H}$ or $\bar{B}$ like we have represented either in $\bar{E}$ or $\bar{D}$ in electrostatic fields. Consider a differential volume in a magnetic field $\bar{B}$ as shown in Fig.2.25.

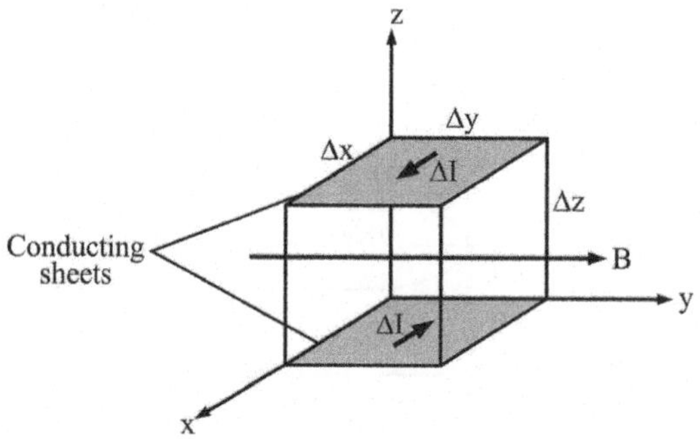

Fig. 2.25 Differential volume

Top and bottom faces of volume are carrying current ΔI

The inductance 'L' can be written as $L = \frac{\Delta\psi}{\Delta I}$

Where Δψ = differential flux

= The product of magnetic field and the surface perpendicular to it

$$\Delta\psi = \mu \bar{H}\, \Delta x\, \Delta z$$

Similarly the differential current ΔI in terms of $\bar{H}$ is $\Delta I = \bar{H}\, \Delta y$

∴ differential magnetic energy $\Delta W_m = \frac{1}{2} L\, \Delta I^2$

$$\Delta W_m = \frac{1}{2} \frac{\Delta\psi}{\Delta I} \Delta I^2$$

$$= \frac{1}{2} \mu \bar{H}\, \Delta x\, \Delta z \cdot \bar{H}\, \Delta y$$

$$= \frac{1}{2} \mu H^2 \Delta x\, \Delta y\, \Delta z$$

$$= \frac{1}{2} \mu H^2 \Delta v$$

Magnetic energy $W_m = \frac{\mu}{2} \int_v H^2 dv$

$$= \frac{\mu}{2} \int_v \bar{H} \cdot \bar{H}\, dv = \frac{1}{2} \int_v \frac{B^2}{\mu} dv \qquad \text{.....(2.9.7)}$$

Magnetic energy density is

$$\frac{\Delta W_m}{\Delta V} = \frac{1}{2} \mu H^2 \ \text{J/m}^3 \qquad \text{.....(2.9.8)}$$

Problem 2.29

Determine the self inductance of a co-axial cable of inner radius 'a' and outer radius 'b' and length of the co-axial cable is '*l*'.

Solution

Assume inner conductor carries a current I in Z-direction and outer conductor carries a current I in opposite direction as shown in Fig.2.26.

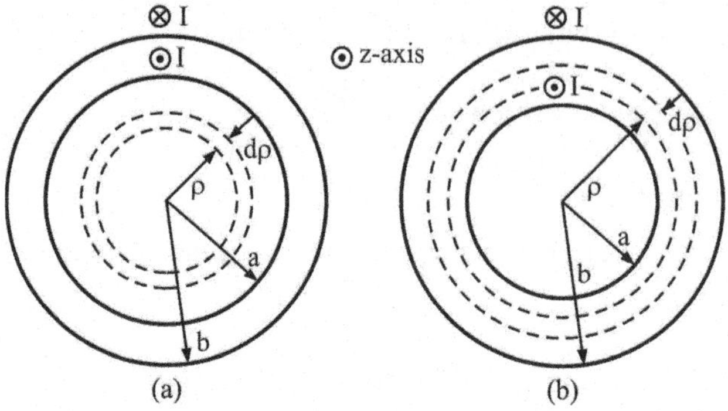

Fig. 2.26 (a) For finding L_{in} **(b)** For finding L_{out}

Inner Conductor

We know for inner conductor $\bar{H} = \dfrac{I\rho}{2\pi a^2}\bar{a}_\phi$

Since it is a cylinder, we can use the cylindrical coordinate system to solve the problem

We have

$$W_m = \frac{1}{2}\int_v \mu H^2 dv$$

$$W_m = \frac{1}{2}\int_v \mu \frac{I^2\rho^2}{4\pi^2 a^4}\rho\, d\rho\, d\phi\, dz$$

$$W_m = \frac{1}{2}\int_{z=0}^{l}\int_{\rho=0}^{a}\int_{\phi=0}^{2\pi} \mu \frac{I^2\rho^2}{4\pi^2 a^4}\rho d\rho\, d\phi\, dz$$

$$L_{in} = \frac{2W_m}{I^2} = \frac{\mu}{4\pi^2}\int_{z=0}^{l}\int_{\phi=0}^{2\pi} \frac{1}{a^4}\frac{a^4}{4} d\phi\, dz$$

$$L_{in} = \frac{\mu}{4\pi^2} 2\pi l = \frac{\mu l}{8\pi}$$

Outer conductor

We know for outer conductor $\bar{H} = \dfrac{I}{2\pi\rho}\bar{a}_\phi$

$$L_{out} = \frac{2W_m}{I^2} = \frac{2}{I^2}\frac{1}{2}\int_{z=0}^{l}\int_{\rho=a}^{b}\int_{\phi=0}^{2\pi}\mu\left(\frac{I}{2\pi\rho}\right)^2 \rho\, d\rho\, d\phi\, dz$$

$$L_{out} = \frac{\mu}{4\pi^2}\times 2\pi\times l\ln\frac{b}{a} = \frac{\mu l}{2\pi}\ln\frac{b}{a}$$

The self inductance of co-axial cable is

$$L = L_{in} + L_{out} = \frac{\mu l}{2\pi}\left(\frac{1}{4} + \ln\frac{b}{a}\right)$$

Review Questions and Answers

1. **What is the force on a charge, moving in a uniform magnetic field?**

   The force on a charge Q moving in a uniform magnetic field B with velocity *v* is given by

   $$F = Q(v \times B) \text{ N}$$

   $$F = BQv\sin\theta \text{ N}$$

   where, θ is the angle between the direction of B and the direction in which the charge moves.

2. **What is the force experienced by a current carrying element in a uniform magnetic field?**

   The force experienced by a current carrying element *Idl* in a uniform magnetic field B is given by

   $$F = I(dl \times B) \text{ N}$$

   $$F = BIl\sin\theta \text{ N}$$

   where, θ is the angle between the direction of B and the direction of current in the conductor.

3. **Give the Lorentz force equation.**

   The Lorentz force equation gives the force on a charge Q moving in a region where both the electric field E and magnetic field B are present.

   $$F = Q(E + v \times B) \text{ N}$$

   where v is the velocity with which the charge moves in the field.

4. Define magnetic flux density.

The magnetic flux per unit area is called the magnetic flux density B.

$$B = \frac{\phi}{A} \text{ Wb/m}^2 \text{ (or) Tesla}$$

5. State Biot-Savart's law.

The Biot-Savart law states that at any point P the magnitude of the magnetic field produced by the differential element dl is

(a) Proportional to the product of current, the magnitude of the differential length, and the sine of the angle lying between the element and a line connecting the element to the point P where the field is desired,

(b) Inversely proportional to the square of the distance from the differential element to the point P.

(c) Directly proportional to the constant of the medium (μ) and

(d) Directed normal to the plane containing the differential element and the line drawn from the filament to the point P.

$$B = \int \frac{\mu I dl \sin\theta}{4\pi r^2} \text{ Wb/m}^2$$

6. State Ampere's law for a magnetic field.

The Ampere's law states that the line integral of H around a single closed path is equal to the current enclosed.

It can also be stated as the line integral of B around a single closed path is equal to the permeability of the medium times the current enclosed.

$$\int H \cdot dl = I$$

$$\int B \cdot dl = \mu I$$

7. What is the force between two current carrying conductors?

The force between the two conductors carrying current I_1 and I_2 separated by a distance r is given by

$$F = \frac{\mu_0 I_1 I_2}{2\pi r} \text{ N}$$

8. Give the relation between magnetic flux density and magnetic field intensity.

The permeability of any medium is the ratio of the magnetic flux density B to the magnetic field intensity H.

$$\mu = \frac{B}{H} \text{ H/m}$$

9. State the Gauss's law for magnetic fields.

The integral of the magnetic flux density B over a closed surface is zero. This is called the Gauss's law for magnetic fields.

$$\oint B \cdot dS = 0$$

where, dS is the normal component of the surface.

10. What is the torque on a current carrying loop?

The torque, or moment, of a force is a vector whose magnitude is the product of the magnitudes of the vector force, the vector lever arm, and the sine of the angle between these two vectors. The direction of the vector torque is normal to both the force and lever arm.

$$T = R \times F \quad \text{Nm}$$

where, R is the vector lever arm, F is the force vector.

11. What is the torque on a planar coil?

The torque on a planar coil of any size in a uniform magnetic field is the product of the magnitudes of magnetic moment 'm', magnetic flux density B and the sine of the angle between these two. It is given by

$$T = m \times B \quad \text{Nm}$$

12. Define magnetic moment.

The magnetic moment 'm' is defined as the product of the loop current and the vector area of the loop. It is given by,

$$m = IA \quad \text{Am}^2$$

where A is the vector area.

13. What is current density.

The current flowing through unit surface area is called the current density.

$$J = \frac{dI}{dS} a_r \quad \text{H/m}$$

The direction of J is the same as the direction of motion of positive charges.

14. Give the expressions relating B and H with the current density J.

$$B = \frac{\mu_0}{4\pi} \iiint \frac{J \times a_r}{r^2} dv$$

$$curl B = \mu_0 J$$

$$curl H = J$$

15. Give the expressions relating magnetic vector potential with the current density J.

$$-\nabla^2 A = \mu J$$

$$A = \frac{\mu}{4\pi}\iiint \frac{J}{r} dr \text{ Wb/m}$$

16. Give the relation between B and magnetic vector potential.

$$B = Curl A$$

17. What is the magnetic field at any point due to a infinitely long conductor carrying current?

$$B = \frac{\mu_0 I}{2\pi d} \quad \text{Wb/m}^2$$

$$H = \frac{I}{2\pi d} \quad \text{A/m}$$

where, d is distance between the conductor and the point where the field is required.

18. What is the magnetic field at any point due to a finite length conductor carrying current?

$$B = \frac{\mu_0 I}{4\pi d}[\cos\theta_1 - \cos\theta_2] \quad \text{Wb/m}^2$$

$$H = \frac{I}{4\pi d}[\cos\theta_1 - \cos\theta_2] \quad \text{A/m}$$

19. What is the magnetic field at any point on the axis of a circular coil carrying current?

$$B = \frac{\mu_0 I a^2}{2(a^2 + d^2)^{3/2}} \quad \text{Wb/m}^2$$

$$H = \frac{I a^2}{2(a^2 + d^2)^{3/2}} \quad \text{A/m}$$

where 'a' is the radius of the circle.

20. What is the magnetic field at the centre of the circular coil carrying current?

$$B = \frac{\mu_0 I}{2a} \quad \text{Wb/m}^2$$

$$H = \frac{I}{2a} \quad \text{A/m}$$

21. What is the magnetic field at any point on the axis of a solenoid carrying current?

$$B = \frac{\mu_0 NI}{2l}[\cos\theta_1 - \cos\theta_2] \quad \text{Wb/m}^2$$

$$H = \frac{I}{2l}[\cos\theta_1 - \cos\theta_2] \quad \text{A/m}$$

22. What is the magnetic field at a point midway on the axis of the solenoid carrying current?

$$B = \frac{\mu_0 NI}{2\sqrt{(l/2)^2 + a^2}} \quad \text{Wb/m}^2$$

$$H = \frac{NI}{2\sqrt{(l/2)^2 + a^2}} \quad \text{A/m}$$

23. What is the magnetic field at any end of the axis of the solenoid carrying current?

$$B = \frac{\mu_0 NI}{2\sqrt{l^2 + a^2}} \quad \text{Wb/m}^2$$

$$H = \frac{NI}{2\sqrt{l^2 + a^2}} \quad \text{A/m}$$

Multiple Choice Questions

1. Determine the magnetic field $\Delta H(P)$ due to a very short current element at the origin of the coordinate system. The current element is oriented along the positive x-axis, its current is I = 1000 A, and its length is $\Delta L = 10^{-3} m$. The observation point P is with coordinates (0, 2, 0) m.

   (a) $\overline{E} = -\nabla V \ \Delta\overline{H} = 1/(8\pi)\overline{a}_x$ A/m
   (b) $\Delta\overline{H} = 1/(16\pi)\overline{a}_z$ A/m
   (c) $\Delta\overline{H} = 1000\overline{a}_y$ A/m
   (d) $\Delta\overline{H} = 0.25\overline{a}_y$ A/m

2. The magnetic field intensity inside a cylindrical wire with uniformly distributed current density depends on the distance ρ from the center of the wire as

   (a) $H \sim \rho$
   (b) $H \sim 1/\rho$
   (c) $H \sim 1/\rho^2$
   (d) $H \sim const.$

3. At a point P, the static magnetic field is given by $\bar{H}(P) = 4z\bar{a}_y - 4y\bar{a}_z$ A/m. Find the electric current density **J**(P) (in A/m^2) at P.
   (a) $\bar{J} = 0$ A/m
   (b) $\bar{J} = -4\bar{a}_y$ A/m
   (c) $\bar{J} = 8\bar{a}_x$ A/m
   (d) $\bar{J} = -8\bar{a}_x$ A/m

4. The magnetic vector potential **A** in a given region of space is given as the gradient of a scalar function: $\bar{A} = \nabla\psi$, where $\psi(x,y,z) = xyz$. Find the magnetic flux density **B**.
   (a) $\bar{B} = yz\bar{a}_x + xz\bar{a}_y + xy\bar{a}_z$ T
   (b) $\bar{B} = (y-z)\bar{a}_x - (x-z)\bar{a}_y + (x-y)\bar{a}_z$ T
   (c) $\bar{B} = const$ T
   (d) $\bar{B} = 0$ T

5. The magnetic vector potential is known to be $\bar{A} = (3x + 10xy - 13xz)\bar{a}_x$ Wb / m in a uniform nonmagnetic (i.e., $\mu = \mu_0$) isotropic region. What is the current density **J** in this region?
   (a) $\bar{J} = 0$ A/m^2
   (b) $\bar{J} = -(3 + 10y - 13x)/\mu_0\bar{a}_x$ A/m^2
   (c) $\bar{J} = (3 + 10y - 13x)\bar{a}_x$ A/m^2
   (d) $\bar{J} = 3\bar{a}_x$ A/m^2

6. The equivalent of Ohm's Law for a magnetic circuit is
   (a) $\bar{J} = \sigma\bar{E}$
   (b) $\bar{B} = \mu\bar{H}$
   (c) $\bar{H} = -\nabla V_m$
   (d) $\bar{E} = -\nabla V$

7. The inductance of a coil with no magnetic core does NOT depend on
   (a) Number of turns
   (b) Current
   (c) Ratio of flux to current
   (d) Depends on all of these

8. An electron moves through a uniform magnetic field. The force on the electron is greatest when it
   (a) moves perpendicular to the magnetic field at a fast velocity.
   (b) moves perpendicular to the magnetic field at a slow velocity
   (c) moves parallel to the magnetic field at a fast velocity
   (d) moves parallel to the magnetic field at a slow velocity.
   (e) does not move.

Explanation: **(a)** The magnitude of the force is equal to the product of the charge, speed, magnetic field, and the sine of the angle between the velocity and the magnetic field, which is θ

Questions 2.9-2.11

An electron moving with velocity 4.0×10^6 m/s enters a uniform magnetic field of 0.5 T as pictured below.

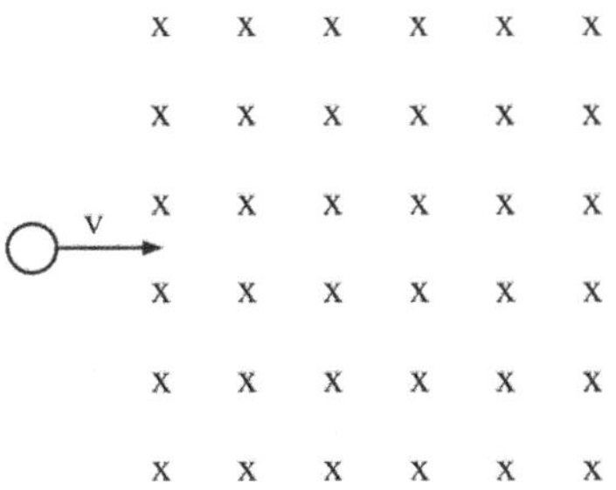

9. The electron will experience a force which is initially
   (a) into the page.
   (b) out of the page.
   (c) toward the top of the page
   (d) toward the bottom of the page.
   (e) to the left.

   Explanation: **d** We use the left-hand rule, since the electron is a negative charge, placing our fingers into the page, and the thumb to the right, in the direction of the velocity which crosses the magnetic field lines. The force, then, comes out of the palm and toward the bottom of the page.

10. The magnitude of the force acting on the electron is
    (a) 8.00×10^{-26} N
    (b) 1.82×10^{-24} N
    (c) 3.64×10^{-24} N
    (d) 3.20×10^{-13} N
    (e) 1.28×10^{-12} N

Explanation:**d** $F = qvB = \left(1.6\times10^{-19}\,C\right)\left(4.0\times10^{6}\text{ m/s}\right)\left(0.5\text{ T}\right) = 3.20\times10^{-13}\,\text{N}$

11. The resulting path of the electron is a circle of radius 4.6×10^{-5}m. The work done by the magnetic field on the electron is
    (a) 2.31×10^{-29} J
    (b) 5.26×10^{-28} J
    (c) 1.05×10^{-27} J
    (d) 9.25×10^{-17} J
    (e) zero

Explanation: **e** The magnetic force on a moving charge is always perpendicular to the magnetic field and hence no work is done by the field. The kinetic energy and the speed of the particle remain unchanged.

Questions 2.12-2.13

A stream of electrons are flowing from the cathode to the anode across a cathode ray tube (left to right) producing a current of 2 A. The strength of the magnetic field between the north pole and the south pole is 0.2 T.

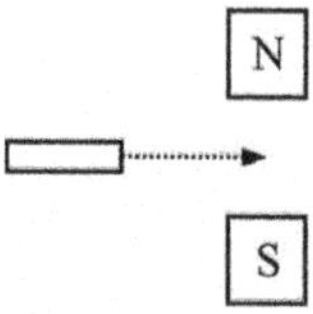

12. With a north pole and south pole as shown above, which direction will the beam deflect?
    (a) up toward the top of the page
    (b) down towards the bottom of the page
    (c) out of the page
    (d) into the page
    (e) The beam of electrons will not be deflected and will travel in a straight path.

Explanation: **c** Using the left hand rule for moving negative charges moving in a magnetic field, we can show that the force on the electron beam is out of the page toward you. According to left-hand rule, the fingers point toward the bottom of the page in the direction of the magnetic field, the thumb points toward the right of the page in the direction of the current, and the force comes out of the palm and out of the page.

13. The length of the beam of electrons in the magnetic field is 0.15 m. The force on the beam of electrons is
    (a) 0.015 N
    (b) 0.06 N
    (c) 0.15 N
    (d) 2.0 N
    (e) 6.7 N

Explanation: **b** $F = ILB = (2\text{ A})(0.2\text{ T})(0.15\text{ m}) = 0.06\text{ N}$

14. All of the following will affect the force acting on a current carrying wire, EXCEPT
    (a) the amount of current in the wire
    (b) the length of the wire.
    (c) the strength of the magnetic field
    (d) the amount of charge per unit time in the wire.
    (e) the magnitude of the charges in the wire.

Explanation: **e** $F = ILB$

15. A wire on the x-axis of a coordinate system has a current I in the +x direction as shown below. What is the direction of the magnetic field due to the wire at point P?

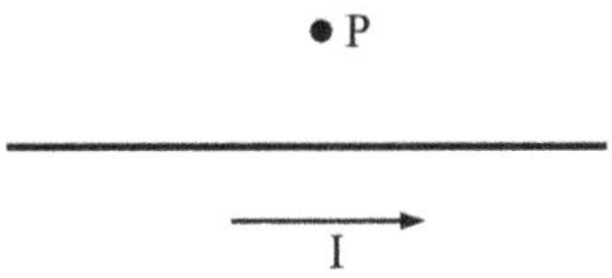

(a) to the left

(b) to the right

(c) down into the page and perpendicular to the page

(d) up out of the page and perpendicular to the page

(e) toward the bottom of the page

Explanation: **d** By the right hand rule, a current flowing to the right in the wire will produce a magnetic field which is into the page below the wire and coming out of the page above the wire.

16. A wire carrying a current I directed toward the bottom of the page is placed in a magnetic field B as shown below. The force on the wire is directed

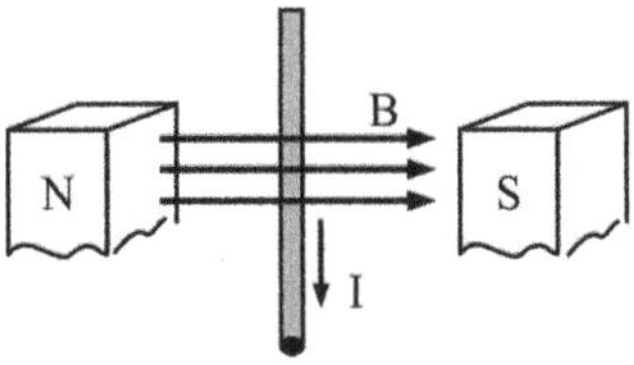

(a) into the page

(b) out of the page

(c) toward the top of the page

(d) toward the bottom of the page

(e) to the left

Explanation: **b** According to right-hand rule, the fingers point to the right of the page in the direction of the magnetic field, the thumb points toward the bottom of the page in the direction of the current, and the force comes out of the palm and out of the page.

17. Two negatively charged particles enter a uniform magnetic field, as pictured below. The particles have the same charge but different masses. Ignoring the gravitational force, the trajectory of the less massive particle is best represented by path

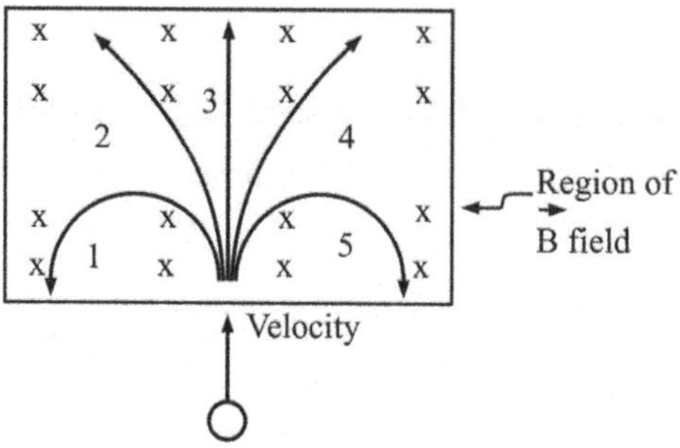

(a) 1 (b) 2 (c) 3 (d) 4

(e) 5

Explanation: **e** The left hand rule indicates the force on the particle is to the right. Since the particles have the same charge and the same velocity, the less massive particle is deflected more than the more massive particle since it has less inertia.

$$r = \frac{mv}{qB}$$

18. What is the force between two current carrying conductors?

(a) $F = \frac{\mu_0 I_1 I_2}{2\pi r}$ N (b) $F = \frac{I_1 I_2}{2\pi r}$ N

(c) $F = \frac{\mu_0 I_1 I_2}{2\pi}$ N (d) $F = \frac{\mu_0 I_1 I_2}{2r}$ N

19. What is the magnetic flux density at the centre of the circular coil carrying current?

(a) $B = \frac{\mu_0 I}{2a}$ Wb / m^2 (b) $B = \frac{I}{2a}$ Wb / m^2

(c) $B = \frac{\mu_0 I}{a}$ Wb / m^2 (d) $B = \frac{\mu_0 I}{4a}$ Wb / m^2

20. What is the magnetic flux density at any point on the axis of a circular coil carrying current?

(a) $B = \frac{Ia^2}{2(a^2 + d^2)^{3/2}}$ Wb/m^2 (b) $B = \frac{\mu_0 Ia^2}{2(a^2 + d^2)^{1/2}}$ Wb/m^2

(c) $B = \frac{\mu_0 Ia^2}{(a^2 + d^2)^{3/2}}$ Wb/m^2 (d) $B = \frac{\mu_0 Ia^2}{2(a^2 + d^2)^{3/2}}$ Wb/m^2

Answers

1.	(b)	2.	(a)	3.	(d)	4.	(d)
5.	(a)	6.	(b)	7.	(b)	8.	(a)
9.	(d)	10.	(d)	11.	(e)	12.	(c)
13.	(b)	14.	(e)	15.	(d)	16.	(b)
17.	(e)	18.	(a)	19.	(a)	20.	(d)

Exercise Questions

1. Define Ampere's Force Law and establish the associated relations.
2. Using the relation for $\bar{H}$ of a finite straight wire, obtain the expressions for the fields due to a semi infinite wire and infinite wire located on Z-axis, carrying current I.
3. State Maxwell's equations for magneto static fields.
4. State Ampere's circuital law. Specify the conditions to be met for determining magnetic field strength H, based on Ampere's circuital law.
5. Define Magnetic flux density and vector magnetic potential.
6. Derive an expression for magnetic field strength H due to a finite filamentary conductor carrying a current I and placed along Z-axis at a point 'p' on Y-axis. Hence deduce the magnetic field strength for the length of the conductor extending from $-\infty$ to ∞.
7. Find the magnetic field strength H at the centre of a square conducting loop of side '2a' in Z=0 plane if the loop is carrying current I in anti-clockwise direction.
8. Derive an equation of continuity for static magnetic fields.
9. Derive an expression for force between two straight long parallel current carrying conductors. What will be the nature of force if he currents are in the same and opposite directions?
10. Derive an expression for energy density in a magnetic field.

Chapter 3

Maxwell's Equations for Time Varying Fields

3.1 Introduction

The static electric fields are represented by $\bar{E}(x,y,z)$ and static magnetic fields are represented by $\bar{H}(x,y,z)$. We know that the static electric fields are generated by static charges and static magnetic fields are generated by moving charges or steady currents. In static electromagnetic fields, the static electric field and static magnetic field are independent. The time varying electric field can be represented with $\bar{E}(x,y,z,t)$ and time varying magnetic field can be represented with $\bar{H}(x,y,z,t)$. In time varying electromagnetic fields, the electric and magnetic fields are interdependent. These fields are generated by accelerated charges or the time varying current waveforms that are shown in Fig.3.1.

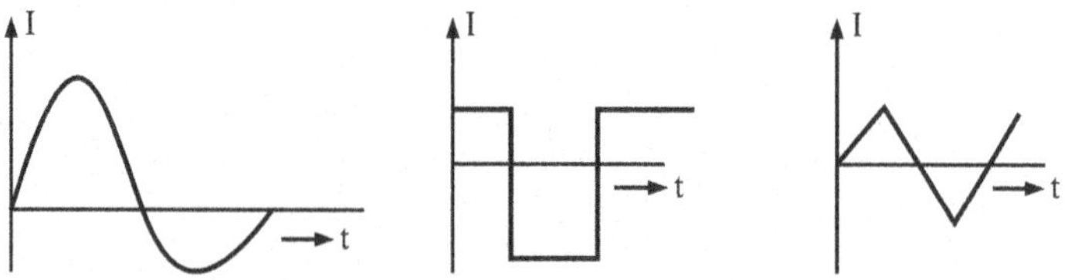

Fig. 3.1 Current waveforms

We can also call time varying electromagnetic fields as electromagnetic waves as they are produced by time varying currents.

3.2 Faraday's Law and Transformer EMF

The two basic concepts that we are going to study in time varying electromagnetic waves are (i) induced emf according to Faraday's law (ii) displacement current.

According to Faraday's law in a closed circuit the induced emf is equal to the time rate of change of magnetic flux linkage

i.e., $$V_{emf} = -\frac{d\lambda}{dt} \quad .....(3.2.1)$$

If each turn of the circuit carries flux 'ψ' then flux linkage $\lambda = N\psi$, then induced emf is

$$V_{emf} = -\frac{Nd\psi}{dt} \quad .....(3.2.2)$$

Where –ve sign indicates, the induced emf opposes the magnetic flux linkage. This is called Lenz's law.

So far we know that the electric fields are generated by static charges. But there are other sources that generate electric fields, they are called emf produced fields the sources of generating emf produced fields are Electric generator, Batteries and Fuel cells etc. All these convert non electrical energy into electrical energy. Let us consider a circuit which has battery as shown in Fig.3.2.

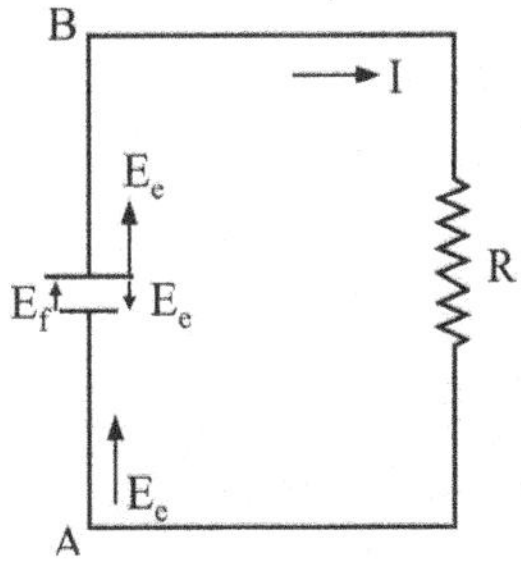

Fig. 3.2 Circuit which has battery

Due to electromechanical action of the battery emf produced field exists E_f inside the battery in the direction shown in Fig.3.2. Due to accumulation of charges on the plates of the battery there will be static electric field in opposite to the E_f in side the battery. The direction of E_e outside the battery is in opposite to the inside the battery.

∴ The total electric field $\bar{E} = \bar{E}_f + \bar{E}_e$

Note that $\bar{E}_f$ outside the battery is '0'.

Let us take the contour line integral of $\bar{E}$

i.e., $$\oint \bar{E} \cdot d\bar{l} = \int_A^B \bar{E}_f \cdot d\bar{l} + 0 = IR = V_{emf}$$

Consider a circuit which has a single turn, then the induced emf according to Faraday's law is $V_{emf} = -\frac{d\psi}{dt}$. We know that the magnetic flux $\psi = \int \overline{B} \cdot \overline{ds}$

$$\therefore \qquad V_{emf} = -\frac{d}{dt}\int \overline{B} \cdot \overline{ds}$$

Consider a constant loop circuit which carries current I and time varying magnetic field $\overline{B}(t)$ as shown in Fig.3.3.

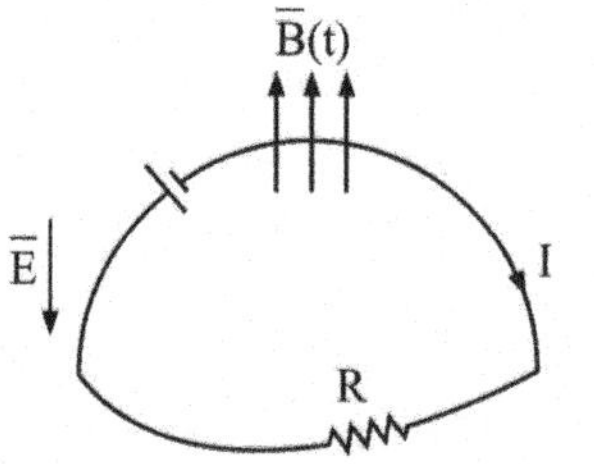

Fig. 3.3 Constant loop circuit whi carries current I and time varyin magnetic fields $\overline{B}(t)$

Then
$$V_{emf} = -\int \frac{\partial \overline{B}}{\partial t} \cdot \overline{ds} = \oint \overline{E} \cdot \overline{dl}$$

here V_{emf} is called transformer emf.

By applying the Stoke's theorem to the last term in the above equation, we get

$$-\int \frac{\partial \overline{B}}{\partial t} \cdot \overline{ds} = \int \nabla \times \overline{E} \cdot \overline{ds}$$

i.e.,
$$\nabla \times \overline{E} = -\frac{\partial \overline{B}}{\partial t} \qquad .....(3.2.3)$$

which is one of the **Maxwell's equation** for time varying fields. From the above equation we can say that both electric field and magnetic field are interdependent in time varying fields.

Problem 3.1

A conducting bar can slide freely over two conducting rails as shown in Fig.3.4. Calculate the induced voltage in bar, if the bar is stationed at y = 8 cm and $\overline{B} = 4\cos\left(10^6 t\right)\overline{a}_z$ mWb/m^2.

Solution

$$\frac{\partial \overline{B}}{\partial t} = -4\sin\left(10^6 t\right)10^6\,\overline{a}_z\, m\,\text{Wb/m}^2 \quad \text{here } \overline{ds} = dx\,dy\,\overline{a}_z$$

$$V_{emf} = -\int \frac{\partial \overline{B}}{\partial t} \cdot \overline{ds}$$

$$= -\int_{x=0}^{0.06}\int_{y=0}^{0.08} -4\times 10^6 \sin\left(10^6 t\right) dx\,dy$$

$$= 19.2 \sin 10^6 \text{ t V}$$

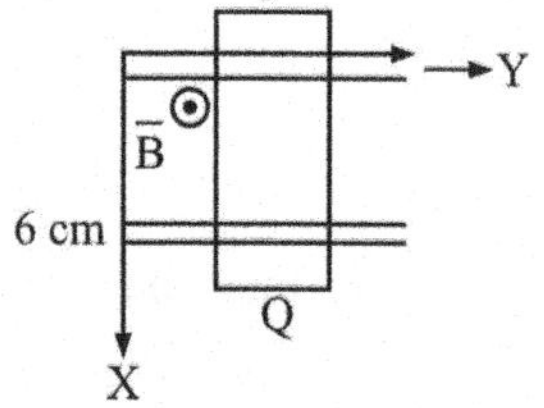

Fig. 3.4

*Problem 3.2

In figure let $\bar{B} = 0.2\cos 120\ \pi t\ T$, and assume that the conductor joining the two ends of the resistor is perfect. It may be assumed that the magnetic field produced by *I*(*t*) is negligible. Find (a) $V_{ab}(t)$ (b) $I(t)$.

Solution

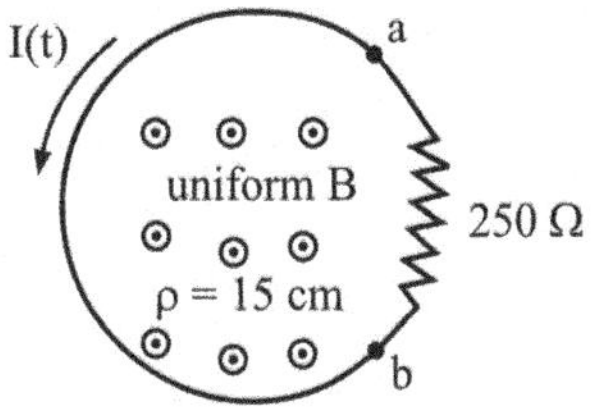

Fig. 3.5

(a) We have $V_{emf} = -\dfrac{d\psi}{dt}$

Here $V_{ab}(t) = V_{emf} = -\dfrac{d\psi}{dt}$

$\psi = Ba$, where 'a' is the area of cross-section of the loop $= \pi\rho^2$

$$\therefore \psi = 0.2\cos 120\pi t \times \pi \times (15\times 10^{-2})^2 = 0.0141\cos 120\pi t$$

$$\text{Hence } V_{ab}(t) = -\frac{d\psi}{dt} = -\frac{d(0.0141\cos 120\pi t)}{dt} = 5.326\sin 120\pi t \text{ V}$$

(b) From Fig.3.5, $I(t) = \dfrac{V_{ba}(t)}{R} = \dfrac{-V_{ab}(t)}{R} = \dfrac{-5.326\sin 120\pi t}{250} = -0.0213\sin 120\pi t$ A

*Problem 3.3

A circular loop conductor of radius 0.1 m lies in the Z = 0 plane and has a resistance of 5Ω given $\bar{B} = 0.20\sin 10^3 t\bar{a}_z\ T$. Determine the current.

Solution

We have $\psi = \int_s \bar{B}.\bar{ds}$, here $\bar{ds} = \rho d\phi\ d\rho\ \bar{a}_z$

$$\psi = \int_s 0.20\sin 10^3 t\bar{a}_z.\rho d\phi\ d\rho\ \bar{a}_z = 0.20\sin 10^3 t \int_{\phi=0}^{2\pi} d\phi \int_{\rho=0}^{0.1} \rho\ d\rho = \frac{2\pi\sin 10^3 t}{10^3}$$

We also have $V_{emf} = -\dfrac{d\psi}{dt}$

$$V_{emf} = -\frac{d\left(\dfrac{2\pi\sin 10^3 t}{10^3}\right)}{dt} = -2\pi\cos 10^3 t\ \text{ V}$$

The current in the loop $I = \dfrac{V_{emf}}{R} = \dfrac{2\pi\cos 10^3 t}{5} = 1.26\cos 10^3 t$ A

3.3 In-Consistency of Ampere's Law and Displacement Current Density

We know the differential from of Ampere's law i.e., $\nabla \times \bar{H} = \bar{J}$.

Where $\bar{J}$ is the conduction current density

By taking divergence on both sides of the above equation

$$\nabla \cdot \left(\nabla \times \bar{H}\right) = \nabla \cdot \bar{J}$$

We know the identity $\nabla \cdot \left(\nabla \times \bar{A}\right) = 0,$

where $\bar{A}$ is any vector

$$\therefore \qquad \nabla \cdot \bar{J} = 0 \qquad \text{.....(3.3.1)}$$

According to Continuity equation for time varying fields

$$\nabla \cdot \bar{J} = \frac{-\partial \rho_V}{\partial t} \qquad \text{.....(3.3.2)}$$

For time varying fields equations (3.3.1) and (3.3.2) are not comparable. Hence Ampere's law can not be used directly for time varying fields. To overcome inconsistency of Ampere's law, modify the differential form of Ampere's law as $\nabla \times \bar{H} = \bar{J} + \bar{J}_D$

where $\bar{J}_D$ is the displacement current density.

Apply divergence on both sides of the above equation

$$\nabla \cdot \left(\nabla \times \bar{H}\right) = \nabla \cdot \bar{J} + \nabla \cdot \bar{J}_D = 0$$

$$\nabla \cdot \bar{J}_D = -\nabla \cdot \bar{J}$$

$$\nabla \cdot \bar{J}_D = \frac{\partial \rho_V}{\partial t}$$

We know from Maxwell's first equation $\rho_V = \nabla \cdot \bar{D}$

$$\therefore \qquad \nabla \cdot \bar{J}_D = \frac{\partial}{\partial t}\left(\nabla \cdot \bar{D}\right) = \nabla \cdot \frac{\partial \bar{D}}{\partial t}$$

From the above equation $\bar{J}_D = \dfrac{\partial \bar{D}}{\partial t}$ .....(3.3.3)

Now the modified Maxwell's equation for time varying fields is

$$\nabla \times \bar{H} = \bar{J} + \frac{\partial \bar{D}}{\partial t} \quad .....(3.3.4)$$

The displacement current $I_D = \int \bar{J}_D \cdot \bar{ds}$ .....(3.3.5)

***Problem 3.4**

The electric field intensity in the region $0 < x < 5,\ 0 < y < \pi/12,\ 0 < z < 0.06$ m in free space is given by $\bar{E} = c\sin 12y \sin az \cos 2\times 10t\bar{a}_x$ V/m. Beginning with the $\nabla \times \bar{E}$ relationship, use Maxwell's equations to find a numerical value for a, if it is known that a is greater than '0'.

Solution

Given $\bar{E} = c\sin 12y \sin az \cos 2\times 10t\bar{a}_x$ V / m for $a > 0$

Consider $$\nabla \times \bar{E} = \begin{vmatrix} \bar{a}_x & \bar{a}_y & \bar{a}_z \\ \frac{\partial}{\partial x} & \frac{\partial}{\partial y} & \frac{\partial}{\partial z} \\ E_x & E_y & E_z \end{vmatrix} = \begin{vmatrix} \bar{a}_x & \bar{a}_y & \bar{a}_z \\ \frac{\partial}{\partial x} & \frac{\partial}{\partial y} & \frac{\partial}{\partial z} \\ c\sin 12y \sin az \cos 2\times 10t & 0 & 0 \end{vmatrix}$$

$$\nabla \times \bar{E} = ac\sin 12y \cos az \cos 2\times 10t\bar{a}_y - 12c\cos 12y \sin az \cos 2\times 10t\bar{a}_z$$

We have $\nabla \times \bar{E} = \frac{-\partial \bar{B}}{\partial t}$

$$\frac{-\partial \bar{B}}{\partial t} = ac\sin 12y \cos az \cos 2\times 10t\bar{a}_y - 12c\cos 12y \sin az \cos 2\times 10t\bar{a}_z$$

Integrating

$$\bar{B} = \left(\frac{-ac\sin 12y \cos az \sin 2\times 10t\bar{a}_y}{20} + \frac{12c\cos 12y \sin az \sin 2\times 10t\bar{a}_z}{20} \right)$$

But $\bar{B} = \mu_0 \bar{H}$

$$\bar{H} = \left(\frac{-ac\sin 12y \cos az \sin 2\times 10t\bar{a}_y}{20\mu_0} + \frac{12c\cos 12y \sin az \sin 2\times 10t\bar{a}_z}{20\mu_0} \right)$$

$$\nabla\times\bar{H}=\begin{vmatrix}\bar{a}_x & \bar{a}_y & \bar{a}_z\\ \frac{\partial}{\partial x} & \frac{\partial}{\partial y} & \frac{\partial}{\partial z}\\ H_x & H_y & H_z\end{vmatrix}=\begin{vmatrix}\bar{a}_x & \bar{a}_y & \bar{a}_z\\ \frac{\partial}{\partial x} & \frac{\partial}{\partial y} & \frac{\partial}{\partial z}\\ 0 & \frac{-ac\sin 12y\cos az\sin 2\times 10t}{20\mu_0} & \frac{12c\cos 12y\sin az\sin 2\times 10t}{20\mu_0}\end{vmatrix}$$

$$\nabla\times\bar{H}=-\left(\frac{12^2+a^2}{20\mu_0}\right)c\sin 12y\sin az\sin 2\times 10t\bar{a}_x$$

But for free space($\rho_v=0$), we have $\nabla\times\bar{H}=\frac{\partial\bar{D}}{\partial t}=\in_0\frac{\partial\bar{E}}{\partial t}$

$$-\left(\frac{12^2+a^2}{20\mu_0}\right)c\sin 12y\sin az\sin 2\times 10t\bar{a}_x=\in_0\frac{\partial(c\sin 12y\sin az\cos 2\times 10t)\bar{a}_x}{\partial t}$$

$$-\left(\frac{12^2+a^2}{20\mu_0}\right)c\sin 12y\sin az\sin 2\times 10t\bar{a}_x=-20\in_0 c\sin 12y\sin az\sin 2\times 10t\bar{a}_x$$

$$\left(\frac{12^2+a^2}{20\mu_0}\right)=20\in_0\Rightarrow 12^2+a^2=20^2\in_0\mu_0$$

$$a^2=20^2\times 8.854\times 10^{-12}\times 4\pi\times 10^{-7}-12^2\Rightarrow a=j12$$

***Problem 3.5**

A certain material has $\sigma=0$ *and* $\in_R=1$. If $\bar{H}=4\sin(10^6t-0.01z)\bar{a}_y$ A/m make use of Maxwell's equations to find μ_r.

Solution

$$\nabla\times\bar{H}=\begin{vmatrix}\bar{a}_x & \bar{a}_y & \bar{a}_z\\ \frac{\partial}{\partial x} & \frac{\partial}{\partial y} & \frac{\partial}{\partial z}\\ H_x & H_y & H_z\end{vmatrix}=\begin{vmatrix}\bar{a}_x & \bar{a}_y & \bar{a}_z\\ \frac{\partial}{\partial x} & \frac{\partial}{\partial y} & \frac{\partial}{\partial z}\\ 0 & 4\sin(10^6t-0.01z) & 0\end{vmatrix}$$

$$\nabla \times \bar{H} = -\frac{\partial[4\sin(10^6 t - 0.01z)\bar{a}_x]}{\partial z} = 0.04\cos(10^6 t - 0.01z)\bar{a}_x$$

Since $\sigma = 0$, we have $\nabla \times \bar{H} = \frac{\partial \bar{D}}{\partial t} = \epsilon_0 \epsilon_r \frac{\partial \bar{E}}{\partial t} = \epsilon_0 \frac{\partial \bar{E}}{\partial t}$

$$\bar{E} = \frac{1}{\epsilon_0}\int 0.04\cos(10^6 t - 0.01z)\bar{a}_x dt$$

$$\bar{E} = \frac{0.04\sin(10^6 t - 0.01z)\bar{a}_x}{10^6\,\epsilon_0}$$

$$\nabla \times \bar{E} = \begin{vmatrix} \bar{a}_x & \bar{a}_y & \bar{a}_z \\ \frac{\partial}{\partial x} & \frac{\partial}{\partial y} & \frac{\partial}{\partial z} \\ E_x & E_y & E_z \end{vmatrix} = \begin{vmatrix} \bar{a}_x & \bar{a}_y & \bar{a}_z \\ \frac{\partial}{\partial x} & \frac{\partial}{\partial y} & \frac{\partial}{\partial z} \\ \frac{0.04\sin(10^6 t - 0.01z)}{10^6\,\epsilon_0} & 0 & 0 \end{vmatrix}$$

$$\nabla \times \bar{E} = -\frac{4\times10^{-4}\cos(10^6 t - 0.01z)}{10^6 \times 8.854\times10^{-12}}\bar{a}_y$$

$$\nabla \times \bar{E} = -45.2\cos(10^6 t - 0.01z)\bar{a}_y$$

We have $\nabla \times \bar{E} = \frac{-\partial \bar{B}}{\partial t} = -\mu_0 \mu_r \frac{\partial \bar{H}}{\partial t}$

$$-45.2\cos(10^6 t - 0.01z)\bar{a}_y = -\mu_0 \mu_r \frac{\partial \bar{H}}{\partial t}$$

$$45.2\cos(10^6 t - 0.01z)\bar{a}_y = 4\times10^6\,\mu_0\mu_r\cos(10^6 t - 0.01z)\bar{a}_y$$

$$\mu_r = \frac{45.2}{4\times10^6\,\mu_0} = \frac{45.2}{4\times10^6\times4\pi\times10^{-7}} = 8.99$$

3.3.1 Ratio of Conduction Current and Displacement Current

If the current is passing through a conductor whose conductivity is 'σ', then let us find the ratio of J_D to J_c. We know the conduction current density $J_c = \sigma E$. For time varying fields assume $E = E_0\, e^{j\omega t}$.

$\therefore \quad J_c = \sigma E_0\, e^{j\omega t}$.

Displacement current density $J_D = \dfrac{\partial D}{\partial t}$

$$= \in \frac{\partial E}{\partial t} = \in j\omega E_0\, e^{j\omega t}$$

$$\therefore \quad \frac{J_D}{J_C} = \frac{\in j\omega E_0\, e^{j\omega t}}{E_0\, e^{j\omega t}\, \sigma}$$

$$\left|\frac{J_D}{J_C}\right| = \frac{\omega \in}{\sigma}$$

$$\Rightarrow \quad \left|\frac{I_D / A}{I_C / A}\right| = \frac{\omega \in}{\sigma} \Rightarrow \frac{I_D}{I_C} = \frac{\omega \in}{\sigma} \qquad \text{.....(3.3.6)}$$

Problems 3.6

Show that the displacement current through the capacitor is equal to the conduction current.

Solution

We know the conduction current $I = C\dfrac{dv}{dt}$. Assume that the capacitor plates are having area 'A' and are separated by a distance 'd' as shown in Fig. 3.6.

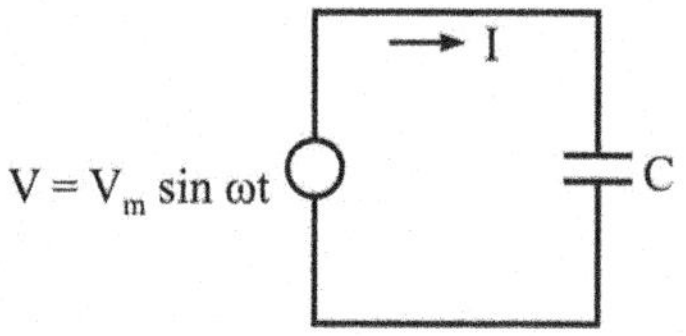

Fig. 3.6

Then $$C = \frac{\in A}{d}$$

$$I = \frac{\in A}{d}\frac{d}{dt}(V_m \sin \omega t)$$

$$= \frac{A \in}{d} V_m \omega \cos \omega t$$

We have

$$J_D = \frac{\partial D}{\partial t}$$

But $$J_D = \frac{I_D}{A}$$

$$\frac{I_D}{A} = \frac{\partial D}{\partial t} \qquad \text{where} \qquad D = \epsilon E$$

$$\frac{I_D}{A} = \epsilon \frac{\partial E}{\partial t} \qquad \text{where} \qquad E = \frac{V}{d}$$

$$\frac{I_D}{A} = \frac{\epsilon}{d}\frac{\partial V}{\partial t} = \frac{\epsilon}{d}\frac{\partial}{\partial t}(V_m \sin \omega t) = \frac{\epsilon}{d} V_m \omega \cos \omega t$$

$$I_D = \frac{A\epsilon}{d} V_m \omega \cos \omega t$$

∴ conduction current = displacement current

Problem 3.7

A parallel plate capacitor with plate area of 5 cm^2 and plate separation of 3 mm has a voltage 50 sin ($10^3\, t$) V applied to it's plates. Calculate the displacement current assuming $\epsilon = 2\epsilon_0$.

Solution

$$I_D = \frac{A\epsilon}{d} V_m \omega \cos \omega t$$

$$= \frac{5\times10^{-4}\times 2\epsilon_0}{3\times10^{-3}} 50\times10^3 \cos 10^3 t$$

$$= 10^3 \times 16.65\, \epsilon_0 \cos 10^3 t$$

$$= 16.65\epsilon_0 \cos 10^3 t \text{ kA}$$

$$= 147.4 \cos 10^3 t \text{ n A}$$

Problem 3.8

A 'Cu' wire carries a conduction current of 1A at 60 Hz. What is the displacement current in the wire. For 'Cu' wire $\epsilon = \epsilon_0$, $\sigma = 5.8 \times 10^7$ and $\mu = \mu_0$.

Solution:

$$I_D = I_C\left(\frac{\omega\epsilon}{\sigma}\right) = 1\left(\frac{2\pi\times60\times8.854\times10^{-12}}{5.8\times10^7}\right)$$

$$= 575.2 \times 10^{-19} \text{ A}$$

Problem 3.9

In free space $\bar{E} = 20\cos(\omega t - 50x)\bar{a}_y$ V/m. Calculate $\bar{J}_D$.

Solution

$$\bar{D} = \epsilon_0 \bar{E} = 20 \epsilon_0 \cos(\omega t - 50x)\bar{a}_y$$

$$\bar{J}_D = \frac{\partial \bar{D}}{\partial t} = -20\omega \epsilon_0 \sin(\omega t - 50x)\bar{a}_y \quad \text{A/m}^2$$

3.4 Maxwell's Equations in different Final Forms and Word Statements

S.No.	Differential form (or) point form	Integral form	Remarks
1	$\nabla \times \bar{H} = \bar{J} + \frac{\partial \bar{D}}{\partial t}$	$\oint_L \bar{H} \cdot d\bar{l} = \int_s \left(\bar{J} + \frac{\partial \bar{D}}{dt}\right) \cdot d\bar{s}$	Ampere's circuit Law
2	$\nabla \times \bar{E} = -\frac{\partial \bar{B}}{\partial t}$	$\oint_L \bar{E} \cdot d\bar{l} = -\int_s \frac{\partial \bar{B}}{\partial t} \cdot d\bar{s}$	Faraday's law
3	$\nabla \cdot \bar{D} = \rho_V$	$\oint_s \bar{D} \cdot d\bar{s} = \int_v \rho_V dv$	Gauss's law
4	$\nabla \cdot \bar{B} = 0$	$\oint_s \bar{B} \cdot d\bar{s} = 0$	Gauss's Law for magnetic fields

***Word statements for Maxwell's equations given in the table are as follows*:**

1. The magneto motive force around a closed path is equal to the sum of conduction current density and the time derivative of electric flux density through a surface i.e., bounded by path.
2. The electromotive force around a closed path is equal to the time derivative of magnetic flux density through a surface i.e., bounded by path.
3. The electric flux through a closed surface is equal to the charge enclosed by that surface.
4. The magnetic flux through a closed surface is zero. Which also says Nonexistence of isolated magnetic charge.

Problem 3.10

A two dimensional Electric field is given by $\bar{E} = x^2\bar{a}_x + x\bar{a}_y$ V/m. Show that this electric field can not arise from a static distribution of charge.

Solution

For static Electric field, we have $\nabla \times \bar{E} = 0$

$$\nabla \times \bar{E} = \begin{vmatrix} \bar{a}_x & \bar{a}_y & \bar{a}_z \\ \frac{\partial}{\partial x} & \frac{\partial}{\partial y} & \frac{\partial}{\partial z} \\ x^2 & x & 0 \end{vmatrix} = \bar{a}_x(0-0) - \bar{a}_y(0-0) + \bar{a}_z(1) = \bar{a}_z \neq 0$$

$\therefore$ The given $\bar{E}$ is not due to static distribution of charge

Problem 3.11

A conductor carries a steady current of 'I' amp. The components of current density vector $\bar{J}$ are $J_x = 2ax$ and $J_y = 2ay$. Find the third component J_z.

Solution

Given steady current, that indicates static fields. For static fields, we have

$$\nabla \cdot \bar{J} = \frac{\partial \rho_V}{\partial t} = 0$$

$$\Rightarrow \quad \frac{\partial}{\partial x}2ax + \frac{\partial}{\partial y}2ay + \frac{\partial}{\partial z}J_z = 0$$

$$\Rightarrow \quad 2a + 2a + \frac{\partial}{\partial z}J_z = 0$$

$$\frac{\partial}{\partial z}J_z = -4a$$

Integrating, we get third component as

$J_z = -4az + c$, where c is constant of integration.

Problem 3.12

Do the fields $\bar{E} = E_m \sin x \sin t\, \bar{a}_y$

$\bar{H} = \frac{E_m}{\mu_0} \cos x \cos t\, \bar{a}_z$ satisfy Maxwell's equations.

Solution

We have

$$\nabla \times \bar{E} = -\frac{\partial}{\partial t}\bar{B}$$

$$\nabla \times \bar{E} = -\mu_0 \frac{\partial}{\partial t}\bar{H}$$

Assume the electromagnetic wave is traveling along x-direction, then the components of $\bar{E}$ and $\bar{H}$ along x-direction will be zero. Variations along y and z directions are zero i.e., $\frac{\partial}{\partial y}$ and $\frac{\partial}{\partial z}$ are zero.

$$\nabla \times \bar{E} = \begin{vmatrix} \bar{a}_x & \bar{a}_y & \bar{a}_z \\ \frac{\partial}{\partial x} & 0 & 0 \\ 0 & E_m \sin x \sin t & 0 \end{vmatrix} = \bar{a}_z E_m \cos x \sin t$$

$$-\mu_0 \frac{\partial}{\partial t}\bar{H} = -E_m \cos x \sin t\, \bar{a}_z$$

$$\therefore \quad \nabla \times \bar{E} = -\frac{\partial}{\partial t}\bar{B}$$

$\therefore$ $\bar{\text{E}}$ and $\bar{\text{H}}$ satisfy the Maxwell's equations.

Problem 3.13

Find the frequency at which conduction current density and displacement current density are equal in a medium with $\sigma = 2 \times 10^{-4}$ ℧/m and $\epsilon_r = 81$.

Solution

We know

$$\frac{J_D}{J_C} = \frac{\omega \epsilon}{\sigma}$$

$$\Rightarrow \quad 1 = \frac{2\pi f \times 8.854 \times 10^{-12} \times 81}{2 \times 10^{-4}}$$

f = 44.384 kHz

Problem 3.14

Calculate the ratio J_D/J_C for 'Al' at frequencies of 50 Hz and 50 MHz, given $\sigma = 10^5$ ℧/m and $\epsilon_r = 1$.

Solution

For f = 50 Hz

$$\frac{J_D}{J_C} = \frac{2\pi 50 \times 8.854 \times 10^{-12} \times 1}{10^5}$$
$$= 2.782 \times 10^{-14}$$

and for f = 50 MHz

$$\frac{J_D}{J_C} = 2.782 \times 10^{-8}$$

3.5 Boundary Conditions for Electric Fields

So far we have seen an electric field on a single homogeneous medium. Now let us have an electric field in two different media. The conditions that the electric field satisfies at the interface of two media are called boundary conditions.

Boundary conditions between dielectric-dielectric, conductor–dielectric and conductor-free space are determined as:

To find boundary conditions we must use the Maxwell's equations $\oint \bar{E} \cdot d\bar{l} = 0$ and $\oint_s \bar{D} \cdot d\bar{s} = Q_{enc}$

Also we have to decompose electric field strength, $\bar{E}$ into tangential component ($\bar{E}_t$) and normal component ($\bar{E}_n$).

$$\therefore \quad \bar{E} = \bar{E}_t + \bar{E}_n \qquad \text{.....(3.5.1)}$$

Similarly $\bar{D}$ can be decomposed

3.5.1 Dielectric – Dielectric

Let us consider two dielectric media with ϵ_1 and ϵ_2 as shown in the Fig.3.7

where ϵ_1 is the permittivity in medium 1 and ϵ_2 is the permittivity in medium 2.

and $\epsilon_1 = \epsilon_0 \epsilon_{r1}$, $\epsilon_2 = \epsilon_0 \epsilon_{r2}$.

Apply $\oint \bar{E} \cdot d\bar{l} = 0$ around abcda loop

$$0 = \bar{E}_{1t}\Delta W - \bar{E}_{1n}\frac{\Delta h}{2} - \bar{E}_{2n}\frac{\Delta h}{2} - \bar{E}_{2t}\Delta W +$$

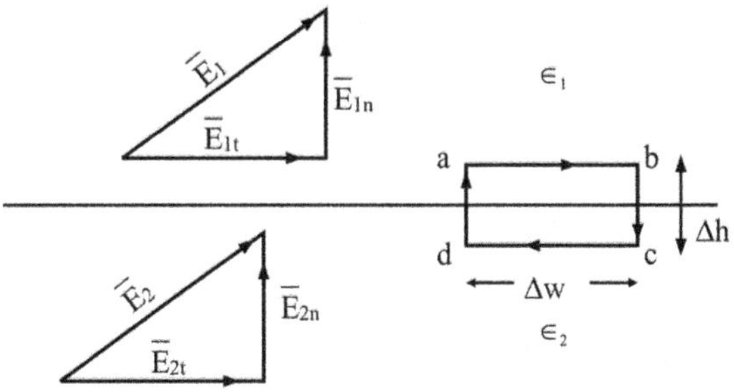

Fig. 3.7 Dielectric – Dielectric boundary for establishing relation between tangential components

$\bar{E}_{2n}\frac{\Delta h}{2}+\bar{E}_{1n}\frac{\Delta h}{2}$ as Δh tends to '0' the above equation becomes

$$\bar{E}_{1t}=\bar{E}_{2t} \qquad .....(3.5.2)$$

i.e., the tangential component in medium 1 is equal to the tangential component in medium 2. Hence the tangential component of electric field intensity is continuous at the interface.

We know $\bar{E}=\frac{\bar{D}}{\epsilon}$ based on this, the above equation becomes

$$\frac{\bar{D}_{1t}}{\epsilon_1}=\frac{\bar{D}_{2t}}{\epsilon_2} \qquad .....(3.5.3)$$

Hence the electric flux density is discontinuous.

To find the relation between normal components consider a small cylinder which intersects both the media as shown in the Fig.3.8.

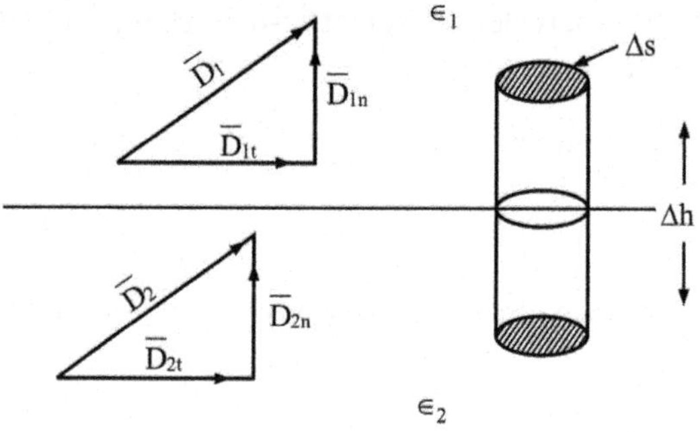

Fig. 3.8 Dielectric – Dielectric boundary for establishing relation between normal components

Apply $\int_s \bar{D}\cdot d\bar{s}=Q=\int \rho_s ds$ to the considered cylindrical surface.

Then $\Delta Q=\rho_s\,\Delta s$

$$\therefore \quad \bar{D}_{1n}\Delta S-\bar{D}_{2n}\Delta S=\rho_s\Delta S$$

$$\bar{D}_{1n}-\bar{D}_{2n}=\rho_s$$

Where ρ_s is the surface charge distribution at the interface.

If $\rho_s=0$ then $\bar{D}_{1n}=\bar{D}_{2n}$ .....(3.5.4)

i.e., the normal component of electric flux density is continuous.

Since $\bar{D}=\epsilon\bar{E}$

$$\Rightarrow \quad \epsilon_1\bar{E}_{1n}=\epsilon_2\bar{E}_{2n} \qquad .....(3.5.5)$$

i.e., the normal component of electric field intensity is discontinuous.

3.5.2 Law of Refraction for Electric Fields

Angle of refraction can be determined with the boundary conditions. Consider the Fig.3.9.

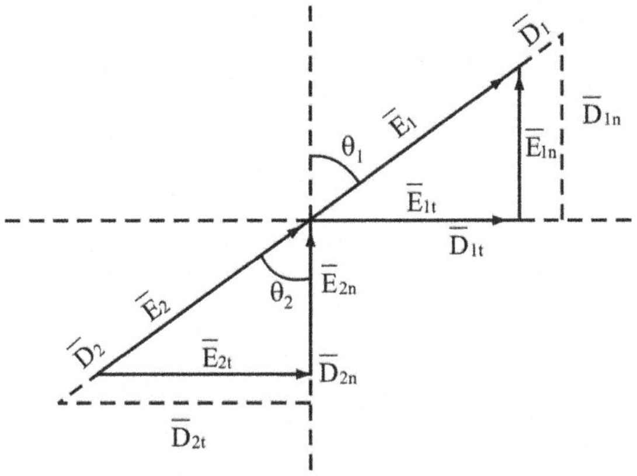

Fig. 3.9 Evaluation of angle of refraction

Where θ_1 is the angle made by $\bar{E}_1$ with the normal component to the interface. Similarly θ_2 is the angle made by $\bar{E}_2$ with the normal component to the interface. From Fig.3.9, $\bar{E}_{1t} = \bar{E}_1 \sin\theta_1$ and $\bar{E}_{2t} = \bar{E}_2 \sin\theta_2$. We know $\bar{E}_{1t} = \bar{E}_{2t}$

$$\therefore \quad \bar{E}_1 \sin\theta_1 = \bar{E}_2 \sin\theta_2 \qquad \text{.....(3.5.6a)}$$

and $\quad \bar{D}_{1n} = \bar{D}_1 \cos\theta_1$. Similarly $\bar{D}_{2n} = \bar{D}_2 \cos\theta_2$

we know $\bar{D}_{1n} = \bar{D}_{2n}$

$$\therefore \quad \bar{D}_1 \cos\theta_1 = \bar{D}_2 \cos\theta_2$$

$$\epsilon_1 \bar{E}_1 \cos\theta_1 = \epsilon_2 \bar{E}_2 \cos\theta_2 \qquad \text{..... (3.5.6b)}$$

$$\frac{(3.5.6a)}{(3.5.6b)} \Rightarrow \frac{\sin\theta_1}{\cos\theta_1 \epsilon_1} = \frac{\sin\theta_2}{\cos\theta_2 \epsilon_2} = \frac{\tan\theta_1}{\tan\theta_2} = \frac{\epsilon_1}{\epsilon_2} \qquad \text{..... (3.5.6c)}$$

which is called law of refraction.

3.5.3 Conductor – Dielectric

Consider two media out of which one is conductor and the other one is dielectric medium, which are meeting at certain point as shown in Fig.3.10. σ = ∞ for a perfect conductor. So electric field does not exist in the perfect conductor.

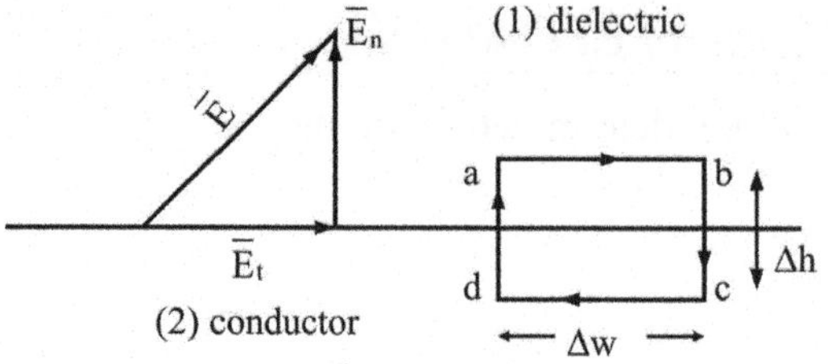

Fig. 3.10 Conductor – Dielectric boundary for establishing relation between tangential components

On applying Maxwell's equation for the closed path

$$0 = \bar{E}_t \Delta W - \bar{E}_n \frac{\Delta h}{2} - 0\left(\frac{\Delta h}{2}\right) - 0(\Delta W) + 0\frac{\Delta h}{2} + \bar{E}_n \frac{\Delta h}{2}$$

as Δh tends to zero.

$$\Rightarrow \qquad E_t = 0 \Rightarrow D_t = 0 \qquad \text{.....(3.5.7)}$$

i.e., the tangential component of electric field will be '0' when a conductor and dielectric are meeting.

To find the expression for normal component consider a small cylinder which intersects both the dielectric medium and conductor as shown in Fig.3.11.

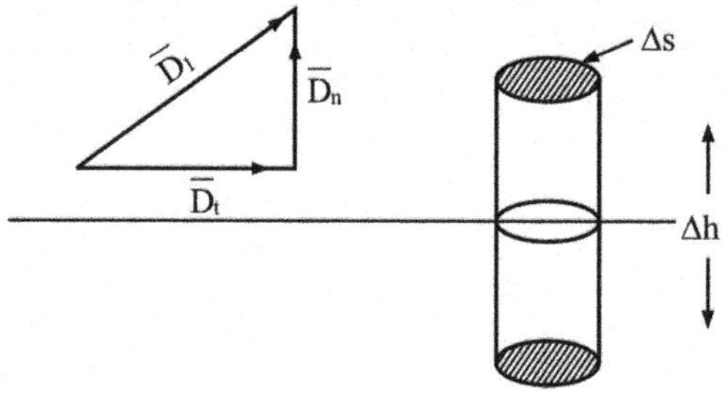

Fig. 3.11 Conductor – Dielectric boundary for establishing relation between normal components

Apply $\int_s \bar{D} \cdot d\bar{s} = Q = \int \rho_s ds$ to the considered cylindrical surface.

Then $\Delta Q = \rho_s$ ds

$$\therefore \qquad \bar{D}_n \Delta s - 0 = \rho_s \Delta s$$

$$\Rightarrow \qquad \bar{D}_n = \rho_s$$

$$\epsilon_1 \bar{E}_n = \rho_s \qquad \text{.....(3.5.8)}$$

3.5.4 Conductor – Free Space

Consider the Fig.3.12.

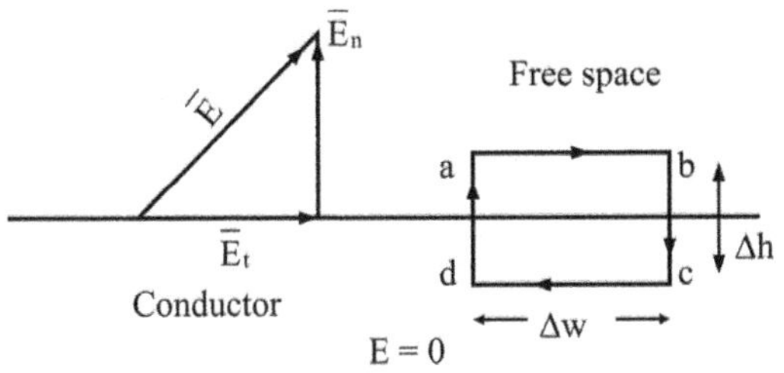

Fig. 3.12 Conductor – free space boundary for establishing relation between tangential components

$$0 = \bar{E}_t(\Delta W) - \bar{E}_n\left(\frac{\Delta h}{2}\right) + \bar{E}_n\left(\frac{\Delta h}{2}\right)$$

at the interface $\bar{E}_t = 0 \Rightarrow \bar{D}_t = 0$ .....(3.5.9)

i.e., the tangential component is '0'.

The normal component is

$$\bar{D}_n = \rho_S$$

$$\epsilon_0\,\bar{E}_n = \rho_S \quad .....(3.5.10)$$

Problem 3.15

Two extensive homogeneous isotropic dielectrics meet on plane z = 0. For z ≥ 0, $\epsilon_{r1} = 4$ and for z ≤ 0, $\epsilon_{r2} = 3$. A uniform electric field $\bar{E}_1 = 5\bar{a}_x - 2\bar{a}_y + 3\bar{a}_z$ kV/m exists for z ≥ 0. Find:

(a) $\bar{E}_2$ for z ≤ 0

(b) the angles $\bar{E}_1$ and $\bar{E}_2$ make with the interface

(c) the energy densities in J/m^3 in both dielectrics

(d) the energy with in a cube of side 2 m centered at (3, 4, –5).

Solution

Consider the Fig.3.13.

(a) Given $\bar{E}_1 = 5\bar{a}_x - 2\bar{a}_y + 3\bar{a}_z$ kV/m

We have

$\bar{E}_1 = \bar{E}_{1t} + \bar{E}_{1n}$, comparing the above two equations

The normal component is $3\,\bar{a}_z$

$\therefore \quad \bar{E}_{1n} = 3\bar{a}_z$

$\therefore \quad \bar{E}_{1n} = 3\bar{a}_z = 5\bar{a}_x - 2\bar{a}_y$

We know $\bar{E}_{1t} = \bar{E}_{2t}$

$\therefore \quad \bar{E}_{2t} = 5\bar{a}_x - 2\bar{a}_y$

Also known $\epsilon_1 \bar{E}_{1n} = \epsilon_2 \bar{E}_{2n}$

$\Rightarrow \quad \bar{E}_{2n} = 4\bar{a}_z$

$\therefore \quad \bar{E}_2 = \bar{E}_{2t} + \bar{E}_{2n} = 5\bar{a}_x - 2\bar{a}_y + 4\bar{a}_z$

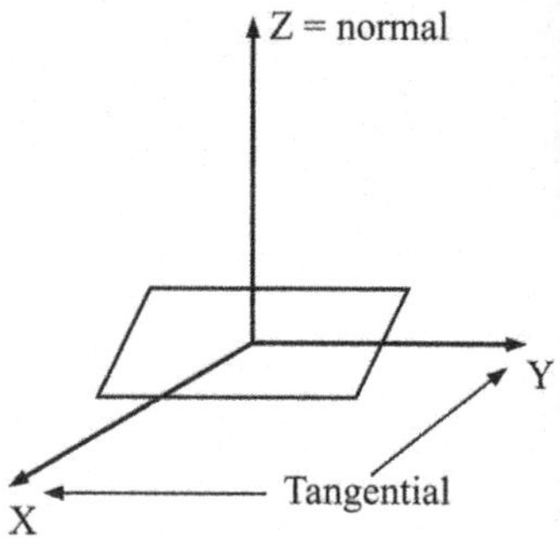

Fig. 3.13

(b) Consider the Fig.3.14.

Here $\alpha_1 = 90 - \theta_1$ and $\alpha_2 = 90 - \theta_2$

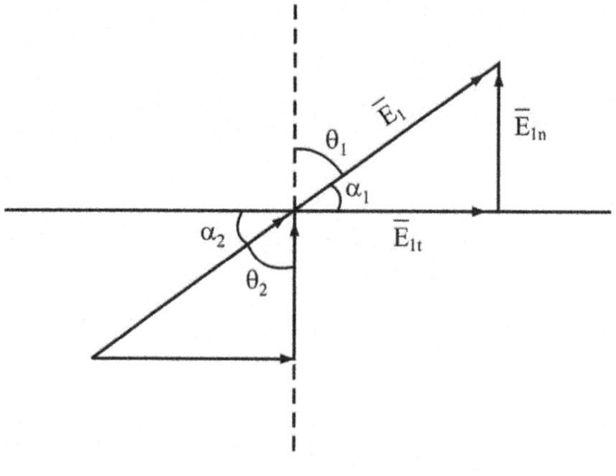

Fig. 3.14

$\bar{E}_{1t} = \bar{E}_1 \sin\theta_1$

$$\sin\theta_1 = \frac{\bar{E}_{1t}}{\bar{E}_1} = \frac{|5\bar{a}_x - 2\bar{a}_y|}{|5\bar{a}_x - 2\bar{a}_y + 3\bar{a}_z|} = \frac{\sqrt{29}}{\sqrt{38}} \Rightarrow \theta_1 = 60.88^\circ$$

$\alpha_1 = 90 - \theta_1 = 29.12^\circ$

and $$\sin\theta_2 = \frac{|\bar{E}_{2t}|}{|\bar{E}_2|} = \frac{\sqrt{29}}{\sqrt{45}} \Rightarrow \theta_2 = 53.4^\circ$$

$\therefore \quad \alpha_2 = 36.6^\circ$

(c) Energy density in first medium is

$$W_1 = \frac{1}{2}\epsilon_1 |E_1|^2$$

$$= \frac{1}{2}\epsilon_0 \epsilon_{r1} (38)$$

$$= 672.9 \ \mu\text{J/m}^3$$

And Energy density in second medium is

$$W_2 = \frac{1}{2}\epsilon_0 \epsilon_{r2} (45)$$

$$= 597.6 \ \mu\text{J/m}^3$$

(d) Given cube of side 2 m centered at (3, 4, –5)

i.e., $z = -5 < 0$, we have to consider the second medium

Limits are $2 \le x \le 4$

$3 \le y \le 5$

$-6 \le z \le -4$

The energy in a given cube is

$$w_2 = \int W_2 dv \quad \text{where } dv = dx\, dy\, dz$$

$$= \int_V 597.6 \, dx\, dy\, dz \ \mu\text{J / m}^3$$

$$= 4.776 \text{ mJ}$$

***Problem 3.16**

X-Z plane is a boundary between two dielectrics. Region y < 0 contains dielectric material $\epsilon_{r1} = 2.5$ while region $y > 0$ has dielectric with $\epsilon_{r2} = 4.0$. If $\bar{E} = -30\bar{a}_x + 50\bar{a}_y + 70\bar{a}_z$ V / m, find normal and tangential components of the E field on both sides of the boundary.

Solution

Here y < 0 is medium-1 and y > 0 is medium-2

Assume given $\bar{E}$ belongs to medium-1

$$\bar{E} = -30\bar{a}_x + 50\bar{a}_y + 70\bar{a}_z \quad \text{V/m}$$

x–z components are tangential and y component is normal

$$\bar{E}_{1t} = -30\bar{a}_x + 70\bar{a}_z \quad \text{and} \quad \bar{E}_{1n} = 50\bar{a}_y$$

The boundary condition on tangential component of $\bar{E}$ is $\bar{E}_{1t} = \bar{E}_{2t}$

$$\therefore \qquad \bar{E}_{2t} = -30\bar{a}_x + 70\bar{a}_z \ \text{V/m}$$

The boundary condition on normal component of $\bar{E}$ is $\bar{E}_{2n} = \dfrac{\epsilon_{r1}}{\epsilon_{r2}}\bar{E}_{1n}$

$$\therefore \qquad \bar{E}_{2n} = \frac{2.5}{4}50\bar{a}_y = 31.25\bar{a}_y$$

3.6 Boundary Conditions for Magnetic Field

Consider two magnetic media having μ_1 in one medium and μ_2 in the other medium. We need to use the following laws

$$\oint \bar{H} \cdot d\bar{l} = I_{enc} \quad \text{(Amperes circuit law)}$$

$$\oint \bar{B} \cdot d\bar{s} = 0 \quad \text{(Gauss's law for magnetic field)}$$

For tangential components

Apply $\oint \bar{H} \cdot d\bar{l} = I_{enc}$ to the closed path shown in Fig.3.15

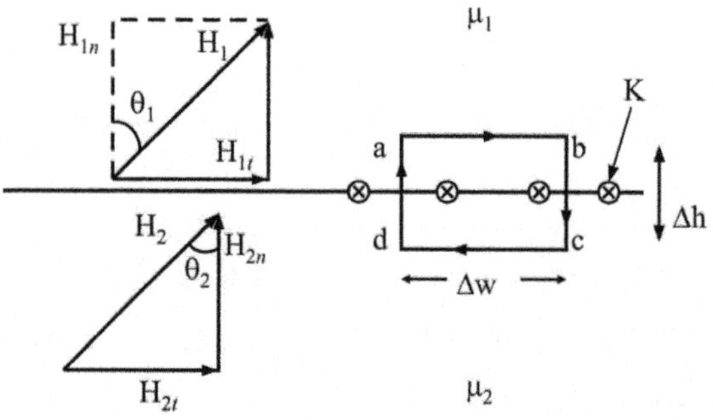

Fig. 3.15 Boundary conditions between two magnetic media for $\bar{H}$

$$\bar{K}\Delta W = \bar{H}_{1t}\Delta W - \bar{H}_{1n}\frac{\Delta h}{2} - \bar{H}_{2n}\frac{\Delta h}{2} + \bar{H}_{2t}\Delta W + \bar{H}_{2n}\frac{\Delta h}{2} + \bar{H}_{1n}\frac{\Delta h}{2}$$

as Δh tends to 0 and if we consider the surface current density distribution is zero, then we can say that

$$\bar{H}_{1t} = \bar{H}_{2t} \quad \text{(Continuous)}$$

$$\frac{\bar{B}_{1t}}{\mu_1} = \frac{\bar{B}_{2t}}{\mu_2} \quad \text{(Discontinuous)}$$

For normal components

Apply $\oint \bar{B}\cdot d\bar{s} = 0$ to the cylinder shown in the Fig.3.16

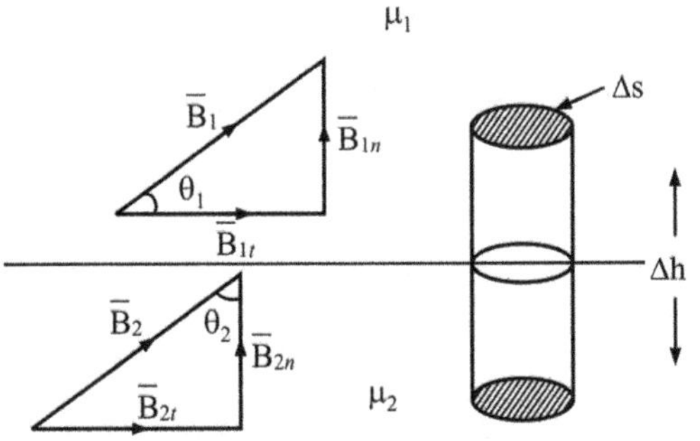

Fig. 3.16 Boundary conditions between two magnetic media for $\bar{B}$

$$\bar{B}_{1n}\Delta S - \bar{B}_{2n}\Delta S = 0$$

$$\bar{B}_{1n} = \bar{B}_{2n} \quad \text{(Continuous)} \qquad \text{.....(3.6.1)}$$

$$\Rightarrow \quad \mu_1\bar{H}_{1n} = \mu_2\bar{H}_{2n} \quad \text{(Discontinuous)} \qquad \text{.....(3.6.2)}$$

3.6.1 Law of Refraction for Magnetic Fields

Consider the Fig.3.15, where θ_1 is the angle made by $\bar{H}_1$ with the normal component to the interface. Similarly θ_2 is the angle made by $\bar{H}_2$ with the normal component to the interface.

From Fig.3.15, $\bar{H}_{1t} = \bar{H}_1 \sin\theta_1$ and

$$\bar{H}_{2t} = \bar{H}_2 \sin\theta_2$$

We know that $\bar{H}_{1t} = \bar{H}_{2t}$

$$\bar{H}_1 \sin\theta_1 = \bar{H}_2 \sin\theta_2 \quad \text{.....(3.6.3a)}$$

$$\bar{B}_{1n} = \bar{B}_1 \cos\theta_1, \quad \bar{B}_{2n} = \bar{B}_2 \cos\theta_2$$

we know that $\bar{B}_{1n} = \bar{B}_{2n} \Rightarrow \bar{B}_1 \cos\theta_1 = \bar{B}_2 \cos\theta_2$ .....(3.6.3b)

$$\frac{(3.6.3\ (a))}{(3.6.3\ (b))} = \frac{\tan\theta_1}{\tan\theta_2} = \frac{\mu_1}{\mu_2} \quad \text{.....(3.6.3c)}$$

which is law of refraction for magnetic fields.

***Problem 3.17**

Region 1, for which $\mu_{r1} = 3$ is defined by x < 0 and region 2, x > 0 has $\mu_{r1} = 3$ given $\bar{H}_1 = 4\bar{a}_x + 3\bar{a}_y + 6\bar{a}_z$ A/m. Determine $\bar{H}_2$ for x > 0 and the angles that $\bar{H}_1$ and $\bar{H}_2$ make with the interface.

Solution

Consider the Fig. 3.17

Given $\bar{H}_1 = 4\bar{a}_x + 3\bar{a}_y + 6\bar{a}_z$ A/m

We have

$\bar{H}_1 = \bar{H}_{1t} + \bar{H}_{1n}$, comparing the above two equations

The normal component is $4\bar{a}_x$

$$\therefore \quad \bar{H}_{1n} = 4\bar{a}_x$$

$$\bar{H}_{1t} = \bar{H}_1 - \bar{H}_{1n} = 3\bar{a}_y + 6\bar{a}_z$$

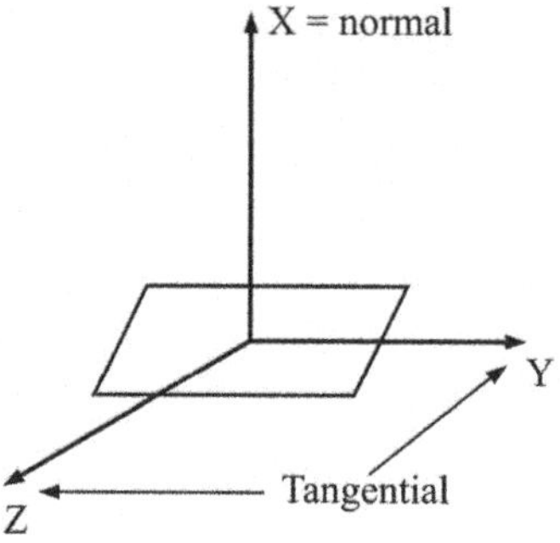

Fig. 3.17

We know $\bar{H}_{1t} = \bar{H}_{2t}$

$$\therefore \quad \bar{H}_{2t} = 3\bar{a}_y + 6\bar{a}_z$$

Also known $\mu_1 \bar{H}_{1n} = \mu_2 \bar{H}_{2n}$

$$\Rightarrow \quad \bar{H}_{2n} = \frac{\mu_1}{\mu_2} 4\bar{a}_x = \frac{\mu_{r1}\mu_0}{\mu_{r2}\mu_0} 4\bar{a}_x = \frac{3}{5} 4\bar{a}_x = 2.4\bar{a}_x$$

$$\therefore \quad \bar{H}_2 = \bar{H}_{2t} + \bar{H}_{2n} = 2.4\bar{a}_x + 3\bar{a}_y + 6\bar{a}_z$$

Consider the Fig.3.18.

Here $\alpha_1 = 90 - \theta_1$ and $\alpha_2 = 90 - \theta_2$

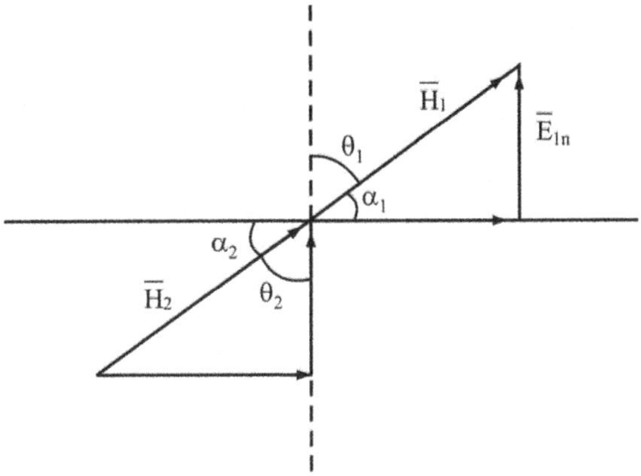

Fig. 3.18

From Fig. 3.18, $\bar{H}_{1t} = \bar{H}_1 \sin\theta_1$

$$\sin\theta_1 = \frac{\bar{H}_{1t}}{\bar{H}_1} = \frac{|3\bar{a}_y + 6\bar{a}_z|}{|4\bar{a}_x + 3\bar{a}_y + 6\bar{a}_z|} = \frac{\sqrt{45}}{\sqrt{61}} \Rightarrow \theta_1 = 59.19^\circ$$

$$\alpha_1 = 90 - \theta_1 = 30.8^\circ$$

And
$$\sin\theta_2 = \frac{|\bar{H}_{2t}|}{|\bar{H}_2|} = \frac{\sqrt{45}}{\sqrt{50.76}} \Rightarrow \theta_2 = 70.313^\circ$$

$\therefore \quad \alpha_2 = 19.686^\circ$

Review Questions and Answers

1. **Write the boundary conditions at the interface between two perfect dielectrics.**

   (i) The tangential component of electric field is continuous i.e., $E_{t1} = E_{t2}$

   (ii) The normal component of electric flux density is continuous i.e., $D_{n1} = D_{n2}$

2. **Write down the magnetic boundary conditions.**

   (i) The normal components of flux density B is continuous across the boundary.

   (ii) The tangential component of field intensity is continuous across the boundary.

3. **Name the two basic concepts involved in EM waves.**

   The two basic concepts that we are going to study in time varying electromagnetic waves are

   (i) induced emf according to Faraday's law

   (ii) displacement current.

4. What is Faradays law?

According to Faraday's law in a closed circuit the induced emf is equal to the time rate of change of magnetic flux linkage i.e., $V_{emf} = -\frac{d\lambda}{dt}$

If each turn of the circuit carries flux 'ψ' then flux linkage $\lambda = N\psi$, then induced emf is $V_{emf} = -\frac{Nd\psi}{dt}$

5. What is Lenz's law?

The induced emf is equal to the time rate of change of magnetic flux linkage i.e., $V_{emf} = -\frac{d\lambda}{dt}$

If each turn of the circuit carries flux 'ψ' then flux linkage $\lambda = N\psi$, then induced emf is $V_{emf} = -\frac{Nd\psi}{dt}$

where –ve sign indicates the induced emf opposes the magnetic flux linkage. This is called Lenz's law.

6. Write Maxwell's equation for time varying fields.

$$\nabla \times \bar{E} = -\frac{\partial \bar{B}}{\partial t}$$

7. Name the sources of generating emf produced fields.

Electric generator, Batteries and Fuel cells etc.

8. What is the In-Consistency of Ampere's Law?

We know the differential form of Ampere's law i.e., $\nabla \times \bar{H} = \bar{J}$.

where $\bar{J}$ is the conduction current density

By taking divergence on both sides of the above equation

$$\nabla \cdot \left(\nabla \times \bar{H}\right) = \nabla \cdot \bar{J}$$

We know the identity $\nabla \cdot \left(\nabla \times \bar{A}\right) = 0$,

where $\bar{A}$ is any vector

$$\therefore \nabla \cdot \bar{J} = 0 \qquad \text{.....(a)}$$

According to Continuity equation for time varying fields

$$\nabla \cdot \bar{J} = \frac{-\partial \rho_V}{\partial t} \qquad \text{...... (b)}$$

For time varying fields equations (a) and (b) are not comparable. Hence Ampere's law can not be used directly for time varying fields. To overcome inconsistency of Ampere's law, modify the differential form of Ampere's law as $\nabla \times \bar{H} = \bar{J} + \bar{J}_D$

where $\bar{J}_D$ is the displacement current density.

9. What is the modified Maxwell's equation for time varying fields?

$$\nabla \times \bar{H} = \bar{J} + \frac{\partial \bar{D}}{\partial t}$$

10. What is the ratio of Displacement to Current Conduction Current?

$$\frac{\textit{Displacement}\ \text{current}}{\textit{Conduction}\ \text{current}} = \frac{I_D}{I_C} = \frac{\omega \in}{\sigma}$$

11. Write word form of Ampere's circuit Law.

The magneto motive force around a closed path is equal to the sum of conduction current density and the time derivative of electric flux density through a surface i.e., bounded by path.

12. Write word form of Faraday's Law.

The electromotive force around a closed path is equal to the time derivative of magnetic flux density through a surface i.e., bounded by path.

13. Write word form of Gauss's Law.

The electric flux through a closed surface is equal to the charge enclosed by that surface.

14. Write word form of Gauss's Law for magnetic fields.

The magnetic flux through a closed surface is zero. Which also says Non existence of isolated magnetic charge.

15. What is law of refraction for Electric fields?

$$\frac{\tan\theta_1}{\tan\theta_2} = \frac{\in_1}{\in_2}$$

16. What is law of refraction for Magnetic fields?

$$\frac{\tan\theta_1}{\tan\theta_2} = \frac{\mu_1}{\mu_2}$$

17. Write integral form of Ampere's circuit Law.

$$\oint_L \bar{H}\cdot d\bar{l} = \int_s \left(\bar{J} + \frac{\partial \bar{D}}{dt}\right)\cdot d\bar{s}$$

18. Write integral form of Faraday's Law.

$$\oint_L \bar{E}\cdot d\bar{l} = -\int_s \frac{\partial \bar{B}}{\partial t}\cdot d\bar{s}$$

19. Write integral form of Gauss's Law.

$$\oint_s \bar{D}\cdot d\bar{s} = \int_v \rho_v dv$$

20. Write integral form of Gauss's Law for magnetic fields.

$$\int_s \bar{B}\cdot d\bar{s} = 0$$

21. What is the significance of displacement current?

The concept of displacement current was introduced to justify the production of magnetic field in empty space. It signifies that a changing electric field induces a magnetic field. In empty space the conduction current is zero and the magnetic fields are entirely due to displacement current.

22. Distinguish between conduction and displacement currents.

The current through a resistive element is termed as conduction current whereas the current through a capacitive element is termed as displacement current.

Multiple Choice Questions

1. The magnetic intensity of a time-varying field is given by $\bar{H} = \pi(-z\bar{a}_y + y\bar{a}_z)\cos(2\pi 10^9 t)\,\text{A/m}$. What is the electric flux density vector $D(x, y, z, t)$?

(a) $\bar{D} = 10^{-9}\sin(2\pi 10^9 t)\bar{a}_x\ \text{C/m}^2$

(b) $\bar{D} = -2\pi\sin(2\pi 10^9 t)\bar{a}_x\ \text{C/m}^2$

(c) $\bar{D} = 2\pi\cos(2\pi 10^9 t)\bar{a}_z\ \text{C/m}^2$

(d) $\bar{D} = -10^{-9}\cos(2\pi 10^9 t)\bar{a}_y\ \text{C/m}^2$

2. Is it possible to derive the divergence Maxwell equation $\nabla.\bar{B}=0$ from any other Maxwell equation? If yes, which one?

(a) No
(b) $yes:\nabla\times\bar{E}=-\partial\bar{B}/\partial t$
(c) $yes:\nabla\times\bar{H}=\partial\bar{D}/\partial t+\bar{J}$
(d) $yes:\nabla.\bar{D}=\rho_v$

3. Is it possible to derive the divergence Maxwell equation $\nabla.\bar{D}=\rho_v$ from any other Maxwell equation and the continuity law, $\nabla.J=-\partial\rho_v/\partial t$? If yes, which one?

(a) No
(b) $yes:\nabla\times\bar{E}=-\partial\bar{B}/\partial t$
(c) $yes:\nabla\times\bar{H}=\partial\bar{D}/\partial t+\bar{J}$
(d) $yes:\nabla.\bar{B}=0$

4. The flux through each turn of a 100-turn coil is (t^3 – 2t) mWb, where t is in seconds. The induced emf at $t = 2s$ is

(a) 1 V
(b) –1 V
(c) 4 mV
(d) 0.4 V

5. The concept of displacement current was a major contribution to

(a) Faraday
(b) Lenz
(c) Maxwell
(d) Lorentz

6. Maxwell's equations depend on ______ Law(s)

(a) Gauss's
(b) Faraday's
(c) Ampere's
(d) All of the above

7. According to Ampere's circuit law Maxwell's equation is

(a) $\nabla\times\bar{H}=\bar{J}+\frac{\partial\bar{D}}{\partial t}$
(b) $\nabla\times\bar{E}=-\frac{\partial\bar{B}}{\partial t}$
(c) $\nabla\cdot\bar{D}=\rho_V$
(d) $\nabla\cdot\bar{B}=0$

8. According to Faraday's law Maxwell's equation is

(a) $\nabla\times\bar{H}=\bar{J}+\frac{\partial\bar{D}}{\partial t}$
(b) $\nabla\times\bar{E}=-\frac{\partial\bar{B}}{\partial t}$
(c) $\nabla\cdot\bar{D}=\rho_v$
(d) $\nabla\cdot\bar{B}=0$

9. According to Gauss's law Maxwell's equation is

(a) $\nabla\times\bar{H}=\bar{J}+\frac{\partial\bar{D}}{\partial t}$
(b) $\nabla\times\bar{E}=-\frac{\partial\bar{B}}{\partial t}$
(c) $\nabla\cdot\bar{D}=\rho_v$
(d) $\nabla\cdot\bar{B}=0$

10. According to Gauss's law for Magnetic fields Maxwell's equation is

(a) $\nabla \times \bar{H} = \bar{J} + \frac{\partial \bar{D}}{\partial t}$ (b) $\nabla \times \bar{E} = -\frac{\partial \bar{B}}{\partial t}$

(c) $\nabla \cdot \bar{D} = \rho_v$ (d) $\nabla \cdot \bar{B} = 0$

11. What does a changing electric field induce?

(a) Charges (b) Magnetic field

(c) Light (d) Electrons

12. Which of Maxwell's equations can be used, to calculate the magnetic field lines form closed loops?

(a) $\oint_L \bar{H} \cdot d\bar{l} = \int_s \left(\bar{J} + \frac{\partial \bar{D}}{dt} \right) \cdot d\bar{s}$ (b) $\oint_L \bar{E} \cdot d\bar{l} = -\int_s \frac{\partial \bar{B}}{\partial t} \cdot d\bar{s}$

(c) $\oint_s \bar{D} \cdot d\bar{s} = -\int_v \rho_v d_v$ (d) $\oint_s \bar{B} . d\bar{s} = 0$

13. Which of the law can be used, to calculate the magnetic field of a long straight current carrying wire?

(a) Ampere's circuit law (b) Faraday's law

(c) Gauss law (d) Coulomb's law

14. Which of Maxwell's equations can be used, to calculate electric field produced by a uniform time-varying magnetic field?

(a) $\oint_L \bar{H} \cdot d\bar{l} = \int_s \left(\bar{J} + \frac{\partial \bar{D}}{dt} \right) \cdot d\bar{s}$ (b) $\oint_L \bar{E} \cdot d\bar{l} = -\int_s \frac{\partial \bar{B}}{\partial t} \cdot d\bar{s}$

(c) $\oint_s \bar{D} \cdot d\bar{s} = -\int_v \rho_v d_v$ (d) $\oint_s \bar{B} . d\bar{s} = 0$

15. Induced electric currents can be explained using which of the following laws?

(a) Ampere's circuit law (b) Faraday's law

(c) Gauss law (d) Coulomb's law

16. An infinitely long wire carries a current of three amps. The magnetic field outside the wire:

(a) points radially away from the wire (b) points radially inward

(c) circles the wire (d) is zero

17. The equation $\nabla \times \bar{E} = -\frac{\partial \bar{B}}{\partial t}$ is the generalization of

(a) Ampere's Law (b) Faraday's Law

(c) Gauss's Law (d) Biot-Savert's Law

18. Law of refraction for Electric fields is

(a) $\frac{\tan\theta_1}{\tan\theta_2} = \frac{\epsilon_1}{\epsilon_2}$ (b) $\frac{\tan\theta_1}{\tan\theta_2} = \frac{\mu_1}{\mu_2}$

(c) $\frac{\tan\theta_1}{\tan\theta_2} = \frac{\epsilon_2}{\epsilon_1}$ (d) $\frac{\tan\theta_1}{\tan\theta_2} = \frac{\mu_2}{\mu_1}$

19. Law of refraction for magnetic fields is

(a) $\frac{\tan\theta_1}{\tan\theta_2} = \frac{\epsilon_1}{\epsilon_2}$ (b) $\frac{\tan\theta_1}{\tan\theta_2} = \frac{\mu_1}{\mu_2}$

(c) $\frac{\tan\theta_1}{\tan\theta_2} = \frac{\epsilon_2}{\epsilon_1}$ (d) $\frac{\tan\theta_1}{\tan\theta_2} = \frac{\mu_2}{\mu_1}$

20. The tangential component of electric field intensity is_______ at the interface between dielectric-dielectric.

(a) Continuous (b) Discontinuous

(c) Zero (d) Infinity

Answers

1.	(a)	2.	(b)	3.	(c)	4.	(b)	5.	(c)
6.	(d)	7.	(a)	8.	(b)	9.	(c)	10.	(d)
11.	(b)	12.	(d)	13.	(a)	14.	(b)	15.	(b)
16.	(c)	17.	(b)	18.	(a)	19.	(b)	20.	(a)

Exercise Questions

1. In free space $\bar{D} = D_m \sin(\omega t + \beta z)\bar{a}_x$. Determine $\bar{B}$ and displacement current density.
2. What is the inconsistency in Ampere's law? How it is rectified by Maxwell?

3. Show that the total displacement current between the condenser plates connected to an alternating voltage sources is exactly the same as the value of charging current (conduction current).
4. Derive the boundary conditions for the tangential and normal components of Electrostatic fields at the boundary between two perfect dielectrics.
5. Derive Maxwell's equations in integral form and differential form for time varying fields.
6. Explain how the concept of Displacement current was introduced by Maxwell to account for the production of Magnetic fields in the empty space.
7. Write down the Maxwell's equations for Harmonically varying fields.
8. Let the internal dimensions of coaxial capacitor be a = 1.2 cm, b = 4 cm and l = 40 cm, the Homogeneous material inside the capacitor has the parameters $\in = 10^{-11}$ F/m, $\mu = 10^{-5}$ H/m and σ = 10^{-5} S/m. If the electric field intensity is $\bar{E} = (10^6/\rho)\cos 10^5 t\bar{a}_\rho$ V/m find

   (a) $\bar{J}$
   (b) The total conduction current I_c through the capacitor
   (c) The total displacement current I_d through the capacitor
   (d) The ratio of the amplitude of I_d to that of I_c, the quality factor of the capacitor.

   (*Hint*: $I_c = C\dfrac{dV}{dt}$, where capacitance of the coaxial cable $C = \dfrac{2\pi \in l}{ln(b/a)}$ and

   $I_d = J_d(2\pi a l)$ where $J_d = \dfrac{\partial D}{\partial t}$)

Chapter

4

EM Wave Characteristics

4.1 Introduction

Here we apply Maxwell's equations to obtain Electromagnetic (EM) wave equations, where waves are means of transporting energy or information. Typical examples of EM waves are radio waves, TV signals, radar beams and light rays.

We have the Maxwell's equations for time varying fields

$$\nabla \times \bar{E} = -\frac{\partial \bar{B}}{\partial t}$$

$$\nabla \times \bar{H} = \bar{J} + \frac{\partial \bar{D}}{\partial t}$$

$$\nabla \cdot \bar{D} = \rho_v$$

$$\nabla \cdot \bar{B} = 0$$

From the first curl equation we can conclude that time varying magnetic field gives rise the electric field that varies with the space and time, and from the second curl equation we can say that, the time varying electric field gives rise of the magnetic field that varies with space and time.

For sinusoidal variations we can write $\bar{B} = \bar{B}_0 e^{j\omega t}$ and $\bar{D} = \bar{D}_0 e^{j\omega t}$.

$$\therefore \qquad \frac{\partial \bar{B}}{\partial t} = j\omega \bar{B} \text{ and } \frac{\partial \bar{D}}{\partial t} = j\omega \bar{D}$$

Hence the Maxwell's equations can be written in phasor form or complex form as

$$\nabla \times \bar{E} = -j\omega \bar{B}$$

$$\nabla \times \bar{H} = \bar{J} + j\omega \bar{D}$$

$$\nabla \cdot \bar{D} = \rho_v$$

$$\nabla \cdot \bar{B} = 0$$

4.2 EM Wave Equations

Wave equation in terms of $\bar{E}$:

To find EM wave equation in terms of only $\bar{E}$

Consider $$\nabla \times \bar{E} = -\frac{\partial \bar{B}}{\partial t}$$

We know $$\bar{B} = \mu \bar{H}$$

$\therefore$ $$\nabla \times \bar{E} = -\mu \frac{\partial \bar{H}}{\partial t}$$

Take curl on both sides, we get

$$\nabla \times \nabla \times \bar{E} = -\mu \frac{\partial}{\partial t}\left(\nabla \times \bar{H}\right)$$

According to an identity $\nabla \times \nabla \times \bar{A} = \nabla\left(\nabla \cdot \bar{A}\right) - \nabla^2 \bar{A}$

where $\bar{A}$ is a general vector

$\therefore$ $$\nabla \times \nabla \times \bar{E} = \nabla\left(\nabla \cdot \bar{E}\right) - \nabla^2 \bar{E}$$

According to Maxwell's equation $\nabla \cdot \bar{D} = \rho_v \Rightarrow \nabla . \bar{E} = \frac{\rho_v}{\in}$

$\therefore$ $$\nabla \times \nabla \times \bar{E} = \nabla\left(\frac{\rho_v}{\in}\right) - \nabla^2 \bar{E} \qquad \text{.....(4.2.1)}$$

$$-\mu \frac{\partial}{\partial t}\left(\nabla \times \bar{H}\right) = -\mu \frac{\partial}{\partial t}\left(\bar{J} + \frac{\partial \bar{D}}{\partial t}\right) = -\mu \frac{\partial \bar{J}}{\partial t} - \mu \in \frac{\partial^2 \bar{E}}{\partial t^2}$$

$$\nabla\left(\frac{\rho_v}{\in}\right) - \nabla^2 \bar{E} = -\mu \frac{\partial \bar{J}}{\partial t} - \mu \in \frac{\partial^2 \bar{E}}{\partial t^2}$$

$$\Rightarrow \quad \nabla^2 \bar{E} - \mu \in \frac{\partial^2 \bar{E}}{\partial t^2} = \nabla\left(\frac{\rho_v}{\in}\right) + \mu\left(\frac{\partial \bar{J}}{\partial t}\right) \qquad .....(4.2.2)$$

For charge free medium $\nabla^2 E - \mu \in \frac{\partial^2 \bar{E}}{\partial t^2} = 0$ .....(4.2.3)

The generalized wave equation which travels with speed 'v' is

$$\nabla^2 \bar{F} - \frac{1}{v^2}\frac{\partial^2 \bar{F}}{\partial t^2} = 0 \qquad .....(4.2.4)$$

on comparing (4.2.3) & (4.2.4), we get $v = \frac{1}{\sqrt{\mu \in}}$

$\therefore$ In free space equation (4.2.3) is a wave which travels with the speed

$$v = \frac{1}{\sqrt{\mu_0 \in_0}} = 3 \times 10^8 \text{ m/s.}$$

Wave equation in terms of $\bar{\mathbf{H}}$:

To get EM wave equation in terms of only $\bar{H}$

Consider $\nabla \times \bar{H} = \bar{J} + \frac{\partial \bar{D}}{\partial t}$

Take curl on both sides

$$\nabla \times \nabla \times \bar{H} = \nabla \times \bar{J} + \frac{\partial}{\partial t}(\nabla \times \bar{D})$$

$$\Rightarrow \quad \nabla(\nabla \cdot \bar{H}) - \nabla^2 \bar{H} = \nabla \times \bar{J} + \in \frac{\partial}{\partial t}(\nabla \times \bar{E})$$

$$\Rightarrow \quad \nabla\left(\nabla \cdot \frac{\bar{B}}{\mu}\right) - \nabla^2 \bar{H} = \nabla \times \bar{J} + \in \frac{\partial}{\partial t}(\nabla \times \bar{E})$$

$$\Rightarrow \quad 0 - \nabla^2 \bar{H} = \nabla \times \bar{J} + \in \frac{\partial}{\partial t}(\nabla \times \bar{E}) \qquad \because \nabla.\bar{B} = 0$$

$$\Rightarrow \quad -\nabla^2 \bar{H} = \nabla \times \bar{J} + \in \frac{\partial}{\partial t}\left(-\frac{\partial \bar{B}}{\partial t}\right)$$

$$\Rightarrow \quad -\nabla^2 \bar{H} = \nabla \times \bar{J} - \in \frac{\partial^2 \bar{B}}{\partial t^2}$$

$$\Rightarrow \qquad -\nabla^2 \bar{H} = \nabla \times \bar{J} - \in \mu \frac{\partial^2 \bar{H}}{\partial t^2}$$

$$\Rightarrow \qquad \nabla^2 \bar{H} - \in \mu \frac{\partial^2 \bar{H}}{\partial t^2} = -\nabla \times \bar{J} \qquad .....(4.2.5)$$

For charge free medium $\nabla^2 \bar{H} - \in \mu \frac{\partial^2 \bar{H}}{\partial t^2} = 0$ .....(4.2.6)

Compare this equation with the generalized one.

Then in free space $v = \frac{1}{\sqrt{\mu_0 \in_0}} = 3 \times 10^8$ m/s.

Equations (4.2.3) and (4.2.6) are the EM wave equations in terms of $\bar{E}$ *and* $\bar{H}$ respectively, which travel with the speed $v = \frac{1}{\sqrt{\mu \in}}$ m/s.

4.3 Transverse Electromagnetic Wave

Consider an electromagnetic wave and if $\bar{a}_E$, $\bar{a}_H$ *and* $\bar{a}_d$ are the unit vectors along the $\bar{E}$ field, the $\bar{H}$ field and the direction of wave propagation respectively, then it can be shown that

$$\bar{a}_d = \bar{a}_E \times \bar{a}_H \qquad .....(4.3.1)$$

i.e., electric and magnetic fields are everywhere normal to the direction of propagation, both the fields lie in a plane that is transverse or orthogonal to the direction of wave propagation. They form an EM wave that has no electric or magnetic field components along the direction of propagation, such a wave is called a transverse electromagnetic (TEM) wave and such a wave with same magnitude throughout any transverse plane is also called Uniform plane wave. The direction in which the electric field points is the **polarization** of a TEM wave.

4.4 Uniform Plane Wave

Uniform plane wave is the wave that will have variation only in the direction of propagation and it's characteristics remain constant across the planes normal to the direction of propagation.

Here we study EM wave propagation in the following media:

(i) Free space $(\sigma = 0,\ \epsilon = \epsilon_0,\ \mu = \mu_0)$

(ii) Lossless dielectrics $\left(\sigma = 0, \in = \in_0 \in_r, \mu = \mu_0 \mu_r \ll \omega \in\right)$

(iii) Lossy dielectrics $\left(\sigma = 0, \ \in = \in_0 \in_r, \ \mu = \mu_0 \mu_r\right)$

(iv) Good conductors $\left(\sigma = \infty, \ \in = \in_0, \ \mu = \mu_0 \ \mu_r \text{ or } \sigma >> \omega \in\right)$

We study about Uniform plane waves and then we discuss on EM wave propagation through Lossy dielectric medium, as it is a general case, from which by changing the values of σ, $\in$, and μ, we can obtain the solutions for other media.

4.4.1 Propagation of Uniform Plane Wave

We know the wave equation (EM wave) which travels in the free space as

$$\nabla^2 \bar{E} - \mu_0 \in_0 \frac{\partial^2 \bar{E}}{\partial t^2} = 0$$

Assume uniform plane wave which is traveling along X-direction. When a wave is traveling along X-direction it's E_x or H_x will be zero and the partial differentiations w.r.t. 'y' and 'z' will be '0'.

$\therefore$ The above equation becomes $\dfrac{\partial^2 \bar{E}}{\partial x^2} = \mu_0 \in_0 \dfrac{\partial^2 \bar{E}}{\partial t^2}$

where $\bar{E}$ contains the y and z components only.

Consider the y component of the above equation

$$\frac{\partial^2 E_y}{\partial x^2} = \mu_0 \in_0 \frac{\partial^2 E_y}{\partial t^2}$$

For a wave traveling along +ve X-direction, the solution for the above equation is $E_{y1} = f_1(x - vt)$. Where 'v' is the velocity with which wave travels, where $v = \dfrac{1}{\sqrt{\mu_0 \in_0}}$

and for the wave that is traveling along –ve X-direction the solution for the above equation is

$$E_{y2} = f_2(x + vt).$$

The generalized solution is $E_y = f_1(x - vt) + f_2(x + vt)$

4.4.2 Relation between $\bar{E}$ and $\bar{H}$ for Uniform Plane Wave

To find the relation between $\bar{E}$ and $\bar{H}$, consider the Maxwell's equation $\nabla \times \bar{H} = \bar{J} + \dfrac{\partial \bar{D}}{\partial t}$

Assume it is traveling in free space i.e., which has no free charges and current then the above equation becomes

$$\nabla \times \bar{H} = \in \frac{\partial \bar{E}}{\partial t} \qquad \because \ \bar{D} = \in \bar{E}$$

Assume the uniform plane wave is traveling along X-direction. Then E_x and H_x are '0' and.

$$\therefore \qquad \bar{E} = E_y \bar{a}_y + E_z \bar{a}_z$$

$$\begin{bmatrix} \bar{a}_x & \bar{a}_y & \bar{a}_z \\ \frac{\partial}{\partial x} & \frac{\partial}{\partial y} & \frac{\partial}{\partial z} \\ 0 & H_y & H_z \end{bmatrix} = \in \left[\frac{\partial E_y \bar{a}_y}{\partial t} + \frac{\partial E_z \bar{a}_z}{\partial t} \right]$$

$$\Rightarrow \bar{a}_x \left[\frac{\partial}{\partial y}(H_z) - \frac{\partial}{\partial z}(H_y) \right] - \bar{a}_y \left[\frac{\partial}{\partial x}(H_z) - 0 \right] + \bar{a}_z \left[\frac{\partial}{\partial x}(H_y) - 0 \right] = \in \left(\frac{\partial E_y \bar{a}_y}{\partial t} + \frac{\partial E_z \bar{a}_z}{\partial t} \right)$$

Since $\qquad \frac{\partial H_z}{\partial y} = 0 \ \& \ \frac{\partial H_y}{\partial z} = 0$

$$\Rightarrow \qquad -\bar{a}_y \left(\frac{\partial}{\partial x} H_z \right) + \bar{a}_z \left(\frac{\partial}{\partial x} H_y \right) = \in \left(\frac{\partial E_y \bar{a}_y}{\partial t} + \frac{\partial E_z \bar{a}_z}{\partial t} \right)$$

By equating $\bar{a}_y$ components

$$\frac{\partial H_z}{\partial x} = - \in \frac{\partial E_y}{\partial t} \qquad \text{.....(4.4.1)}$$

And equating $\bar{a}_z$ components

$$\frac{\partial H_y}{\partial x} = \in \frac{\partial E_z}{\partial t} \qquad \text{.....(4.4.2)}$$

We know the solution for the wave equation that travels along X-direction as $E_y = f_1\,(x - vt)$ where $v = \frac{1}{\sqrt{\mu \in}}$

Assume $x - vt = u$

Then $\qquad E_y = f_1(u)$

Differentiate w.r.t 't' on both sides

$$\frac{\partial E_y}{\partial t} = \frac{\partial f_1(u)}{\partial t} = \frac{\partial f_1(u)}{\partial u} \cdot \frac{\partial u}{\partial t}$$

But $$\frac{\partial u}{\partial t} = -v$$

$$\therefore \quad \frac{\partial E_y}{\partial t} = \frac{\partial f_1(u)}{\partial u} \cdot (-v) = -vf_1^1$$

where $$f_1^1 = \frac{\partial f_1(u)}{\partial u}$$

From equation (4.4.1)

$$\frac{\partial H_z}{\partial x} = \in vf_1^1$$

$$\partial H_z = \in vf_1^1 \partial x$$

Integrating on both sides

$$H_z = \in v\int f_1^1 dx \qquad \text{.....(4.4.3)}$$

Take $$\frac{\partial E_y}{\partial x} = \frac{\partial f_1(u)}{\partial x} = \frac{\partial f_1(u)}{\partial u}\frac{\partial u}{\partial x} = f_1^1$$

$$\because \quad \frac{\partial u}{\partial x} = 1 \quad \& \; f_1^1 = \frac{\partial f_1(u)}{\partial u}$$

from equation (4.4.3)

$$H_z = \in v\int \frac{\partial E_y}{\partial x} \cdot dx \qquad \text{.....(4.4.4)}$$

$$H_z = \in vE_y$$

$$H_z = \in \frac{1}{\sqrt{\mu \in}} E_y = \sqrt{\frac{\in}{\mu}} E_y$$

$$E_y = \sqrt{\frac{\mu}{\in}} H_z \qquad \text{.....(4.4.5)}$$

Another solution for the wave equation that travels along X-direction is $E_z = f_1\,(x - vt)$

Assume $x - vt = u$

$$E_z = f_1(u)$$

$$\frac{\partial E_z}{\partial t} = \frac{\partial f_1(u)}{\partial t} = \frac{\partial f_1(u)}{\partial u}\frac{\partial u}{\partial t} = -vf_1^1$$

from equation (4.4.2)

$$\frac{\partial H_y}{\partial x} = -\in v f_1^1$$

$$\Rightarrow \qquad H_y = -\in v \int f_1^1 dx \qquad \text{.....(4.4.6)}$$

$$\frac{\partial E_z}{\partial x} = \frac{\partial f_1(u)}{\partial x} = \frac{\partial f_1(u)}{\partial u}\frac{\partial u}{\partial x} = f_1^1$$

Substitute this in equation (4.4.6)

$$H_y = -\in v \int \frac{\partial E_z}{\partial x} dx$$

$$H_y = -\in v E_z$$

$$H_y = -\in \frac{1}{\sqrt{\mu \in}} E_z$$

$$H_y = -\sqrt{\frac{\in}{\mu}} E_z$$

$$\Rightarrow \qquad E_z = -\sqrt{\frac{\mu}{\in}} H_y \qquad \text{.....(4.4.7)}$$

On squaring equations (4.4.5) and (4.4.7), we get

$$E_y^2 = \frac{\mu}{\in} H_z^2 \quad \text{and} \quad E_z^2 = \frac{\mu}{\in} H_y^2$$

$$\bar{E} = E_y \bar{a}_y + E_z \bar{a}_z$$

$$|\bar{E}| = \sqrt{E_y^2 + E_z^2} = \sqrt{\frac{\mu}{\in}\left(H_z^2 + H_y^2\right)} = \sqrt{\frac{\mu}{\in}} \cdot \sqrt{H_z^2 + H_y^2}$$

$$|\bar{E}| = \sqrt{\frac{\mu}{\in}} |\bar{H}| \Rightarrow \frac{E}{H} = \sqrt{\frac{\mu}{\in}}$$

$$\bar{E} \cdot \bar{H} = \left(E_y \bar{a}_y + E_z \bar{a}_z\right) \cdot \left(H_y \bar{a}_y + H_z \bar{a}_z\right)$$

$$= E_y H_y + E_z H_z$$

$$= \sqrt{\frac{\mu}{\in}} H_z H_y - \sqrt{\frac{\mu}{\in}} H_z H_y = 0$$

$\bar{E}$ and $\bar{H}$ will be perpendicular to each other.

The ratio $\dfrac{E}{H}$ is denoted as 'η' which is called as intrinsic impedance or characteristic impedance. For free space $\eta = \sqrt{\dfrac{\mu_0}{\epsilon_0}} = 377\ \Omega$ .....(4.4.8)

4.5 Wave Propagation in a Conducting Medium or Lossy Medium

A lossy dielectric is a medium in which an EM wave loses it's power, as it propagates through the medium or lossy dielectric is a partially conducting medium (imperfect dielectric or imperfect conductor) with $\sigma \neq 0$.

From equation (4.2.2), we have

$$\nabla^2 \bar{E} - \mu \epsilon \frac{\partial^2 \bar{E}}{\partial t^2} = \nabla\left(\frac{\rho_v}{\epsilon}\right) + \mu \frac{\partial \bar{J}}{\partial t}$$

Assume a linear, isotropic, homogeneous, lossy dielectric medium that has some conductivity 'σ' and there are no free charges.

i.e., $\rho_v = 0$ and $\bar{J} = \sigma \bar{E}$

$$\therefore \quad \nabla^2 \bar{E} - \mu \epsilon \frac{\partial^2 \bar{E}}{\partial t^2} = \sigma \mu \frac{\partial \bar{E}}{\partial t}$$

$$\Rightarrow \quad \nabla^2 \bar{E} - \mu \epsilon \frac{\partial^2 \bar{E}}{\partial t^2} - \sigma \mu \frac{\partial \bar{E}}{\partial t} = 0 \qquad .....(4.5.1)$$

which is the EM wave equation in conducting medium in terms of $\bar{E}$

To find the solution for a wave equation in a conducting medium consider sinusoidal variations i.e.,

$$\bar{E} = \bar{E}_0 e^{j\omega t}$$

$$\frac{\partial \bar{E}}{dt} = \bar{E}_0 j\omega e^{j\omega t} = j\omega \bar{E}\ ; \quad \frac{\partial^2 \bar{E}}{\partial t^2} = j^2 \omega^2 \bar{E}$$

∴ The equation (4.5.1) becomes

$$\nabla^2 \bar{E} - \mu \epsilon j^2 \omega^2 \bar{E} - \sigma \mu j \omega \bar{E} = 0$$

$$\nabla^2 \bar{E} - j\omega\mu(\sigma + j\omega \epsilon)\bar{E} = 0$$

Replace $j\omega\mu(\sigma + j\omega\in) = \gamma^2$, where γ is the **propagation constant** of the medium measured in terms of **reciprocal meters**.

∴ The above equation becomes

$$\nabla^2\bar{E} - \gamma^2\bar{E} = 0 \qquad .....(4.5.2)$$

Similar procedure can be applied to equation (4.2.5) and we can obtain EM wave equation in conducting medium in terms of $\bar{H}$

$$\nabla^2\bar{H} - \gamma^2\bar{H} = 0 \qquad .....(4.5.3)$$

Equations (4.5.2) and (4.5.3) are known as homogeneous vector **Helmholtz's** equations or vector wave equations in conducting medium.

If the wave is traveling along X-direction it's E_x component is '0' and

$$\frac{\partial^2\bar{E}}{\partial z^2} \text{ and } \frac{\partial^2\bar{E}}{\partial y^2} = 0$$

Then equation (4.5.2) becomes

$$\frac{\partial^2\bar{E}}{\partial x^2} - \gamma^2\bar{E} = 0$$

Consider only the 'y' component of $\bar{E}$

i.e., $\bar{E} = E_y\bar{a}_y$

Then $$\frac{\partial^2 E_y}{\partial x^2} - \gamma^2 E_y = 0$$

The solution for the above Differential equation is

$$E_y = C_1e^{-\gamma x} + C_2e^{\gamma x}$$

The displacement $E_y(t)$ at any time in Y-direction can be written as

$$E_y(t) = E_ye^{j\omega t}$$

$$= \left[C_1e^{-\gamma x} + C_2e^{\gamma x}\right]e^{j\omega t}$$

$$= C_1e^{j\omega t - \gamma x} + C_2e^{j\omega t + \gamma x}$$

We know that γ is a complex quantity, so replace γ with $\alpha + j\beta$

$$E_y(t) = C_1e^{(j\omega t - \alpha x - j\beta x)} + C_2e^{(j\omega t + \alpha x + j\beta x)}$$

Consider the first term i.e., $C_1 e^{(j\omega t - j\beta x)} e^{-\alpha x}$

From $e^{-\alpha x}$ we can say that the amplitude of the wave will be decaying as it travels along X-direction.

α is a measure of spatial rate of decay of the wave in the medium and is called attenuation constant or attenuation coefficient, of the medium, measured in nepers per meter (Np/m).

$$1\text{Np} = 20\log_{10}\text{e} = 8.686 \text{ dB}$$

and β is a measure of the phase shift per unit length and is called phase constant or wave number, measured in radians per meter.

In terms of β, the wave velocity v and wavelength λ are, respectively given by

$$v = \frac{\omega}{\beta}, \quad \lambda = \frac{2\pi}{\beta} \qquad .....(4.5.4)$$

Let us evaluate α and β in terms of μ, ∈, ω and σ.

We know $\gamma^2 = j^2\mu\epsilon\omega^2 + j\omega\mu\sigma$

$$= -\mu\epsilon\omega^2 + j\omega\mu\sigma \qquad .....(4.5.5)$$

and from $\gamma = \alpha + j\beta$

$$\gamma^2 = \alpha^2 - \beta^2 + 2j\alpha\beta \qquad .....(4.5.6)$$

On comparing the equations (4.5.5) & (4.5.6)

$$\alpha^2 - \beta^2 = -\mu\epsilon\omega^2 \quad \text{and} \quad 2\alpha\beta = \omega\mu\sigma$$

for simplicity denote $\mu\epsilon\omega^2 = a^2$ and $\omega\mu\sigma = b^2$

$$\therefore \quad \alpha^2 - \beta^2 = -a^2 \qquad .....(4.5.7)$$

$$\therefore \quad 2\alpha\beta = b^2 \qquad .....(4.5.8)$$

On squaring and adding equations (4.5.7) and (4.5.8), we get

$$(\alpha^2 - \beta^2)^2 + 4\alpha^2\beta^2 = a^4 + b^4$$

$$(\alpha^2 + \beta^2)^2 = a^4 + b^4$$

$$\alpha^2 + \beta^2 = \sqrt{a^4 + b^4} \qquad .....(4.5.9)$$

Add equations (4.5.7) and (4.5.9)

$$2\alpha^2 = -a^2 + \sqrt{a^4 + b^4}$$

$$\alpha^2 = -\frac{a^2}{2} + \frac{1}{2}\sqrt{a^4 + b^4}$$

$$\alpha = \sqrt{-\frac{a^2}{2} + \frac{1}{2}\sqrt{a^4 + b^4}}$$

Substituting a^2 and b^2, we get

$$\alpha = \left[-\frac{\mu \in \omega^2}{2} + \frac{1}{2}\sqrt{\mu^2 \in^2 \omega^4 + \omega^2 \mu^2 \sigma^2}\right]^{1/2}$$

$$\alpha = \omega\sqrt{\mu \in}\left[-\frac{1}{2} + \frac{1}{2}\sqrt{1 + \frac{\sigma^2}{\omega^2 \in^2}}\right]^{1/2}$$

$$\Rightarrow \qquad \alpha = \omega\sqrt{\mu \in}\left[\frac{1}{2}\left(\sqrt{1 + \left(\frac{\sigma}{\omega \in}\right)^2} - 1\right)\right]^{1/2} \qquad .....(4.5.10)$$

Subtracting equations (4.5.7) and (4.5.9)

$$2\beta^2 = a^2 + \sqrt{a^4 + b^4}$$

$$\beta = \sqrt{\frac{a^2}{2} + \frac{1}{2}\sqrt{a^4 + b^4}}$$

Substituting a^2 and b^2, we get

$$\beta = \left[\frac{\mu \in \omega^2}{2} + \frac{1}{2}\sqrt{\mu^2 \in^2 \omega^4 + \omega^2 \mu^2 \sigma^2}\right]^{1/2}$$

$$\beta = \omega\sqrt{\mu \in}\left[\frac{1}{2} + \frac{1}{2}\sqrt{1 + \frac{\sigma^2}{\omega^2 \in^2}}\right]^{1/2}$$

$$\Rightarrow \qquad \beta = \omega\sqrt{\mu \in}\left[\frac{1}{2}\left(\sqrt{1 + \left(\frac{\sigma}{\omega \in}\right)^2} + 1\right)\right]^{1/2} \qquad .....(4.5.11)$$

4.6 Wave Propagation in Free Space

For charge free medium we have from equation (4.2.3)

$$\nabla^2 \overline{E} - \mu \in \frac{\partial^2 \overline{E}}{\partial t^2} = 0$$

Wave equation in terms of $\overline{E}$ in free space is

$$\nabla^2 \overline{E} - \mu_0 \in_0 \frac{\partial^2 \overline{E}}{\partial t^2} = 0 \qquad (4.6.1)$$

For sinusoidal variations $\overline{E}$ can be written as $\overline{E} = \overline{E}_0 e^{j\omega t}$

$$\frac{\partial^2 \overline{E}}{\partial t^2} = -\omega^2 \overline{E}_0 e^{j\omega t} = -\omega^2 \overline{E}$$

from equation (4.6.1) $\quad \nabla^2 \overline{E} + \omega^2 \mu_0 \in_0 \overline{E} = 0 \qquad$ .....(4.6.2)

The above equation is called as **Helmholtz** equation in free space

If wave is traveling along X-direction it's E_x component is '0'

and $$\frac{\partial^2 \overline{E}}{\partial z^2} \;\&\; \frac{\partial^2 \overline{E}}{\partial y^2} = 0$$

∴ The equation (4.6.2) becomes

$$\frac{\partial^2 \overline{E}}{\partial x^2} + \omega^2 \mu_0 \in_0 \overline{E} = 0$$

where $\overline{E} = E_y \overline{a}_y + E_z \overline{a}_z$

Let us consider only the y-component

Then $$\frac{\partial^2 E_y}{\partial x^2} + \omega^2 \mu_0 \in_0 E_y = 0$$

Replace $\omega^2 \mu_0 \in_0$ with β^2

$$\beta = \omega \sqrt{\mu_0 \in_0}$$

where β is the phase constant.

$$\frac{\partial^2 E_y}{\partial x^2} + \beta^2 E_y = 0$$

The solution of above differential equation is

$$E_y = C_1 e^{-j\beta x} + C_2 e^{j\beta x}$$

The displacement $E_y(t)$ at any time in Y-direction can be written as

$$E_y(t) = E_y e^{j\omega t}$$

$$E_y(t) = \left[C_1 e^{-j\beta x} + C_2 e^{j\beta x}\right] e^{j\omega t}$$

$$= C_1 e^{j(\omega t - \beta x)} + C_2 e^{j(\omega t + \beta x)} \qquad \text{.....(4.6.3)}$$

The above equation contains two waves in which the first term is a wave that travels along +ve X-direction and second term is a wave that travels along –ve X-direction.

If we consider the source at origin, the first term will be traveling away from the source and the second term will be traveling towards the source.

Phase Velocity:

To find the phase velocity consider the real part of first term of equation (4.6.3)

$$C_1 \cos(\omega t - \beta x) = C_1 \cos \omega \left(t - \frac{\beta}{\omega} x\right)$$

$$= C_1 \cos \omega \left(t - \sqrt{\mu_0 \in_0} x\right)$$

Equate phase to zero i.e., $\omega\left(t - \sqrt{\mu_0 \in_0}\ x\right)$

$$\Rightarrow \qquad t - x\sqrt{\mu_0 \in_0} = 0$$

Differentiate the above equation

$$\Rightarrow \qquad dt - dx\sqrt{\mu_0 \in_0} = 0$$

$$\frac{dx}{dt} = \frac{1}{\sqrt{\mu_0 \in_0}} \qquad \text{.....(4.6.4)}$$

which is the phase velocity in free space.

Here the phase velocity is equal to the wave velocity '*v*' and it's generalized equation in any medium is

$$v = \frac{1}{\sqrt{\mu \in}} \qquad \text{.....(4.6.5)}$$

Problem 4.1

An uniform plane wave with an intensity of electric field = 1 V/m is traveling in free space. Find the magnitude of associated magnetic field.

Solution

Electric field intensity = 1 V/m

The magnitude of the magnetic field is found by $\frac{|\bar{E}|}{|\bar{H}|} = \sqrt{\frac{\mu_0}{\epsilon_0}}$

$$|\bar{H}| = 2.6 \text{ mA/m}$$

Problem 4.2

Show that electric and magnetic energy densities in a traveling plane wave are equal.

Solution

$$\frac{E}{H} = \sqrt{\frac{\mu}{\epsilon}}$$

$$\Rightarrow \quad \frac{E^2}{H^2} = \frac{\mu}{\epsilon}$$

$$\Rightarrow \quad \epsilon E^2 = \mu H^2$$

$$\Rightarrow \quad \frac{1}{2} \epsilon E^2 = \frac{1}{2} \mu H^2$$

Energy density for electric field = energy density for magnetic field

Problem 4.3

For a uniform plane wave traveling in X-direction in free space E_y=10 sin (2π 10^8 t – βx). Find the phase constant, phase velocity and equation for H_z if $E_z = H_y = 0$.

Solution

Given E_y=10 sin (2π 10^8 t – βx)

$$\omega = 2\pi\ 10^8; \quad \beta^2 = \omega^2 \mu_0 \epsilon_0 = (2\pi 10^8)^2 \times 8.854 \times 10^{-12} \times 4\pi \times 10^{-7}$$

The Phase constant $\beta = 2.096$

The Phase Velocity $v = \frac{\omega}{\beta} = \frac{2\pi \times 10^8}{2.096} = 2.99 \times 10^8$ m/s

We have $\frac{E_y}{H_z} = 377$

$$\Rightarrow \quad H_z = \frac{10}{377} \sin(2\pi 10^8 t - \beta x)$$

4.7 Comparison between Conductors and Dielectrics

When the wave is traveling in a conductor, we know that the amplitude of the wave decreases. When σ is large the decrease in the amplitude of the wave is rapid. When the 'σ' is small the decrease in the amplitude of the wave is less as shown in Fig. 4.1(a).

To see the dependency of the amplitude of the wave on σ

Consider Maxwell's equation

$$\nabla \times \bar{H} = \bar{J} + \frac{\partial \bar{D}}{\partial t}$$

For sinusoidal variations assume $\bar{D} = \bar{D}_0 \, e^{j\omega t}$

$$\therefore \qquad \frac{\partial \bar{D}}{\partial t} = j\omega \bar{D}_0 e^{j\omega t} = j\omega \bar{D}$$

$$\nabla \times \bar{H} = \bar{J} + j\omega \bar{D} \qquad \because \bar{J} = \sigma \bar{E} \ \& \bar{D} = \in \bar{E}$$

$$\nabla \times \bar{H} = \sigma \bar{E} + j\omega \in \bar{E}$$

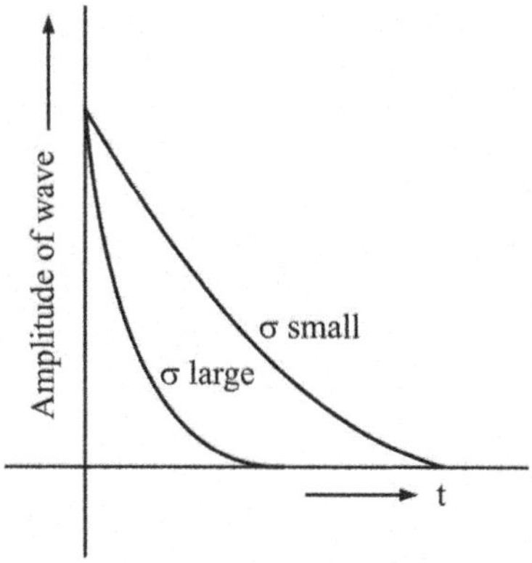

Fig. 4.1(a) Amplitude variation w.r.t. time

Where $\sigma\bar{E}$ is the conduction current density and $j\omega \in \bar{E}$ is the displacement current density. The ratio of magnitudes of conduction current density and displacement current density is $\left|\frac{\sigma \bar{E}}{j\omega \in \bar{E}}\right| = \frac{\sigma}{\omega \in} = \tan\theta$ which serves as a boundary between conductors and dielectrics and is called loss tangent and θ is called loss angle, which is shown in Fig. 4.1(b).

For good conductors $\frac{\sigma}{\omega \in} >> 1$

For perfect dielectrics the ratio tends to 'zero' and for good dielectrics the ratio is <<1.

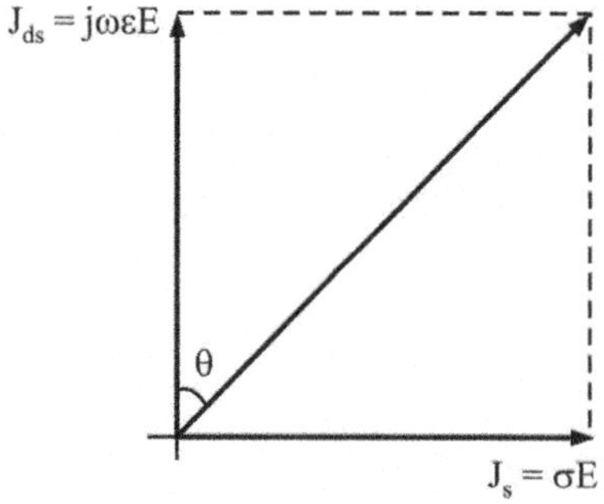

Fig. 4.1(b) Loss angle

Good Conductors

As $\frac{\sigma}{\omega \in} >> 1$ for good conductors, the equations for α and β become as follows:

We know α in a general medium as

$$\alpha = \omega\sqrt{\mu \in}\left[\frac{1}{2}\left(\sqrt{1+\left(\frac{\sigma}{\omega \in}\right)^2} - 1\right)\right]^{1/2} \quad \text{as } \frac{\sigma}{\omega \in} >> 1$$

$$\alpha = \omega\sqrt{\mu \in}\left[\frac{1}{2}\left(\frac{\sigma}{\omega \in}\right)\right]^{1/2}$$

$$\alpha = \frac{1}{\sqrt{2}}\omega\sqrt{\mu}\sqrt{\in}\frac{\sqrt{\sigma}}{\sqrt{\omega}\sqrt{\in}} = \sqrt{\frac{\omega\sigma\mu}{2}} \qquad \text{.....(4.7.1)}$$

and we know

$$\beta = \omega\sqrt{\mu \in}\left[\frac{1}{2}\left(\sqrt{1+\left(\frac{\sigma}{\omega \in}\right)^2}+1\right)\right]^{1/2} \quad \text{as } \frac{\sigma}{\omega \in} >> 1$$

$$\beta = \omega\sqrt{\mu \in}\frac{1}{\sqrt{2}}\frac{\sqrt{\sigma}}{\sqrt{\omega \in}} = \sqrt{\frac{\omega\sigma\mu}{2}} \qquad \text{.....(4.7.2)}$$

$$\gamma = \alpha + j\beta$$

$$= \sqrt{\frac{\omega\sigma\mu}{2}} + j\sqrt{\frac{\omega\sigma\mu}{2}}$$

$$= \sqrt{\frac{\omega\sigma\mu}{2}}(1+j)$$

$$\gamma = \sqrt{\omega\sigma\mu}\ \angle 45^o \qquad \text{.....(4.7.3)}$$

$$v = \text{phase velocity} = \frac{\omega}{\beta}$$

$$v = \frac{\omega}{\sqrt{\frac{\omega\sigma\mu}{2}}} = \sqrt{\frac{2\omega}{\sigma\mu}} \qquad \text{.....(4.7.4)}$$

Perfect Dielectrics or Loss Less Medium

As $\frac{\sigma}{\omega \in} \to 0$ for perfect dielectrics, the equations for α and β become as follows:

$$\alpha = \omega\sqrt{\mu \in}\left[\frac{1}{2}\left(\sqrt{1+\left(\frac{\sigma}{\omega \in}\right)^2}-1\right)\right]^{1/2} \quad \text{as } \frac{\sigma}{\omega \in} \to 0$$

$$\alpha = \omega\sqrt{\mu \in}\,(0) = 0$$

$$\beta = \omega\sqrt{\mu \in}\left[\frac{1}{2}\left(\sqrt{1+\left(\frac{\sigma}{\omega \in}\right)^2}+1\right)\right]^{1/2}$$

$$\beta = \omega\sqrt{\mu \in}\left[\frac{1}{2}\,2\right]^{1/2} = \omega\sqrt{\mu \in} \qquad \text{.....(4.7.5)}$$

$$\gamma = \alpha + j\beta$$

$$\gamma = j\omega\sqrt{\mu \in} \qquad \text{.....(4.7.6)}$$

and the phase velocity $v = \frac{\omega}{\beta} = \frac{\omega}{\omega\sqrt{\mu \in}} = \frac{1}{\sqrt{\mu \in}}$ .....(4.7.7)

Good Dielectrics (or) Low Loss Dielectrics

As $\frac{\sigma}{\omega \in} << 1$ for good dielectrics, the equations for α and β become as follows:

$$\alpha = \omega\sqrt{\mu \in}\left[\frac{1}{2}\left(\sqrt{1+\left(\frac{\sigma}{\omega \in}\right)^2}-1\right)\right]^{1/2} \quad \text{as} \quad \frac{\sigma}{\omega \in} << 1$$

i.e., $(1+a)^n = 1 + na \quad if \quad a << 1$

$$\therefore \qquad \alpha = \omega\sqrt{\mu \in}\left[\frac{1}{2}\left[1+\frac{1}{2}\left(\frac{\sigma}{\omega \in}\right)^2-1\right]\right]^{1/2}$$

$$\alpha = \omega\sqrt{\mu \in}\left[\frac{1}{2}\left(\frac{\sigma}{\omega \in}\right)\right]$$

$$\alpha = \frac{1}{2}\sigma\sqrt{\frac{\mu}{\in}}$$

and

$$\beta = \omega\sqrt{\mu \in}\left[\frac{1}{2}\left(\sqrt{1+\left(\frac{\sigma}{\omega \in}\right)^2}+1\right)\right]^{1/2}$$

$$= \omega\sqrt{\mu \in}\left[\frac{1}{2}\left[1+\frac{1}{2}\left(\frac{\sigma}{\omega \in}\right)^2+1\right]\right]^{1/2}$$

$$= \omega\sqrt{\mu \in}\left[\frac{1}{2}+\frac{1}{2}+\frac{1}{4}\left(\frac{\sigma}{\omega \in}\right)^2\right]^{1/2}$$

$$= \omega\sqrt{\mu \in}\left[1+\frac{1}{4}\left(\frac{\sigma}{\omega \in}\right)^2\right]^{1/2} = \omega\sqrt{\mu \in}\left[1+\frac{1}{4}\left(\frac{\sigma}{\omega \in}\right)^2\frac{1}{2}\right]$$

$$= \omega\sqrt{\mu \in}\left[1+\frac{1}{8}\left(\frac{\sigma}{\omega \in}\right)^2\right] \quad \text{.....(4.7.8)}$$

$$\gamma = \alpha + j\beta$$

$$= \frac{\sigma}{2}\sqrt{\frac{\mu}{\in}} + j\omega\sqrt{\mu \in}\left[1+\frac{1}{8}\left(\frac{\sigma}{\omega \in}\right)^2\right] \quad \text{.....(4.7.9)}$$

The Phase Velocity $$v = \frac{\omega}{\beta} = \frac{\omega}{\omega\sqrt{\mu \in}\left[1+\frac{1}{8}\left(\frac{\sigma}{\omega \in}\right)^2\right]}$$

$$= \frac{1}{\sqrt{\mu \in}\left[1+\frac{1}{8}\left(\frac{\sigma}{\omega \in}\right)^2\right]} \quad \text{..... (4.7.10)}$$

Complex Permittivity

From the view point of wave propagation, the characteristic behavior of a medium depends not only on its parameters $\sigma, \in$ and μ but also on the frequency of operation. A medium that is regarded as a good conductor at low frequencies may be a good dielectric at high frequencies.

We have

$$\nabla \times \bar{H} = \sigma\bar{E} + j\omega \in \bar{E}$$

$$\nabla \times \bar{H} = j\omega \in \left[1-\frac{j\sigma}{\omega \in}\right]\bar{E}$$

$$\Delta \times \bar{H} = j\omega \in_c \bar{E}$$

where $\in_c = \in \left[1 - \dfrac{j\sigma}{\omega \in}\right]$ is the complex permittivity of the medium. .....(4.7.11)

4.8 Skin Effect and Skin Depth

When a wave enters in to a conducting medium, it's amplitude reduces exponentially as shown in Fig.4.2 and it becomes zero after traveling some distance, as a result the induced current by the wave exist near the surface of the conductor. This effect is called skin effect.

The exponential decay of the wave can be represented by $E = E_0\, e^{-\alpha x}$, where α is the attenuation constant and E_0 is the maximum amplitude at $x = 0$. If the wave penetrates by a distance x = δ, the amplitude of the wave at $x = 1/e$ becomes E_0/e, where δ is the depth of penetration or skin depth. It can be defined as depth of a conductor at which the amplitude of traveling wave reduces by $1/e$ times of it's amplitude at the initial time.

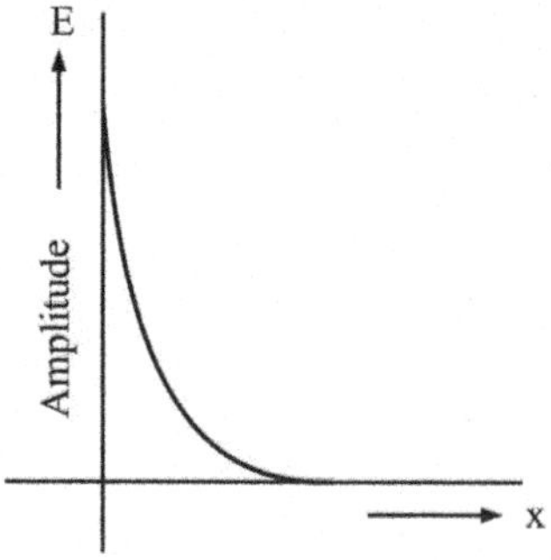

Fig. 4.2 Evaluation of Skin depth

The skin depth $\delta = \dfrac{1}{\alpha}$ .....(4.8.1)

We know α for a good conductor i.e.,

$$\alpha = \sqrt{\frac{\omega\mu\sigma}{2}}$$

$$\therefore \quad \delta = \sqrt{\frac{2}{\omega\mu\sigma}} \qquad .....(4.8.2)$$

From the above equation we can say

(i) The depth of penetration decreases as conductivity increases

(ii) The depth of penetration decreases as frequency of signal or wave increases.

The surface or skin resistance R_s (in Ω) is defined as real part of intrinsic impedance for good conductors.

$$\therefore \quad R_s = \frac{1}{\sigma\delta} = \sqrt{\frac{\omega\mu}{2\sigma}} \qquad .....(4.8.3)$$

which is the real part of intrinsic or surface impedance η or Z_s as in equation (4.9.5)

4.9 Intrinsic Impedance in different Media

To find the generalized equation for intrinsic or characteristic impedance (η) consider the Maxwell's equation $\nabla \times \bar{E} = -\frac{\partial \bar{B}}{\partial t}$

where $\bar{B} = \mu \bar{H}$

$$\nabla \times \bar{E} = -\mu \frac{\partial \bar{H}}{\partial t}$$

$$\begin{vmatrix} \bar{a}_x & \bar{a}_y & \bar{a}_z \\ \frac{\partial}{\partial x} & \frac{\partial}{\partial y} & \frac{\partial}{\partial z} \\ E_x & E_y & E_z \end{vmatrix} = -\mu \left[\frac{\partial H_x}{\partial t} \bar{a}_x + \frac{\partial H_y}{\partial t} \bar{a}_y + \frac{\partial H_z}{\partial t} \bar{a}_z \right]$$

Assume the wave is traveling along X-direction then E_x and H_x are zero and partial differentiations w.r.t. y and z produces '0'.

$$\bar{a}_x (0) - \bar{a}_y \left(\frac{\partial}{\partial x} E_z - 0 \right) + \bar{a}_z \left(\frac{\partial}{\partial x} E_y - 0 \right) = -\mu \left(\frac{\partial H_y}{\partial t} \bar{a}_y + \frac{\partial H_z}{\partial t} \bar{a}_z \right)$$

Equating $\bar{a}_y$ components

$$\frac{\partial E_z}{\partial x} = \mu \frac{\partial H_y}{\partial t}$$

$$\Rightarrow \qquad \frac{\partial E_z}{\partial x} = \mu \frac{\partial H_y}{\partial t}$$

Similarly equating $\bar{a}_z$ components

$$\frac{\partial E_y}{\partial x} = -\mu \frac{\partial H_z}{\partial t} \qquad \text{.....(4.9.1)}$$

Any wave traveling along X-direction can be represented by it's 'y' component of E as

$$E_y = C_1 e^{-\gamma x} e^{j\omega t}$$

Where

$$\gamma = \alpha + j\beta = \sqrt{j\omega\mu(\sigma + j\omega \in)}$$

From above equation $\frac{\partial E_y}{\partial x} = -\gamma C_1 e^{-\gamma x} e^{j\omega t}$

$$= -\gamma E_y$$

Similarly 'z' component of 'H' is $H_z = C_2 e^{-\gamma x} e^{j\omega t}$

$$\frac{\partial H_z}{\partial t} = j\omega H_z$$

Substitute in equation (4.9.1)

$$-\gamma E_y = -\mu j\omega H_z$$

$$\eta = \frac{E_y}{H_z} = \frac{\mu j\omega}{\gamma} = \frac{\mu j\omega}{\sqrt{\mu j\omega(\sigma + j\omega\epsilon)}} = \sqrt{\frac{\mu j\omega}{\sigma + j\omega\epsilon}} \quad \text{.....(4.9.2)}$$

which is the intrinsic impedance in a general medium.

The above expression can also be written as

$$\eta = \sqrt{\frac{j\mu\omega}{j\omega\epsilon\left(1 - \frac{j\sigma}{\omega\epsilon}\right)}} = \sqrt{\frac{\mu/\epsilon}{\left(1 - \frac{j\sigma}{\omega\epsilon}\right)}}$$

$$\eta = \sqrt{\frac{\mu/\epsilon\left(1 + \frac{j\sigma}{\omega\epsilon}\right)}{\left(1 - \frac{j\sigma}{\omega\epsilon}\right)\left(1 + \frac{j\sigma}{\omega\epsilon}\right)}} = \sqrt{\frac{\mu}{\epsilon}}\sqrt{\frac{\left(1 + \frac{j\sigma}{\omega\epsilon}\right)}{\left(1 + \left(\frac{\sigma}{\omega\epsilon}\right)^2\right)}}$$

$$|\eta| = \sqrt{\frac{\mu}{\epsilon}}\sqrt{\left[\frac{1}{\left(1 + \left(\frac{\sigma}{\omega\epsilon}\right)^2\right)^2} + \frac{\left(\frac{\sigma}{\omega\epsilon}\right)^2}{\left(1 + \left(\frac{\sigma}{\omega\epsilon}\right)^2\right)^2}\right]^{1/2}} \quad \text{and} \quad \theta_\eta = \tan^{-1}\left(\sqrt{\frac{\sigma}{\omega\epsilon}}\right)$$

$$|\eta| = \sqrt{\frac{\mu}{\in}} \sqrt{\left[\frac{\left(1+\left(\frac{\sigma}{\omega \in}\right)^2\right)}{\left(1+\left(\frac{\sigma}{\omega \in}\right)^2\right)^2}\right]^{1/2}} \quad \text{and} \quad 2\theta_\eta = \tan^{-1}\left(\frac{\sigma}{\omega \in}\right)$$

$$|\eta| = \frac{\sqrt{\frac{\mu}{\in}}}{\left(1+\left(\frac{\sigma}{\omega \in}\right)^2\right)^{1/4}} \quad \text{and} \quad \tan 2\theta_\eta = \left(\frac{\sigma}{\omega \in}\right) \qquad .....(4.9.3)$$

For a Good Conductor

$$\frac{\sigma}{\omega \in} >> 1 \Rightarrow \sigma >> \omega \in$$

$$\therefore \qquad \eta = \sqrt{\frac{\mu j\omega}{\sigma + j\omega \in}} = \sqrt{\frac{\mu j\omega}{\sigma}} = (j)^{1/2}\sqrt{\frac{\mu\omega}{\sigma}} \qquad .....(4.9.4)$$

$$\eta = (\cos \pi / 2 + j \sin \pi / 2)^{1/2}\sqrt{\frac{\mu\omega}{\sigma}}$$

$$\eta = (e^{j\pi/2})^{1/2}\sqrt{\frac{\mu\omega}{\sigma}} = e^{j\pi/4}\sqrt{\frac{\mu\omega}{\sigma}} = \sqrt{\frac{\mu\omega}{\sigma}}\angle 45^0 \qquad .....(4.9.5)$$

For Perfect dielectrics:

$$\frac{\sigma}{\omega \in} \to 0$$

$$\therefore \qquad \eta = \sqrt{\frac{\mu j\omega}{\omega \in \left(\frac{\sigma}{\omega \in} + j\right)}} = \sqrt{\frac{\mu}{\in}} \qquad .....(4.9.6)$$

For Good dielectrics or low loss medium:

$$\frac{\sigma}{\omega \in} << 1$$

$$\therefore \qquad \eta = \sqrt{\frac{\mu j\omega}{j\omega \in \left(1+\frac{\sigma}{j\omega \in}\right)}} = \sqrt{\frac{\mu}{\in}}\left(1+\frac{\sigma}{j\omega \in}\right)^{-1/2} \quad \text{as} \quad \frac{\sigma}{\omega \in} << 1$$

$$\eta = \sqrt{\frac{\mu}{\in}}\left(1-\frac{1}{2}\frac{\sigma}{j\omega \in}\right)$$

$$\eta = \sqrt{\frac{\mu}{\in}}\left(1+\frac{1}{2}\frac{\sigma j}{\omega \in}\right) \qquad \text{.....(4.9.7)}$$

Problem 4.4

A plane wave traveling in +ve X-direction in a loss less unbounded medium having permeability 4.5 times that of free space and a permittivity twice that of free space. (a) Find the phase velocity of the wave (b) If the electric field $\bar{E}$ has only a 'y' component with an amplitude of 20 V/m. Find the amplitude and direction of magnetic field intensity.

Solution

(a) Phase velocity $v = \frac{1}{\sqrt{\mu \in}} = \frac{1}{\sqrt{4.5\mu_0 \, 2 \in_0}}$

$= 1 \times 10^8$ m/s

(b) $H_z = \sqrt{\frac{\in}{\mu}} E_y = 0.0354 \text{ A/m}$, it's direction is along 'z'.

***Problem 4.5**

A wave propagating in a lossless dielectric has the components

$$\bar{E} = 500\cos\left(10^7 t - \beta z\right)\bar{a}_x \text{ V/m and } \bar{H} = 1.1\cos\left(10^7 t - \beta z\right)\bar{a}_y \text{ A/m}.$$

If the wave is traveling at v = 0.5c. Where 'c' is the velocity in free space. Find (a) μ_r (b) $\in_r$ (c) β (d) λ (e) Z.

Solution

The wave is propagating along Z-direction, $\bar{E}$ is along X and $\bar{H}$ is along 'Y' directions

(a) v = 0.5 c

$$V = 0.5\frac{1}{\sqrt{\mu_0 \in_0}}$$

$$\frac{1}{\sqrt{\mu_0\mu_r \in_0\in_r}} = 0.5\frac{1}{\sqrt{\mu_0 \in_0}}$$

$$\frac{1}{\sqrt{\mu_r \in_r}} = 0.5$$

$$\sqrt{\mu_r \in_r} = \frac{1}{0.5} \Rightarrow \mu_r \in_r = 4$$

$$z = 454.5 = \sqrt{\frac{\mu}{\in}} = \sqrt{\frac{\mu_r \mu_0}{\in_0\in_r}} = (3.767)\times10^{-10}\sqrt{\frac{\mu_r}{\in_r}}$$

$$\Rightarrow \qquad \frac{\mu_r}{\in_r} = \left(\frac{454.5}{3.767\times10^{-10}}\right)^2 = 1.456\times10^{24}$$

$$\Rightarrow \qquad \mu_r = \sqrt{4\times1.456\times10^{24}} = 2.41\times10^{12}$$

(b) $\in_r = \sqrt{4/1.456\times10^{24}} = 1.66\times10^{-12}$

(c) given $\omega = 10^7$, as $\frac{\sigma}{\omega\in} << 1$ for low loss dielectric,

$$\beta = \omega\sqrt{\mu\in} = 10^7\sqrt{8.854\times10^{-12}\times1.66\times10^{-12}\times4\times\pi\times10^{-7}\times2.41\times10^{12}} = 0.0667$$

(d) Wavelength $\lambda = \frac{2\pi}{\beta} = \frac{2\pi}{0.0667} = 94.2$ m

(e) Z = η = intrinsic impedance = $\frac{|E|}{|H|} = \frac{500}{1.1} = 454.5\ \Omega$

***Problem 4.6**

Dry ground has a conductivity of 5×10^{-4} mhos/m and relative dielectric constant of 10 at a frequency of 500 MHz. Compute (i) the intrinsic impedance, (ii) the propagation constant, (iii) the phase velocity.

Solution

Given $\sigma = 5\times10^{-4}$ mhos / m, $\varepsilon_r = 10$ and $f = 500$ MHz

Loss tangent $\dfrac{\sigma}{\omega\varepsilon} = \dfrac{5\times10^{-4}}{2\pi\times50\times10^{6}\times10\times\dfrac{10^{-9}}{36\pi}} = 0.018$

i.e., $\dfrac{\sigma}{\omega\varepsilon} \to 0$, it is perfect dielectric medium

(i) intrinsic impedance $\eta = \sqrt{\dfrac{\mu}{\in}}$

$$\eta = \sqrt{\frac{\mu_0\mu_r}{\in_0\in_r}} = \sqrt{\frac{4\pi\times10^{-7}}{\dfrac{10^{-9}}{36\pi}\times10}} = 119.22\ \Omega$$

(ii) the propagation constant $\gamma = j\omega\sqrt{\mu\in}$

$$\gamma = j2\pi\times500\times10^{6}\sqrt{4\pi\times10^{-7}\times\frac{10^{-9}}{36\pi}\times10} = j33.11$$

(iii) the phase velocity $v_p = \dfrac{1}{\sqrt{\mu\in}}$

$$v_p = \frac{1}{\sqrt{4\pi\times10^{-7}\times\dfrac{10^{-9}}{36\pi}\times10}} = 94.868\times10^{6}\ \text{m/s}$$

***Problem 4.7**

Find α, β, γ and η for ferrite at 10 GHz. $\varepsilon_r = 9, \mu_r = 4, \sigma = 10$ mhos / m

Solution

Loss tangent $\dfrac{\sigma}{\omega\varepsilon} = \dfrac{10}{2\pi\times10\times10^{9}\times9\times\dfrac{10^{-9}}{36\pi}} = 2$

i.e., $\dfrac{\sigma}{\omega\varepsilon} > 1$, it is a good conductor

$$\alpha = \sqrt{\frac{\omega\sigma\mu}{2}}$$

$$\alpha = \sqrt{\frac{2\pi \times 10 \times 10^9 \times 10 \times 4\pi \times 10^{-7} \times 4}{2}} = 1256.63 \text{ Np/m}$$

$$\beta = \alpha = \sqrt{\frac{\omega\sigma\mu}{2}} = 1256.63 \text{ rad/m}$$

$$\gamma = \alpha + j\beta$$

$$\gamma = 1256.63 + j1256.63 \text{ m}^{-1}$$

$$\eta = \sqrt{\frac{\mu\omega}{\sigma}}\angle 45^\circ = \sqrt{\frac{2\pi \times 10 \times 10^9 \times 4\pi \times 10^{-7} \times 4}{10}}\angle 45^\circ = 177.71\angle 45^\circ \,\Omega$$

*Problem 4.8

A non magnetic medium has an intrinsic impedance of $240\angle 30\,°\Omega$. Find its (i) Loss tangent, (ii) Dielectric constant, (iii) Complex permittivity, (iv) Attenuation constant at 1 MHz.

Solution

Given $\eta = 240\angle 30°\,\Omega$

i.e., $|\eta| = 240$ and $\theta_\eta = \angle 30°$

(i) we have

$$\text{loss tangent} = \frac{\sigma}{\omega \in} = \tan 2\theta_\eta = \tan(2 \times 30°) = 1.732$$

(ii) we have

$$|\eta| = \frac{\sqrt{\dfrac{\mu}{\in}}}{\left(1 + \left(\dfrac{\sigma}{\omega \in}\right)^2\right)^{1/4}} = 240$$

$$= \frac{\sqrt{\dfrac{\mu_0 \mu_r}{\in_0 \in_r}}}{\left(1 + (1.732)^2\right)^{1/4}} = 240$$

$$\frac{\sqrt{\dfrac{1\times 4\pi\times 10^{-7}}{\epsilon_r \times \dfrac{10^{-9}}{36\pi}}}}{\left(1+(1.732)^2\right)^{1/4}} = 240$$

$$= \frac{84.853\pi}{\sqrt{\epsilon_r}} = 240$$

Dielectric constant $\epsilon_r = 1.2337$

(iii) we have Complex permittivity

$$\epsilon_c = \epsilon\left[1 - \frac{j\sigma}{\omega\epsilon}\right]$$

$$\epsilon_c = \frac{10^{-9}}{36\pi} \times 1.2337[1 - j1.732] = (1.09 - j1.8893)\times 10^{-11}\,\text{F / maq}$$

***Problem 4.9**

A lossy dielectric has an intrinsic impedance of $200\angle 30\ ^\circ\Omega$ at a particular radian frequency ω. If at that frequency the plane wave propagating through the dielectric has the magnetic field component $\bar{H} = 10e^{-\alpha x}\cos(\omega t - 0.5x)\bar{a}_y$ A / m. Find $\bar{E}$ and α. Determine the skin depth and wave polarization.

Solution

From the given $\bar{H}$, we can say that wave travels along x-axis

We have $\bar{a}_d = \bar{a}_E \times \bar{a}_H$

Here $\bar{a}_d = \bar{a}_x$ and $\bar{a}_H = \bar{a}_y$

$\therefore \quad \bar{a}_E = -\bar{a}_z$

Also we have

$$\eta = \frac{|\bar{E}|}{|\bar{H}|}$$

$\therefore \quad |\bar{E}| = |\bar{H}|\eta = 10 \times 200\angle 30^\circ = 2000e^{j\pi/6}$

$\bar{E}$ and $\bar{H}$ will have the same form except magnitude and phase

$$\therefore \qquad \bar{E} = -2000e^{-\alpha x}\cos(\omega t - 0.5x + \pi/6)\bar{a}_z \text{V/m}$$

From the above expression $\beta = 0.5$, α can be determined as

$$\alpha = \omega\sqrt{\mu \in}\left[\frac{1}{2}\left(\sqrt{1+\left(\frac{\sigma}{\omega \in}\right)^2} - 1\right)\right]^{1/2}$$

$$\beta = \omega\sqrt{\mu \in}\left[\frac{1}{2}\left(\sqrt{1+\left(\frac{\sigma}{\omega \in}\right)^2} + 1\right)\right]^{1/2}$$

$$\frac{\alpha}{\beta} = \left[\frac{\sqrt{1+\left(\frac{\sigma}{\omega \in}\right)^2} - 1}{\sqrt{1+\left(\frac{\sigma}{\omega \in}\right)^2} + 1}\right]^{1/2}$$

But

$$\frac{\sigma}{\omega \in} = \tan 2\theta_\eta = \tan 60^0 = \sqrt{3}$$

$$\frac{\alpha}{\beta} = \left[\frac{2-1}{2+1}\right]^{1/2}$$

$$\Rightarrow \qquad \alpha = \frac{0.5}{\sqrt{3}} = 0.2887 \text{N}_\text{p}/\text{m}$$

skin depth

$$\delta = \frac{1}{\alpha} = 2\sqrt{3} = 3.464 \text{ m}$$

Since Electric field points along z-axis, the polarization of the wave is z-direction.

Problem 4.10

A plane wave of 16 GHz frequency and E = 10 V/m propagates through the body of salt water having constants $\in_r = 100$, $\mu_r = 1$ and $\sigma = 100$ ℧/m. Determine attenuation constant, phase shift constant, phase velocity and intrinsic impedance of medium.

Solution

$$\frac{\sigma}{\omega \in} = 1.12 > 1, \text{ it is a good conductor}$$

Attenuation constant $\alpha = \sqrt{\frac{\omega\mu\sigma}{2}} = 2513.27$

Phase shift constant $\beta = \alpha = 2513.27$

Propagation constant $\gamma = \alpha + j\beta = 2513.27 + j\,2513.27$

$$= 3554.3\ \angle 45^\circ$$

Phase velocity $V = \frac{\omega}{\beta} = 40\ \text{Mm/s}$

intrinsic impedance $\eta = \sqrt{\frac{j\omega\mu}{\sigma}} = 35.543\sqrt{j}$

$$= 35.543\angle 45^\circ$$

Problem 4.11

Determine the propagation constant at 500 KHz for a medium in which $\mu_r = 1$, $\in_r = 15$, $\sigma = 0$, at what velocity will an EM wave travel in this medium.

Solution:

$$\frac{\sigma}{\omega \in} = 0 = \text{Perfect dielectric}$$

$\alpha = 0$

$$\beta = \omega\sqrt{\mu \in} = 0.041$$

$\gamma = \alpha + j\beta = 0.041\ \angle 90^\circ$

Wave travels with velocity $v = \frac{\omega}{\beta} = 76.624\ \text{Mm/s}$

Problem 4.12

For silver the conductivity is $\sigma = 3 \times 10^6$ ℧/m, at what frequency will the wave travels, if the depth of penetration is 1 mm.

Solution

$$\delta = \sqrt{\frac{2}{\mu\omega\sigma}}$$

$$1 \times 10^{-3} = \sqrt{\frac{2}{2\pi f \times 4\pi \times 10^{-7} \times 3 \times 10^6}}$$

$$f = \frac{4}{10^{-16} \times 4\pi \times 10^{-7} \times 3 \times 10^6} = 84.4\ \text{KHz}$$

Problem 4.13

In a medium $\bar{E} = 16e^{-x/20} \sin\left(2\times10^8 t - 2x\right) \bar{a}_z$ V/m. Find the direction of propagation, propagation constant, wave length, speed of wave and skin depth.

Solution

Direction of propagation is +Ve X-direction

$\alpha = 1/20, \ \beta = 2$

$\omega = 2 \times 10^8$

$\gamma = \alpha + j\beta = 2\angle 88.57^\circ$

$$v = \frac{1}{\sqrt{\mu \in}} = 3\times10^8 \text{ m/s}$$

Skin depth $\delta = \dfrac{1}{\alpha} = 20$

Problem 4.14

An EM wave propagated through a material $\mu_r = 5$, $\in_r = 10$. Determine (a) Velocity of propagation (b) Intrinsic impedance in free space and in material (c) wavelength in free space and in material, when f = 1 G Hz.

Solution

(a) $v = \dfrac{1}{\sqrt{\mu_0 \mu_r \in_0 \in_r}} = 42.4$ M m/s.

(b) Intrinsic impedance

in free space $\eta = \sqrt{\dfrac{\mu_0}{\in_0}} = \sqrt{\dfrac{4\pi\times10^{-7}}{8.854\times10^{-12}}} = 377\ \Omega$

in material $\eta = \sqrt{\dfrac{\mu_0\mu_r}{\in_0\in_r}} = \sqrt{\dfrac{4\pi\times10^{-7}\times5}{8.854\times10^{-12}\times10}} = 266.39\ \Omega$

(c) Wavelength

in free space $\lambda = \dfrac{v_0}{f} = \dfrac{3\times10^8}{1\times10^8} = 3$ m

in material $\lambda = \dfrac{v}{f} = \dfrac{42.4\times10^6}{1\times10^8} = 0.424$ m

Problem 4.15

A 'Cu' wire carries a conduction current of 1 A. Determine the displacement current in wire at 100 MHz. For 'Cu' $\sigma = 5.8 \times 10^7$ ℧/m and permittivity is same as that of free space.

Solution

$$\frac{I_D}{I_c} = \frac{\omega \in}{\sigma}$$

$$\Rightarrow \qquad I_D = 1A \times \frac{\omega \in}{\sigma} = \frac{1 \times 2 \times \pi \times 100 \times 10^6 \times 8.854 \times 10^{-12}}{5.8 \times 10^7}$$

$$= 9.59 \times 10^{-11}\,\text{A}$$

4.10 Polarization in EM Waves

The word polarization is a manner in which the variation of magnitude and direction of the (Electromagnetic) EM wave is observed. So based on this observation there are three different types of polarized waves exist namely Linearly polarized wave, Circularly polarized wave and Elliptically polarized wave.

4.10.1 Linearly Polarized Wave

Let us consider the wave is traveling along Z-direction, then it will have E_x, H_x and E_y, H_y components only. Assume only electric field components. The electric field $\overline{E}$ is resultant of $\overline{E}_x$ and $\overline{E}_y$. If the phase difference between $\overline{E}_x$ and $\overline{E}_y$ is '0' as shown in Fig. 4.3.

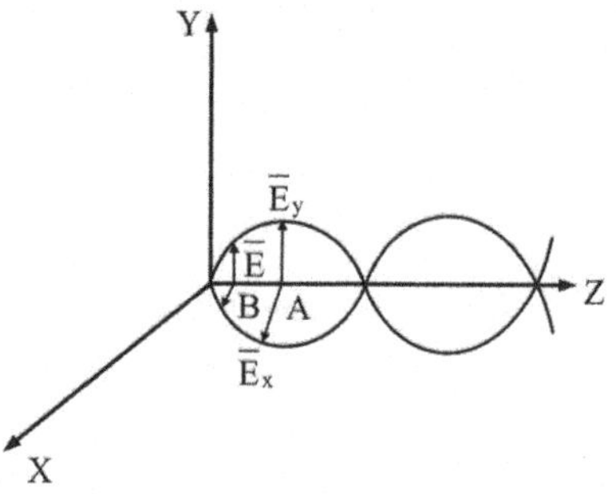

Fig. 4.3 Linear polarization

In Fig. 4.3 $\overline{E}_y$ wave is looking like it exists on YZ plane and $\overline{E}_x$ wave looks like it exists on XZ plane. As both of them are starting at origin the phase difference will be zero.

At 'O' the resultant vector $\overline{E}$ is '0' because $\overline{E}_x$ component and $\overline{E}_y$ component are zero. At 'B' the resultant vector $\overline{E}$ is greater than the value at origin. At 'A' the resultant vector will be maximum as $\overline{E}_x$ and $\overline{E}_y$ components are maximum. Like wise when we draw path formed by $\overline{E}$ on XY co-ordinate system it will be a straight line which makes an angle 'θ' with $\overline{E}_x$ as shown in Fig.4.4. Since the path formed by resultant $\overline{E}$ is a straight line this polarization is called linear polarization and the wave is said to be linearly polarized wave.

The conditions for getting linearly polarized wave are

(i) The phase difference between $\bar{E}_x$ and $\bar{E}_y$ waves must be 0.

(ii) The magnitudes of $\bar{E}_x$ and $\bar{E}_y$ need not be equal.

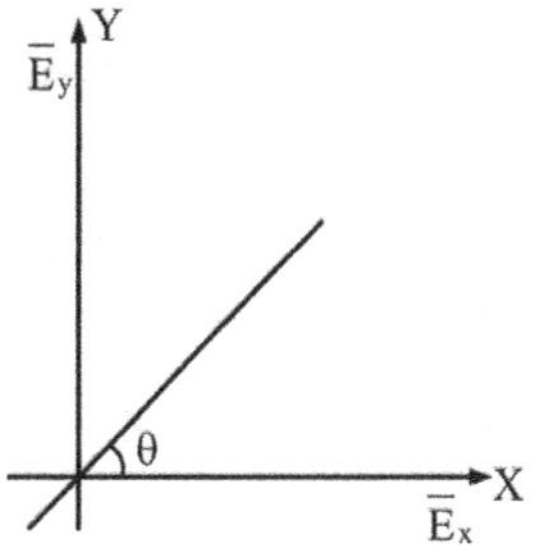

Fig. 4.4

4.10.2 Circularly Polarized Wave

Assume the phase difference between $\bar{E}_x$ and $\bar{E}_y$ as 90° as shown in Fig.4.5a. At 'O' the resultant vector is only $\bar{E}_y$, at 'B' the resultant vector is only $\bar{E}_x$ and at 'C' the resultant vector is $-\bar{E}_y$ and at 'D' the resultant vector is $-\bar{E}_x$ So if we draw path formed by the resultant vector $\bar{E}$ on XY plane, it will be a circle as shown in Fig:4.5b. Since the path formed by the resultant vector $\bar{E}$ is a circle, this type of polarization is called circular polarization and the wave is said to be circularly polarized wave.

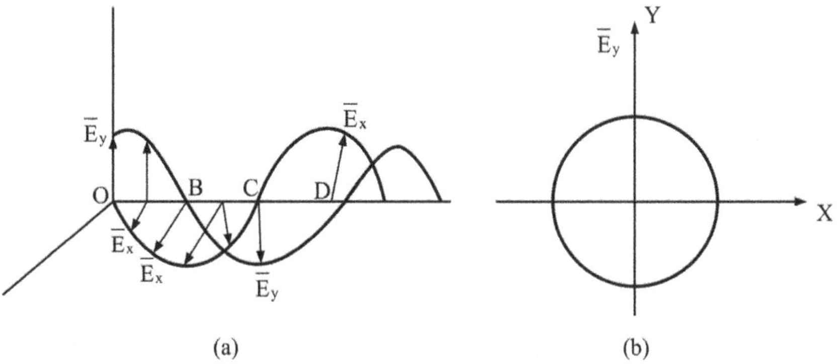

Fig. 4.5 Circular polarization

The conditions for getting circularly polarized wave are

(i) The phase difference between $\bar{E}_x$ and $\bar{E}_y$ is 90°.

(ii) The magnitudes of $\bar{E}_x$ and $\bar{E}_y$ must be equal.

4.10.3 Elliptically Polarized Wave

To get Elliptically polarized wave the conditions are

(i) The phase difference between $\bar{E}_x$ and $\bar{E}_y$ must be other than zero.

(ii) Magnitudes of $\bar{E}_x$ and $\bar{E}_y$ must not be equal.

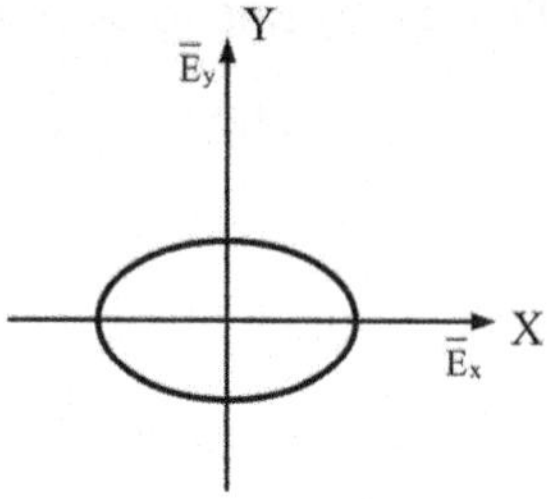

Fig. 4.6

For a phase difference of 90° and $E_x > E_y$ the shape of ellipse will be as shown in Fig.4.6. Here major and minor axises are x and y respectively.

4.10.4 Generalized Equation of Polarized Wave

Assume the wave is travelling along Z-direction with zero attenuation and consider only electric field components. We know electric field wave $\bar{E}$ can be represented as

$$\bar{E} = \bar{E}_x + \bar{E}_y \quad .....(4.10.1)$$

$$\bar{E} = E_x \bar{a}_x + E_y \bar{a}_y \quad .....(4.10.2)$$

For sinusoidal variations, we can write as

$$\bar{E} = \bar{E}_0 e^{-j\beta z} \ e^{j\omega t}$$

By assuming a phase difference of 'δ' between $\bar{E}_x$ and $\bar{E}_y$ the wave equations for $\bar{E}_x$ and $\bar{E}_y$ are

$$\bar{E}_x = \bar{E}_1 e^{j(\omega t - \beta z)} \quad .....(4.10.3)$$

$$\bar{E}_y = \bar{E}_2 e^{j(\omega t - \beta z - \delta)} \quad .....(4.10.4)$$

Substitute (4.10.3) & (4.10.4) in (4.10.1)

$$\bar{E} = \bar{E}_1 e^{j(\omega t - \beta z)} + \bar{E}_2 e^{j(\omega t - \beta z - \delta)}$$

Consider z = 0

$$\bar{E} = \bar{E}_1 e^{j\omega t} + \bar{E}_2 e^{j(\omega t - \delta)}$$

$$= \bar{E}_1 (\cos \omega t + j \sin \omega t) + \bar{E}_2 \left[\cos(\omega t - \delta) + j \sin(\omega t - \delta)\right]$$

$$= E_1 (\cos \omega t + j \sin \omega t)\bar{a}_x + E_2 \left[\cos(\omega t - \delta) + j \sin(\omega t - \delta)\right]\bar{a}_y \quad (4.10.5)$$

compare the real parts of equation (4.10.5) with equation (4.10.2)

$$E_x = E_1 \cos\omega t \quad E_y = E_2 \cos(\omega t - \delta)$$

$$\frac{E_x}{E_1} = \cos\omega t \Rightarrow \sin\omega t = \sqrt{1 - \left(\frac{E_x}{E_1}\right)^2}$$

$$\frac{E_y}{E_2} = \cos(\omega t - \delta)$$

$$\Rightarrow \quad \cos\omega t \cos\delta + \sin\omega t \sin\delta = \frac{E_y}{E_2}$$

$$\Rightarrow \quad \frac{E_x}{E_1}\cos\delta + \sqrt{1 - \left(\frac{E_x}{E_1}\right)^2}\sin\delta = \frac{E_y}{E_2}$$

$$\Rightarrow \quad \frac{E_y}{E_2} - \frac{E_x}{E_1}\cos\delta = \sqrt{1 - \left(\frac{E_x}{E_1}\right)^2}\sin\delta$$

$$\Rightarrow \quad \left(\frac{E_y}{E_2}\right)^2 + \left(\frac{E_x}{E_1}\right)^2 \cos^2\delta - 2\frac{E_y}{E_2}\cdot\frac{E_x}{E_1}\cos\delta = sin^2\delta - \sin^2\delta\left(\frac{E_x}{E_1}\right)^2$$

$$\Rightarrow \quad \left(\frac{E_y}{E_2}\right)^2 + \left(\frac{E_x}{E_1}\right)^2 - 2\frac{E_X}{E_1}\cdot\frac{E_Y}{E_2}\cos\delta = \sin^2\delta \qquad \text{.....(4.10.6)}$$

which is the generalized polarized wave equation.

To get linearly polarized wave the phase difference between $\bar{E}_x$ and $\bar{E}_y$ must be zero and magnitudes of $\bar{E}_x$ and $\bar{E}_y$ need not be equal.

Substitute δ = 0 in the above equation

$$\Rightarrow \quad \left(\frac{E_x}{E_1}\right)^2 + \left(\frac{E_y}{E_2}\right)^2 - 2\frac{E_x}{E_1}\cdot\frac{E_y}{E_2} = 0$$

$$\Rightarrow \quad \left(\frac{E_x}{E_1} - \frac{E_y}{E_2}\right)^2 = 0$$

$$\Rightarrow \qquad \frac{E_x}{E_1} = \frac{E_y}{E_2}$$

$$\Rightarrow \qquad E_x = \left(\frac{E_1}{E_2}\right) E_y \qquad \text{.....(4.10.7)}$$

which is the equation for a straight line.

For circularly polarized wave, substitute $\delta = \frac{\pi}{2}$ and $E_1 = E_2$ in the generalized equation, we get

$$\Rightarrow \qquad \left(\frac{E_x}{E_1}\right)^2 + \left(\frac{E_y}{E_1}\right)^2 = 1 \qquad \text{.....(4.10.8)}$$

which is the equation for circle.

***Problem 4.16**

A traveling wave has two linearly polarized components $E_x = 2\cos\omega t$ and $E_y = 3\cos\left(\omega t + \frac{\pi}{2}\right)$

(a) What is the axial ratio

(b) What is the tilt angle of the major axis of the polarization ellipse.

(c) What is the sense of rotation.

Solution

(a) Ratio of major axis to minor axis = 3/2 = 1.5

(b) Tilt angle is the phase difference between E_x and E_y

$$\therefore \theta = 90^0$$

(c) $\frac{E_x}{2} = \cos\omega t, \qquad \frac{E_y}{3} = \cos\left(\omega t + \frac{\pi}{2}\right) = -\sin\omega t$

$\left(\frac{E_x}{2}\right)^2 + \left(\frac{E_y}{3}\right)^2 = \cos^2\omega t + \sin^2\omega t = 1$, which is an equation for ellipse.

Hence sense of rotation is an ellipse.

4.11 Reflection and Refraction of Plane Waves

When an electromagnetic (EM) wave travels in one medium and encounters at different medium at the boundary between the two different media, the part of the wave will be reflected back and remaining part will be transmitted along second medium. The strength of the reflected and transmitted waves depends upon the type of medium.

4.11.1 Normal Incidence for Perfect Conductor

Let us consider an EM wave is traveling along +ve X-direction and assume that it incidents normally on to a perfect conductor as shown in Fig.4.7. We know that perfect conductor can not sustain either electric or magnetic fields. Therefore the entire incident wave reflects back.

"According to right hand rule thumb indicates the direction of wave propagation, fore finger indicates the direction of electric field component and the middle finger indicates the magnetic field component."

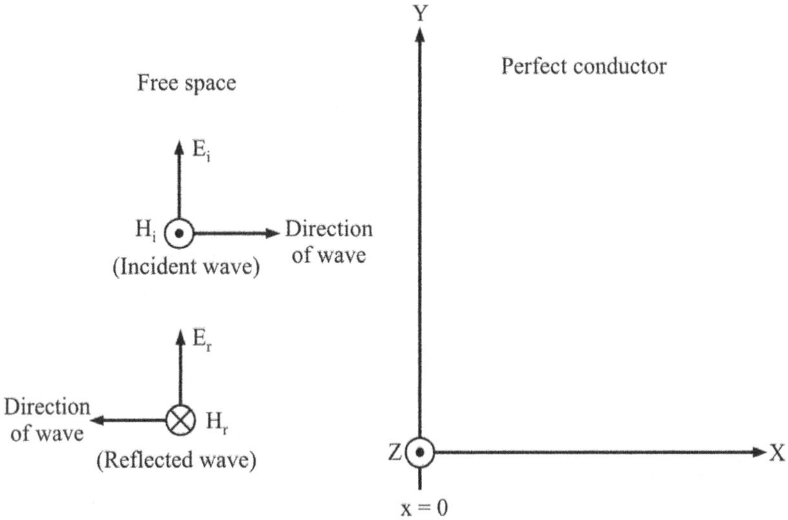

Fig. 4.7 Normal incidence

Let $$E_1 = E_i e^{-j(\omega t - \beta x)} \qquad .....(4.11.1)$$

be the incident wave and

$$E_2 = E_r e^{-j(\omega t + \beta x)} \qquad .....(4.11.2)$$

be the reflected wave.

According to boundary conditions of electric fields, sum of the tangential components in medium1 (Free space) is equal to sum of tangential components in medium 2 (perfect conductor).

Tangential components in free space are E_i and E_r and tangential components in perfect conductor are zero.

$\therefore$ $E_i + E_r = 0$

i.e., $E_i = -E_r$ .....(4.11.3)

The resultant wave E(x, t) due to incident and reflected waves is sum of equations (4.11.1) and (4.11.2).

$$E(x,t) = E_i e^{-j(\omega t-\beta x)} + E_r e^{-j(\omega t+\beta x)}$$

Substituting in equation (4.11.3)

$$E(x,t) = E_i\left[e^{-j(\omega t-\beta x)} - e^{-j(\omega t+\beta x)}\right]$$

By taking real part

$$E(x,t) = E_i\left[\cos(\omega t - \beta x) - \cos(\omega t + \beta x)\right]$$

$$E(x,t) = 2E_i \sin \omega t \sin \beta x \quad .....(4.11.4)$$

which is a standing wave equation in terms of electric field.

Similarly standing wave equation in terms of 'H' can be obtained as follows

Let incident wave be $H_1 = H_i e^{-j(\omega t-\beta x)}$

and reflected wave be $H_2 = H_r e^{-j(\omega t+\beta x)}$

According to boundary conditions of magnetic fields, sum of the tangential components in medium 1 (Free space) is equal to sum of tangential components in medium 2 (perfect conductor).

Tangential components in free space are H_i and H_r but they are in opposite direction and tangential components in perfect conductor are zero.

$\therefore$ $H_i - H_r = 0$

i.e., $H_i = H_r$

The resultant magnetic wave due to incident and reflected waves is

$$H(x,t) = H_i e^{-j(\omega t-\beta x)} + H_r e^{-j(\omega t+\beta x)}$$

Since $H_i = H_r$

$$H(x,t) = H_i\left[e^{-j(\omega t-\beta x)} + e^{-j(\omega t+\beta x)}\right]$$

By taking real part

$$H(x,t) = H_i\left[\cos(\omega t - \beta x) + \cos(\omega t + \beta x)\right]$$

$$H(x,t) = 2H_i \cos\omega t \cos\beta x \qquad .....(4.11.5)$$

which is a standing wave equation in terms of magnetic field.

The standing wave equations for electric and magnetic waves are shown in Fig.4.8.

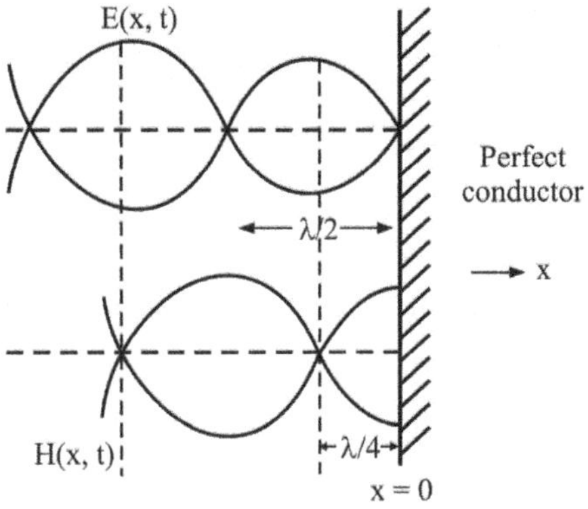

Fig. 4.8 Standing wave

4.11.2 Oblique Incidence for Perfect Conductor

When a wave is traveling along X-direction we can represent the wave as $E = E_0 e^{-j\beta x}$. Consider a plane that is normal to the direction of propagation which is as shown in Fig.4.9.

Consider a point 'p' on the normal plane with radius vector $\bar{r}$, where $\bar{r} = x\bar{a}_x + y\bar{a}_y + z\bar{a}_z$. The unit vector along the direction of propagation is $\bar{a}_x$

$$\therefore \qquad \bar{r}\cdot\bar{a}_x = x$$

The wave equation can be written as

$$E = E_0\, e^{-j\beta(\bar{r}\cdot\bar{a}_x)}$$

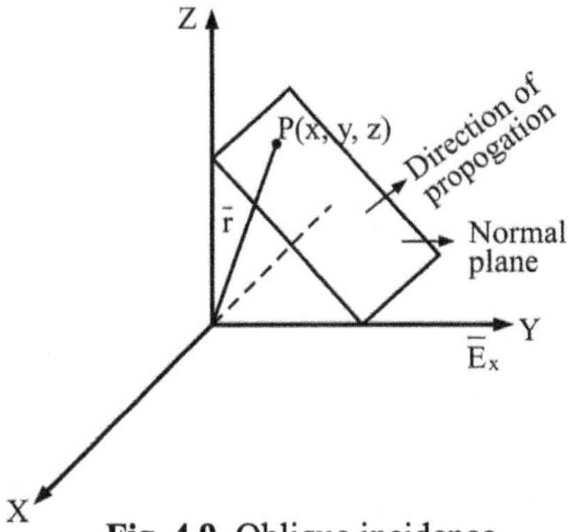

Fig. 4.9 Oblique incidence

Now consider a wave which is traveling along direction OA which makes an angle θ_1 with x-axis and θ_2 with y-axis and θ_3 with Z-axis as shown in Fig.4.10.

Assume the plane is normal to OA and $\bar{n}$ be the unit vector along OA.

From Fig.4.10

$$\bar{n} = \cos\theta_1\bar{a}_x + \cos\theta_2\bar{a}_y + \cos\theta_3\bar{a}_z\,.$$

By assuming any point 'p' on the normal plane at $\bar{r}$, we can write as

$$\bar{r}\cdot\bar{n} = x\cos\theta_1 + y\cos\theta_2 + z\cos\theta_3$$

which will be in the direction of OA.

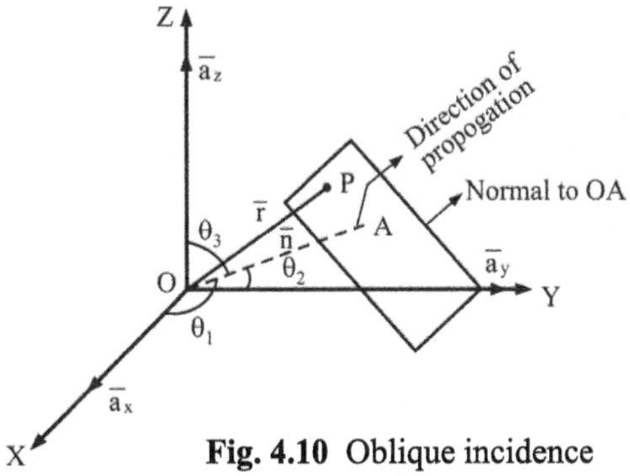

Fig. 4.10 Oblique incidence

∴ The wave equation with an oblique incidence can be written as

$$E = E_0\, e^{-j\beta(\bar{n}\cdot\bar{r})} \qquad \text{.....(4.11.6)}$$

When a wave incidents obliquely at the interface between two media. We have to consider two cases.

(i) Horizontal polarization (perpendicular polarization)

(ii) Vertical polarization (parallel polarization)

Horizontal Polarization

It is obtained when the electric field vector is perpendicular to the plane of incidence (Y-Z plane in this case). Horizontal polarization can also be called as perpendicular polarization.

Let us assume that the EM wave incidents obliquely with an angle θ with Z-axis and which produces horizontal polarization as shown in Fig. 4.11.

Since the second medium is perfect conductor, the entire wave reflects back with an angle θ with the Z- axis.

The equation for incident wave in terms electric field is

$$E_{incident} = E_i\, e^{-j\beta\, \bar{n}_1\cdot\bar{r}} \qquad \text{.....(4.11.7)}$$

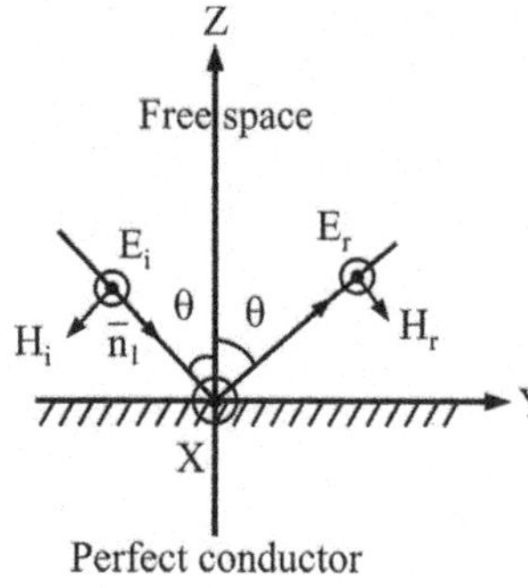

Fig. 4.11 Horizontal polarization

where $\bar{n}_1$ is the unit vector along the direction of propagation of incident wave. Assume that $\bar{n}_1$ makes an angle θ_1, θ_2 and θ_3 with X, Y and Z-axis respectively. To find θ_1, θ_2 and θ_3 in terms of θ draw the diagram as shown in Fig.4.12.

We know $\bar{n}_1 \cdot \bar{r} = x\cos\theta_1 + y\cos\theta_2 + z\cos\theta_3$

$$\bar{n}_1 \cdot \bar{r} = x\cos\frac{\pi}{2} + y\cos\left(\frac{\pi}{2} - \theta\right) + z\cos(\pi - \theta)$$

$$\bar{n}_1 \cdot \bar{r} = 0 + y\sin\theta - z\cos\theta$$

∴ Equation for incident wave is

$$E_{incident} = E_i e^{-j\beta(y\sin\theta - z\cos\theta)} \qquad \text{.....(4.11.8)}$$

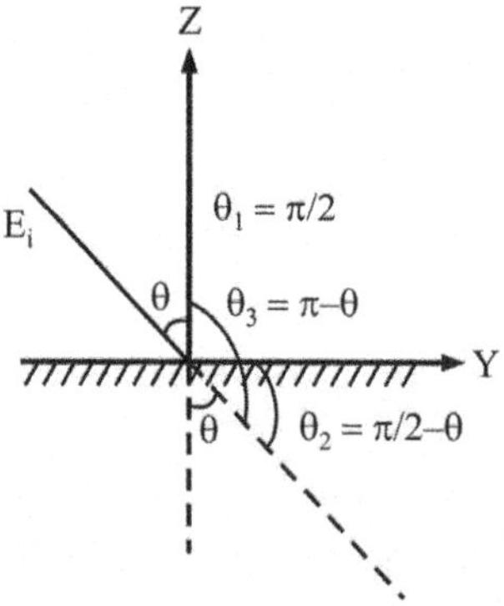

Fig. 4.12 Finding θ_1, θ_2 and θ_3 in terms of θ w.r.t. $\bar{n}_1$

Assume $\bar{n}_2$ is the unit vector along the direction of propagation of reflected wave. Here θ_1, θ_2 and θ_3 are the angles made by $\bar{n}_2$ with X, Y and Z axis respectively.

To find θ_1, θ_2 and θ_3 in terms of 'θ' consider the Fig.4.13.

Equation for reflected wave is

$$E_{reflected} = E_r e^{-j\beta \bar{n}_2 . \bar{r}} \qquad .....(4.11.9)$$

$$\bar{n}_2 \cdot \bar{r} = x\cos 90 + y\cos\left(\frac{\pi}{2} - \theta\right) + z\cos\theta$$

$$= y\sin\theta + z\cos\theta$$

$$E_{reflected} = E_r e^{-j\beta(y\sin\theta + z\cos\theta)} \qquad .....(4.11.10)$$

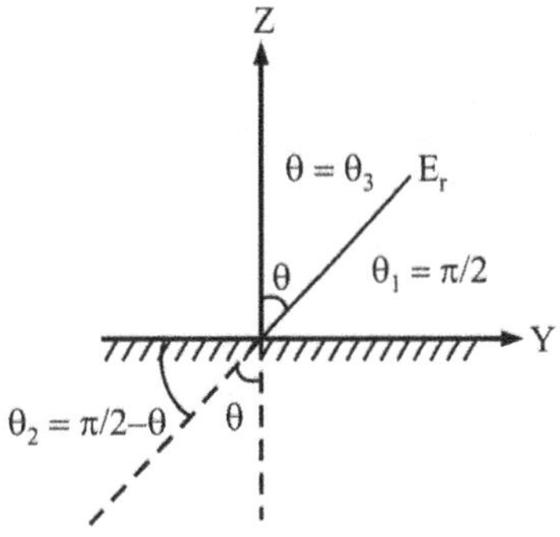

Fig. 4.13 Finding θ_1, θ_2 and θ_3 in terms of θ w.r.t. $\bar{n}_2$

We know at the boundary between dielectric and perfect conductor $E_t = 0$

$\therefore\ E_i = -E_r$

The resultant electric field is given by

$$E(y,z) = E_i\left[e^{-j\beta(y\sin\theta - z\cos\theta)} - e^{-j\beta(y\sin\theta + z\cos\theta)}\right]$$

$$= E_i e^{-j\beta y\sin\theta}\left[e^{+j\beta z\cos\theta} - e^{-j\beta z\cos\theta}\right]$$

$$= E_i e^{-j\beta y\sin\theta}\left[2j\sin(\beta z\cos\theta)\right]$$

$$E(y,z) = 2jE_I\sin(\beta z\cos\theta)_e^{-j\beta y\sin\theta} \qquad .....(4.11.11)$$

We see that the resultant electric field is a standing wave in Z-direction, and a traveling wave in the Y-direction.

The wavelength in the Z-direction is given by

$$\lambda_z = \frac{2\pi}{\beta_z} = \frac{2\pi}{\beta\cos\theta} = \frac{\lambda}{\cos\theta} \qquad .....(4.11.11a)$$

where λ is the wavelength of the incident (and reflected) wave.

In the direction of the Y-axis, the resultant field behaves as a traveling wave, with a phase velocity along the Y-axis

$$v_p = \frac{\omega}{\beta_y} = \frac{\omega}{\beta\sin\theta} = \frac{v}{\sin\theta} \qquad .....(4.11.11b)$$

and wavelength along Y-axis is

$$\lambda_y = \frac{2\pi}{\beta_y} = \frac{2\pi}{\beta\sin\theta} = \frac{\lambda}{\sin\theta} \qquad .....(4.11.11c)$$

Vertical Polarization

Vertical polarization is obtained when the electric field vector is parallel to plane of incidence. Vertical polarization is also called parallel polarization.

Let us consider the boundary between free space and perfect conductor and assume the wave incidents obliquely at the boundary with an angle θ with the Z-axis as shown in Fig. 4.14, where Z-axis is normal to the boundary.

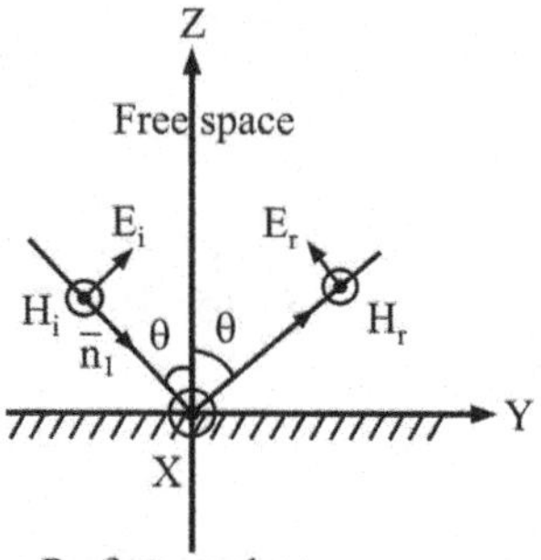

Fig. 4.14 Vertical polarization

Let us consider only the incident wave as shown in Fig. 4.15.

we have $\eta = \dfrac{E_i}{H_i}$

From the Fig. 4.15,

The component of E_i along Z-axis is

$$E_{zi} = E_i \cos(90 - \theta) = \eta\, H_i \sin\theta$$

and the component of E_i along y-axis is

$$E_{yi} = E_i \sin(90 - \theta) = \eta\, H_i \cos\theta$$

The wave equation for the component along Z-axis due to incident wave is

$$E_{zi}\, e^{-j\beta(y\sin\theta - z\cos\theta)} = \eta H_i \sin\theta\, e^{-j\beta(y\sin\theta - z\cos\theta)} \qquad ..(4.11.12)$$

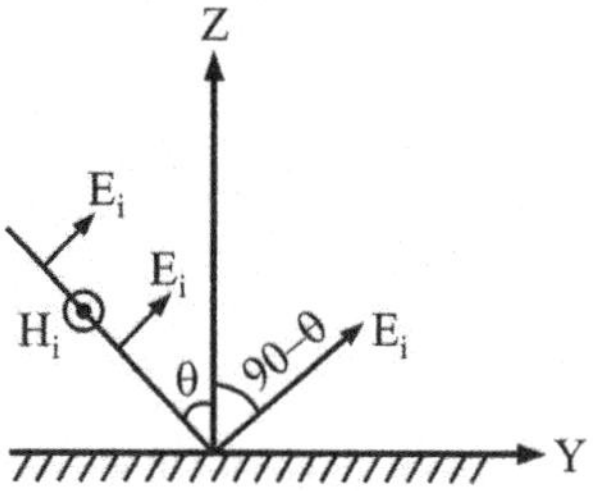

Fig. 4.15 Incident wave with components

The wave equation for the component along Y-axis due to incident wave is

$$E_{yi}\, e^{-j\beta(y\sin\theta - z\cos\theta)} = \eta H_i \cos\theta\, e^{-j\beta(y\sin\theta - z\cos\theta)} \qquad ..(4.11.13)$$

Let us consider only the reflected wave as shown in Fig. 4.16.

We know $\eta = \dfrac{E_r}{H_r}$

From the Fig. 4.16,

The component of E_r along Z-axis is

$$E_{zr} = E_r \cos(90 - \theta) = E_r \sin\theta = \eta\, H_r \sin\theta$$

and the component of E_r along Y-axis is

$$E_{yr} = -E_r \sin(90 - \theta) = -\eta\, H_r \cos\theta$$

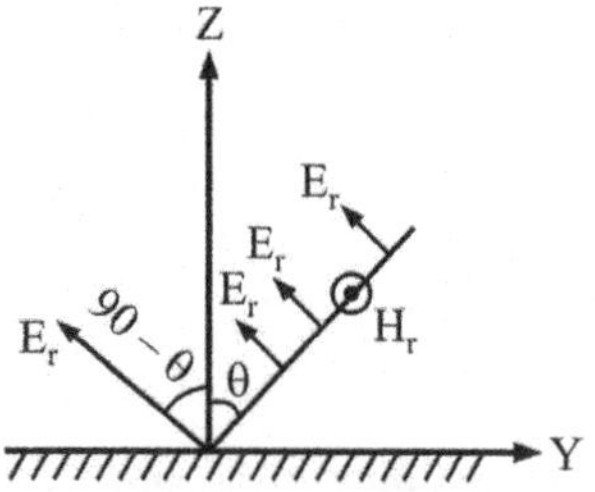

Fig. 4.16 Reflected wave with components

The wave equation for the component along Z-axis due to reflected wave is

$$E_{zr}e^{-j\beta(y\sin\theta+z\cos\theta)} = \eta H_r \sin\theta e^{-j\beta(y\sin\theta+z\cos\theta)} \quad .....(4.11.14)$$

The wave equation for the component along Y-axis due to reflected wave is

$$E_{yr}e^{-j\beta(y\sin\theta+z\cos\theta)} = -\eta H_r \cos\theta e^{-j\beta(y\sin\theta+z\cos\theta)} \quad .....(4.11.15)$$

The resultant wave equation along Z-axis due to incident and reflected wave is sum of equations (4.11.12) and (4.11.14)

$$E_z = \eta H_i \sin\theta e^{-j\beta(y\sin\theta-z\cos\theta)} + \eta H_r \sin\theta e^{-j\beta(y\sin\theta+z\cos\theta)}$$

According boundary conditions $H_i + H_r = 0$

$H_i = -H_r$

$$E_z = \eta H_i \sin\theta e^{-j\beta y\sin\theta}\left[e^{j\beta z\cos\theta} - e^{-j\beta z\cos\theta}\right]$$

$$E_z = 2j\eta H_i \sin\theta e^{-j\beta y\sin\theta}\sin(\beta z\cos\theta) \quad .....(4.11.16)$$

Similarly

$$E_y = \eta H_i \cos\theta e^{-j\beta(y\sin\theta-z\cos\theta)} - \eta H_r \cos\theta e^{-j\beta(y\sin\theta+z\cos\theta)}$$

Since $H_i = -H_r$

$$E_y = \eta H_i \cos\theta e^{-j\beta y\sin\theta}\left[e^{j\beta z\cos\theta} + e^{-j\beta z\cos\theta}\right]$$

$$E_y = 2\eta H_i \cos\theta e^{-j\beta y\sin\theta}\cos(\beta z\cos\theta) \quad .....(4.11.17)$$

4.11.3 Normal Incidence for Perfect Dielectric

Let us assume the wave incidents normally at the interface between two media, whose values are η_1, ϵ_1, μ_1 for medium1 and η_2, ϵ_2, μ_2 for medium 2, as shown in the Fig.4.17.

When a wave incidents normally, part of the wave will be reflected back and the remaining part is transmitted through the second medium (second medium should not be a perfect conductor). Let E_i and H_i are magnitudes of incident electric and magnetic waves, E_r and H_r are magnitudes of reflected electric and magnetic waves and E_t and H_t are magnitudes of transmitted electric and magnetic waves. We know that according to boundary conditions the sum of tangential components in medium 1 is equal to the sum of tangential components in medium 2. In medium 1 the tangential components are E_i, H_i, E_r and H_r because they are parallel to the boundary surface, and in medium2 the tangential components are E_t and H_t.

Here H_i and H_r are in opposite direction.

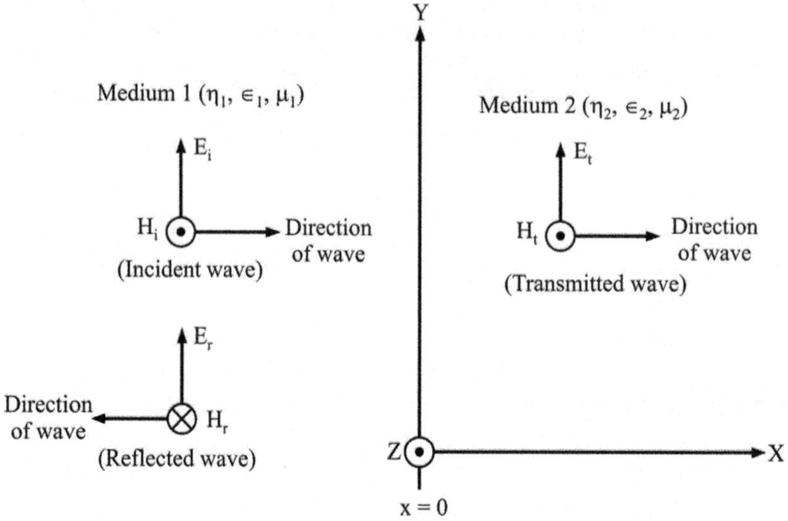

Fig. 4.17 Normal incidence for perfect dielectric

For electric field vectors at boundary $E_i + E_r = E_t$ .....(4.11.18)

Similarly for magnetic field vectors $H_i - H_r = H_t$ .....(4.11.19)

We know $E_i = \eta_1 H_i$ .....(4.11.20)

$E_r = \eta_1 H_r$ .....(4.11.21)

$E_t = \eta_2 H_t$ .....(4.11.22)

Let us find the ratio $\frac{E_r}{E_i}$, known as reflection coefficient

From equations (4.11.20) & (4.11.21)

$$E_i - E_r = \eta_1 (H_i - H_r)$$

$$= \eta_1 H_t \quad \text{from (4.11.19)}$$

$$= \frac{\eta_1}{\eta_2} E_t \quad \text{from (4.11.22)}$$

$$= \frac{\eta_1}{\eta_2} (E_i + E_r) \quad \text{from (4.11.18)}$$

$\therefore$ $$\eta_2 (E_i - E_r) = \eta_1 (E_i + E_r)$$

$$E_i (\eta_2 - \eta_1) = E_r (\eta_1 + \eta_2)$$

$$\frac{E_i}{E_r} = \frac{\eta_1 + \eta_2}{\eta_2 - \eta_1}$$

$$\frac{E_r}{E_i} = \frac{\eta_2 - \eta_1}{\eta_1 + \eta_2} \qquad .....(4.11.23)$$

Let us find the ratio $\frac{E_t}{E_i}$, known as transmission coefficient

$$\frac{E_t}{E_i} = \frac{E_i + E_r}{E_i} \qquad \text{from (4.11.18)}$$

$$= 1 + \frac{\eta_2 - \eta_1}{\eta_1 + \eta_2} = \frac{2\eta_2}{\eta_1 + \eta_2} \qquad .....(4.11.24)$$

From (4.11.20) & (4.11.21)

$$\frac{-H_r}{H_i} = \frac{-E_r}{\eta_1}\frac{\eta_1}{E_i} = -\frac{\eta_2 - \eta_1}{\eta_1 + \eta_2} = \frac{\eta_1 - \eta_2}{\eta_1 + \eta_2} \qquad .....(4.11.25)$$

$$\frac{H_t}{H_i} = \frac{H_i - H_r}{H_i} = 1 - \frac{H_r}{H_i} = \frac{2\eta_1}{\eta_1 + \eta_2} \qquad .....(4.11.26)$$

Let us find out these ratios in terms of '$\in$'. We know $\eta = \sqrt{\frac{\mu}{\in}}$. Practically the permeability values of different dielectric media are equal to free space

$\therefore \qquad \mu_1 = \mu_2 = \mu_0$

$$\eta_1 = \sqrt{\frac{\mu_1}{\in_1}}\,;\ \eta_2 = \sqrt{\frac{\mu_2}{\in_2}}$$

$$\frac{E_r}{E_i} = \frac{\sqrt{\frac{\mu_0}{\in_2}} - \sqrt{\frac{\mu_0}{\in_1}}}{\sqrt{\frac{\mu_0}{\in_2}} + \sqrt{\frac{\mu_0}{\in_1}}}$$

$$= \frac{\sqrt{\in_1 \mu_0} - \sqrt{\in_2 \mu_0}}{\sqrt{\in_1 \mu_0} + \sqrt{\in_2 \mu_0}} = \frac{\sqrt{\in_1} - \sqrt{\in_2}}{\sqrt{\in_1} + \sqrt{\in_2}} \qquad .....(4.11.27)$$

$$\frac{E_t}{E_i} = \frac{2\sqrt{\mu_0}\sqrt{\epsilon_1}\sqrt{\epsilon_2}}{\sqrt{\epsilon_2}\sqrt{\mu_0}\left(\sqrt{\epsilon_1}+\sqrt{\epsilon_2}\right)} = \frac{2\sqrt{\epsilon_1}}{\sqrt{\epsilon_1}+\sqrt{\epsilon_2}} \qquad \text{.....(4.11.28)}$$

$$\frac{-H_r}{H_i} = \frac{\sqrt{\mu_0}\left(\sqrt{\epsilon_2}-\sqrt{\epsilon_1}\right)}{\sqrt{\mu_0}\left(\sqrt{\epsilon_1}+\sqrt{\epsilon_2}\right)} = \frac{\sqrt{\epsilon_2}-\sqrt{\epsilon_1}}{\sqrt{\epsilon_1}+\sqrt{\epsilon_2}} \qquad \text{.....(4.11.29)}$$

$$\frac{H_t}{H_i} = \frac{2\sqrt{\epsilon_2}}{\sqrt{\epsilon_1}+\sqrt{\epsilon_2}} \qquad \text{.....(4.11.30)}$$

4.11.4 Oblique Incidence for Perfect Dielectric

When a wave incidents obliquely with an angle θ_1 with the normal at the interface between two dielectric media, part of the wave will be reflected back with an angle θ_3 with the normal and remaining part will be refracted with an angle θ_2 with the normal as shown in Fig.4.18.

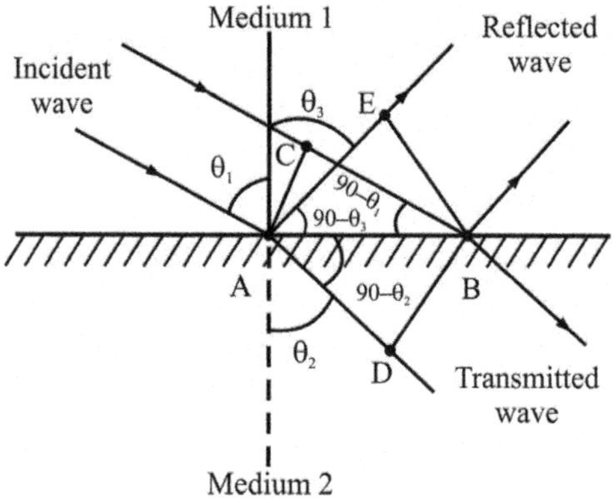

Fig. 4.18 Oblique incidence perfect dielectric

By the time incident wave travels by a distance CB the transmitted wave travels a distance AD. Assuming v_1 is the velocity with which the incident wave travels and v_2 is the velocity with which the transmitted wave travels.

$$\frac{CB}{AD} = \frac{v_1}{v_2}$$

From Triangle ABC $\cos(90 - \theta_1) = \dfrac{CB}{AB}$

$\Rightarrow \qquad CB = AB \sin\theta_1$

From Triangle ABD $\cos(90-\theta_2) = \frac{AD}{AB}$

$\Rightarrow$ $AD = AB \sin\theta_2$

$$\frac{AB \sin\theta_1}{AB \sin\theta_2} = \frac{v_1}{v_2}$$

$$\frac{\sin\theta_1}{\sin\theta_2} = \frac{v_1}{v_2}$$

We know $v = \sqrt{\frac{1}{\mu \in}}$

$\therefore$ $\mu_1 = \mu_2 = \mu_0$

$$\frac{\sin\theta_1}{\sin\theta_2} = \sqrt{\frac{\in_2}{\in_1}} \quad .....(4.11.31)$$

In the Fig.4.18, the distance traveled by incident wave is equal to the distance traveled by reflected wave i.e., CB = AE .....(4.11.32)

From triangle ABC, $CB = AB \sin\theta_1$

and from triangle ABE, $\cos(90-\theta_3) = \frac{AE}{AB}$

$AE = AB \sin\theta_3$

Substituting in equation (4.11.32)

$$\sin\theta_1 = \sin\theta_3 \quad .....(4.11.33)$$

$$\theta_1 = \theta_3$$

which is called as law of Sines or Snell's law.

Reflection Coefficient

The product of electric field intensity and magnetic field intensity gives rise to power per area.

$\therefore$ power / area $= E.\,H$

$$= \frac{E^2}{\eta} \quad \text{where} \quad \eta = \frac{E}{H}$$

The power in the incident wave is $\frac{E_i^2}{\eta_1}$

The component of this along the normal is $\dfrac{E_i^2}{\eta_1}\cos\theta_1$

The power in the reflected wave is $\dfrac{E_r^2}{\eta_1}$

The component of this along the normal is $\dfrac{E_r^2}{\eta_1}\cos\theta_1$ $(\because \theta_3 = \theta_1)$

The power in the transmitted wave is $\dfrac{E_t^2}{\eta_2}$

The component of this along the normal is $\dfrac{E_t^2}{\eta_2}\cos\theta_2$.

According to law of conservation of energy, we can write as

$$\frac{E_i^2}{\eta_1}\cos\theta_1 = \frac{E_r^2}{\eta_1}\cos\theta_1 + \frac{E_i^2}{\eta_1}\cos\theta_2$$

Dividing with E_i^2

$$\frac{\cos\theta_1}{\eta_1} = \frac{E_r^2\cos\theta_1}{E_i^2\eta_1} + \frac{E_t^2\cos\theta_2}{E_i^2\eta_2}$$

$$1 = \frac{E_r^2}{E_i^2} + \frac{\eta_1 E_t^2\cos\theta_2}{\eta_2 E_i^2\cos\theta_1}$$

$$1 - \frac{\eta_1}{\eta_2}\frac{E_t^2\cos\theta_2}{E_i^2\cos\theta_1} = \frac{E_r^2}{E_i^2}$$

$$1 - \frac{\sqrt{\epsilon_2}}{\sqrt{\epsilon_1}}\frac{E_t^2\cos\theta_2}{E_i^2\cos\theta_1} = \frac{E_r^2}{E_i^2} \qquad \text{.....(4.11.34)}$$

$\because \qquad \eta = \sqrt{\dfrac{\mu}{\epsilon}}$ and $\mu_1 = \mu_2 = \mu_0$

Horizontal Polarization

For horizontal polarization the electric field vector is perpendicular to the plane of incidence as shown in the Fig. 4.19.

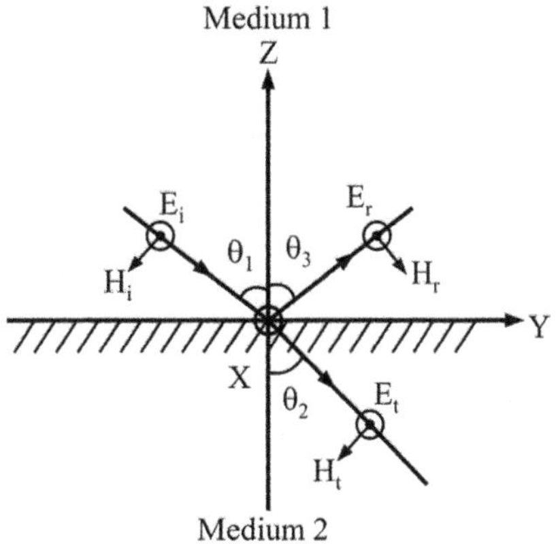

Fig. 4.19 Horizontal polarization

According to boundary conditions, the tangential components in medium 1 = The tangential components in medium 2

i.e., $E_i + E_r = E_t$

Because E_i, E_r, E_t are parallel to boundary surface so they become tangential components

$$\Rightarrow \qquad \frac{E_t}{E_i} = 1 + \frac{E_r}{E_i}$$

Substitute in equation (4.11.34)

$$\Rightarrow \qquad 1 - \frac{\sqrt{\epsilon_2}}{\sqrt{\epsilon_1}}\left(1 + \frac{E_r}{E_i}\right)^2 \frac{\cos\theta_2}{\cos\theta_1} = \frac{E_r^2}{E_i^2}$$

$$\Rightarrow \qquad 1 - \frac{E_r^2}{E_i^2} = \frac{\sqrt{\epsilon_2}}{\sqrt{\epsilon_1}}\left(1 + \frac{E_r}{E_i}\right)^2 \frac{\cos\theta_2}{\cos\theta_1}$$

$$\Rightarrow \qquad \left(1 - \frac{E_r}{E_i}\right)\left(1 + \frac{E_r}{E_i}\right) = \frac{\sqrt{\epsilon_2}}{\sqrt{\epsilon_1}}\left(1 + \frac{E_r}{E_i}\right)^2 \frac{\cos\theta_2}{\cos\theta_1}$$

$$\Rightarrow \qquad \left(1 - \frac{E_r}{E_i}\right) = \frac{\sqrt{\epsilon_2}}{\sqrt{\epsilon_1}}\left(1 + \frac{E_r}{E_i}\right) \frac{\cos\theta_2}{\cos\theta_1}$$

$$\Rightarrow \qquad 1 - \frac{\sqrt{\epsilon_2}}{\sqrt{\epsilon_1}}\frac{\cos\theta_2}{\cos\theta_1} = \frac{E_r}{E_i}\left(\frac{\sqrt{\epsilon_2}}{\sqrt{\epsilon_1}}\frac{\cos\theta_2}{\cos\theta_1} + 1\right)$$

$$\Rightarrow \qquad \frac{E_r}{E_i} = \frac{\sqrt{\epsilon_1}\cos\theta_1 - \sqrt{\epsilon_2}\cos\theta_2}{\sqrt{\epsilon_1}\cos\theta_1 + \sqrt{\epsilon_2}\cos\theta_2} \qquad \text{.....(4.11.35a)}$$

We have $\dfrac{\sin\theta_1}{\sin\theta_2} = \sqrt{\dfrac{\epsilon_2}{\epsilon_1}}$

$$\Rightarrow \qquad \cos\theta_2 = \sqrt{\frac{\epsilon_2 - \epsilon_1 \sin^2\theta_1}{\epsilon_2}}$$

Substitute in (4.11.35a)

$$\frac{E_r}{E_i} = \frac{\cos\theta_1 - \sqrt{\left(\frac{\epsilon_2}{\epsilon_1}\right) - \sin^2\theta_1}}{\cos\theta_1 + \sqrt{\left(\frac{\epsilon_2}{\epsilon_1}\right) - \sin^2\theta_1}} \quad .....(4.11.35b)$$

Equation (4.11.35a) can be further simplified as

$$\Rightarrow \quad \frac{E_r}{E_i} = \frac{\cos\theta_1 - \sqrt{\epsilon_2}/\sqrt{\epsilon_1}\cos\theta_2}{\cos\theta_1 + \sqrt{\epsilon_2}/\sqrt{\epsilon_1}\cos\theta_2}$$

$$\Rightarrow \quad \frac{E_r}{E_i} = \frac{\cos\theta_1 - (\sin\theta_1/\sin\theta_2)\cos\theta_2}{\cos\theta_1 + (\sin\theta_1/\sin\theta_2)\cos\theta_2} \qquad \because \frac{\sin\theta_1}{\sin\theta_2} = \sqrt{\frac{\epsilon_2}{\epsilon_1}}$$

$$\Rightarrow \quad \frac{E_r}{E_i} = \frac{\sin\theta_2\cos\theta_1 - \sin\theta_1\cos\theta_2}{\sin\theta_2\cos\theta_1 + \sin\theta_1\cos\theta_2} = \frac{\sin(\theta_2 - \theta_1)}{\sin(\theta_2 + \theta_1)}$$

$$\Rightarrow \quad \frac{E_r}{E_i} = \frac{\sin\theta_2\cos\theta_1 - \sin\theta_1\cos\theta_2}{\sin\theta_2\cos\theta_1 + \sin\theta_1\cos\theta_2} = \frac{\sin(\theta_2 - \theta_1)}{\sin(\theta_2 + \theta_1)} \quad .....(4.11.35c)$$

which is the reflection coefficient for horizontal polarization.

Vertical Polarization

Consider the Fig. 4.20

The tangential component of E_i is $E_i \cos\theta_1$

The tangential component of E_r is $- E_r \cos\theta_1$

The tangential component of E_t is $E_t \cos\theta_2$

According to boundary condition, the sum of tangential components in medium 1 is equal to the sum of tangential components in medium 2.

$\therefore$ $E_i \cos\theta_1 - E_r \cos\theta_1 = E_t \cos\theta_2$

$$\cos\theta_1 - \frac{E_r}{E_i}\cos\theta_1 = \frac{E_t}{E_i}\cos\theta_2$$

$$\Rightarrow \quad \frac{E_t}{E_i} = \frac{\cos\theta_1}{\cos\theta_2} - \frac{E_r}{E_i}\frac{\cos\theta_1}{\cos\theta_2}$$

$$\frac{E_t}{E_i} = \frac{\cos\theta_1}{\cos\theta_2}\left(1 - \frac{E_r}{E_i}\right)$$

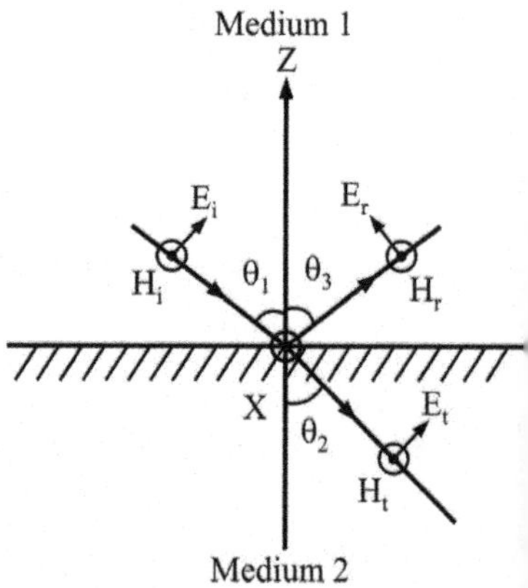

Fig. 4.20 Vertical polarization

substitute in equation (4.11.34)

$$1-\frac{\sqrt{\epsilon_2}}{\sqrt{\epsilon_1}}\frac{\cos^2\theta_1}{\cos^2\theta_2}\frac{\cos\theta_2}{\cos\theta_1}\left(1-\frac{E_r}{E_i}\right)^2=\frac{E_r^2}{E_i^2}$$

$$\Rightarrow \qquad \left(1-\frac{E_r}{E_i}\right)\left(1+\frac{E_r}{E_i}\right)=\frac{\sqrt{\epsilon_2}}{\sqrt{\epsilon_1}}\frac{\cos\ \theta_1}{\cos\ \theta_2}\left(1-\frac{E_r}{E_i}\right)^2$$

$$\Rightarrow \qquad \left(1+\frac{E_r}{E_i}\right)=\frac{\sqrt{\epsilon_2}}{\sqrt{\epsilon_1}}\frac{\cos\ \theta_1}{\cos\ \theta_2}\left(1-\frac{E_r}{E_i}\right)$$

$$\Rightarrow \qquad \left(\frac{E_r}{E_i}\right)\left(\frac{\sqrt{\epsilon_2}}{\sqrt{\epsilon_1}}\frac{\cos\ \theta_1}{\cos\ \theta_2}+1\right)=\frac{\sqrt{\epsilon_2}}{\sqrt{\epsilon_1}}\frac{\cos\ \theta_1}{\cos\ \theta_2}-1$$

$$\Rightarrow \qquad \frac{E_r}{E_i}=\frac{\sqrt{\epsilon_2}\cos\ \theta_1-\sqrt{\epsilon_1}\cos\ \theta_2}{\sqrt{\epsilon_2}\cos\ \theta_1+\sqrt{\epsilon_1}\cos\ \theta_2} \qquad \text{.....(4.11.36a)}$$

We have $\cos\theta_2=\sqrt{1-\frac{\epsilon_1}{\epsilon_2}\sin^2\theta_1}$

Substituting in the above equation

$$\Rightarrow \qquad \frac{E_r}{E_i}=\frac{\sqrt{\epsilon_1}\left(\sqrt{\frac{\epsilon_2}{\epsilon_1}}\cos\ \theta_1-\sqrt{1-\frac{\epsilon_1}{\epsilon_2}\sin^2\theta_1}\right)}{\sqrt{\epsilon_1}\left(\sqrt{\frac{\epsilon_2}{\epsilon_1}}\cos\ \theta_1+\sqrt{1-\frac{\epsilon_1}{\epsilon_2}\sin^2\theta_1}\right)}$$

$$\Rightarrow \qquad \frac{E_r}{E_i}=\frac{\left(\sqrt{\frac{\epsilon_2}{\epsilon_1}}\cos\ \theta_1-\sqrt{1-\frac{\epsilon_1}{\epsilon_2}\sin^2\theta_1}\right)}{\left(\sqrt{\frac{\epsilon_2}{\epsilon_1}}\cos\ \theta_1+\sqrt{1-\frac{\epsilon_1}{\epsilon_2}\sin^2\theta_1}\right)}$$

$$\Rightarrow \qquad \frac{E_r}{E_i}=\frac{\left(\sqrt{\frac{\epsilon_2}{\epsilon_1}}\cos\ \theta_1-\sqrt{\frac{\epsilon_1}{\epsilon_2}}\sqrt{\frac{\epsilon_2}{\epsilon_1}-\sin^2\theta_1}\right)}{\left(\sqrt{\frac{\epsilon_2}{\epsilon_1}}\cos\ \theta_1+\sqrt{\frac{\epsilon_1}{\epsilon_2}}\sqrt{\frac{\epsilon_2}{\epsilon_1}-\sin^2\theta_1}\right)}$$

Multiply both numerator and denominator with $\sqrt{\dfrac{\epsilon_2}{\epsilon_1}}$

$$\frac{E_r}{E_i} = \frac{(\epsilon_2/\epsilon_1)\cos\ \theta_1 - \sqrt{(\epsilon_2/\epsilon_1) - \sin^2\theta_1}}{(\epsilon_2/\epsilon_1)\cos\ \theta_1 + \sqrt{(\epsilon_2/\epsilon_1) - \sin^2\theta_1}} \quad \text{.....(4.11.36b)}$$

Equation (4.11.36a) can be further simplified as

$$\Rightarrow \qquad \frac{E_r}{E_i} = \frac{\sqrt{\dfrac{\epsilon_2}{\epsilon_1}}\cos\ \theta_1 - \cos\ \theta_2}{\sqrt{\dfrac{\epsilon_2}{\epsilon_1}}\cos\ \theta_1 + \cos\ \theta_2}$$

We have $\dfrac{\sin\theta_1}{\sin\theta_2} = \sqrt{\dfrac{\epsilon_2}{\epsilon_1}}$

$$\Rightarrow \qquad \frac{E_r}{E_i} = \frac{\dfrac{\sin\theta_1}{\sin\theta_2}\cos\ \theta_1 - \cos\ \theta_2}{\dfrac{\sin\theta_1}{\sin\theta_2}\cos\ \theta_1 + \cos\ \theta_2}$$

$$\Rightarrow \qquad \frac{E_r}{E_i} = \frac{\sin\theta_1\cos\ \theta_1 - \sin\theta_2\cos\ \theta_2}{\sin\theta_1\cos\ \theta_1 + \sin\theta_2\cos\ \theta_2}$$

Multiplying both numerator and denominator with 2

$$\Rightarrow \qquad \frac{E_r}{E_i} = \frac{2\sin\theta_1\cos\ \theta_1 - 2\sin\theta_2\cos\ \theta_2}{2\sin\theta_1\cos\ \theta_1 + 2\sin\theta_2\cos\ \theta_2}$$

$$\Rightarrow \qquad \frac{E_r}{E_i} = \frac{\sin 2\theta_1 - \sin 2\theta_2}{\sin 2\theta_1 + \sin 2\theta_2} \quad \text{.....(4.11.36c)}$$

which is the reflection co-efficient for vertical polarization.

4.12 Brewster Angle

If any wave incidents at the boundary surface with some an angle and if it is not producing the reflected wave then that angle is called Brewster angle. Brewster angle exists only for vertical polarization. By equating numerator of reflection coefficient equation for vertical polarization to '0', we get Brewster angle as

$$(\epsilon_2/\epsilon_1)\cos\theta_1 - \sqrt{(\epsilon_2/\epsilon_1) - \sin^2\theta_1} = 0$$

$$\frac{\epsilon_2^2}{\epsilon_1^2}\cos^2\theta_1 = \frac{\epsilon_2}{\epsilon_1} - \sin^2\theta_1$$

$$\frac{\epsilon_2^2}{\epsilon_1^2} - \frac{\epsilon_2^2}{\epsilon_1^2}\sin^2\theta_1 = \frac{\epsilon_2}{\epsilon_1} - \sin^2\theta_1$$

$$\frac{\epsilon_2^2}{\epsilon_1^2} - \frac{\epsilon_2}{\epsilon_1} = \sin^2\theta_1\left(\frac{\epsilon_2^2}{\epsilon_1^2} - 1\right)$$

$$\sin^2\theta_1 = \left(\frac{\epsilon_2^2 - \epsilon_1\epsilon_2}{\epsilon_1^2} \times \frac{\epsilon_1^2}{\epsilon_2^2 - \epsilon_1^2}\right)$$

$$= \left(\frac{\epsilon_2^2 - \epsilon_1\epsilon_2}{\epsilon_2^2 - \epsilon_1^2}\right) = \frac{\epsilon_2(\epsilon_2 - \epsilon_1)}{(\epsilon_2 - \epsilon_1)(\epsilon_2 + \epsilon_1)} = \frac{\epsilon_2}{\epsilon_1 + \epsilon_2}$$

$$\cos^2\theta_1 = 1 - \sin^2\theta_1 = \frac{\epsilon_1}{\epsilon_1 + \epsilon_2}$$

$$\tan^2\theta_1 = \frac{\epsilon_2}{\epsilon_1}$$

$$\theta_1 = \tan^{-1}\sqrt{\frac{\epsilon_2}{\epsilon_1}} \qquad \text{.....(4.12.1)}$$

which is the expression for Brewster angle.

4.13 Critical Angle or Total Internal Reflection

The Brewster angle is valid only for vertical polarization and it is valid for either $\epsilon_1 > \epsilon_2$ or $\epsilon_2 > \epsilon_1$.

If the wave incidents at the boundary surface of two dielectric media with some an angle, refracted wave will not be appeared when $\epsilon_1 > \epsilon_2$. In such case the angle made by the incident wave with the interface is called critical angle and the phenomenon is called as total internal reflection.

According to Snell's law we know that

$$\frac{\sin\theta_1}{\sin\theta_2} = \sqrt{\frac{\epsilon_2}{\epsilon_1}}$$

When $\in_1 > \in_2$, $\theta_2 > \theta_1$. The maximum angle θ_2 can have is 90°.

$$\therefore \qquad \sin\theta_1 = \sqrt{\frac{\in_2}{\in_1}} \text{ and}$$

where $\theta_1 = \theta_c$ = critical angle $= \sin^{-1}\sqrt{\frac{\in_2}{\in_1}}$ .....(4.13.1)

∴ If any wave incidents with the critical angle, refraction does not take place.

4.14 Surface Impedance

When an electromagnetic wave encounters a perfect conductor we know that the amplitude of wave decreases exponentially. The current induced by the wave exists only at the surface of the conductor. The surface impedance Z_s is defined as ratio of tangential component of electric field to the conduction current density.

$$\text{i.e.,} \qquad Z_s = \frac{E_{tan}}{J_s} \qquad \text{.....(4.14.1)}$$

The current density distribution is exponentially decaying as shown in Fig. 4.21.

J_s can be obtained by integrating J from 0 to ∞.

$$\therefore \qquad J_s = \int_0^{\infty} J\,dy$$

$$\int_0^{\infty} J_0 e^{-\gamma y}\,dy = \frac{J_0}{\gamma}$$

where J_0 is the conduction current density at the surface and it is equal to σE_{tan}.

γ is the propagation constant.

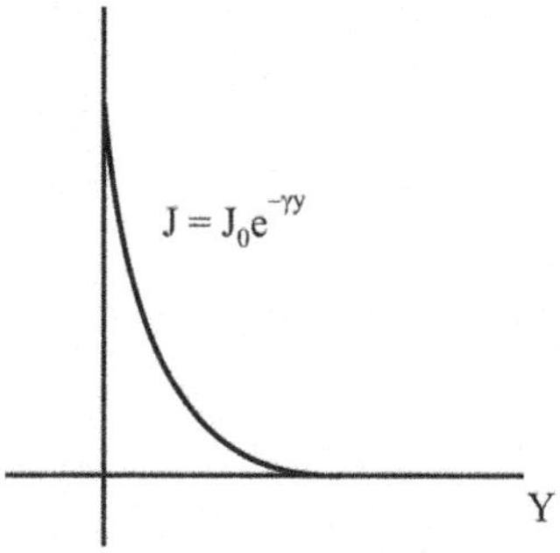

Fig. 4.21 Evaluation of surface impedance

$$\Rightarrow \qquad Z_s = \frac{\gamma}{\sigma} = \frac{\sqrt{j\omega\mu(\sigma + j\omega\in)}}{\sigma}$$

for perfect conductor $\sigma >> j\omega\in$

$$\Rightarrow \quad Z_s = \frac{\sqrt{j\omega\mu\sigma}}{\sigma} = \sqrt{\frac{j\omega\mu}{\sigma}} = \sqrt{\frac{\omega\mu}{\sigma}}\angle 45^0 \qquad \because \quad j = \cos\frac{\pi}{2} + j\sin\frac{\pi}{2} = e^{j\frac{\pi}{2}}$$

$$\Rightarrow \quad \sqrt{j} = j^{1/2} = \left(e^{j\frac{\pi}{2}}\right)^{1/2} = e^{j\frac{\pi}{4}}$$

$$\Rightarrow \quad Z_s = \sqrt{\frac{\omega\mu}{\sigma}} e^{j\frac{\pi}{4}} \qquad \text{.....(4.14.2)}$$

$$\Rightarrow \quad R_s = \sqrt{\frac{\omega\mu}{2\sigma}} \qquad \text{.....(4.14.3)}$$

$$X_s = \sqrt{\frac{\omega\mu}{2\sigma}} \qquad \text{.....(4.14.4)}$$

Problem 4.17

An EM wave traveling in free space incidents on a dielectric medium with relative dielectric constant = 2 at an angle of 45°. Find the angle by which E tilts as the wave crosses the boundary.

Solution

$\in_1 = \in_0 = 8.854 \times 10^{-12}$ Wb/m

$\in_2 = \in_0 \in_r = 2 \times 8.854 \times 10^{-12}$ Wb/m

$$\frac{\sin\theta_1}{\sin\theta_2} = \sqrt{\frac{\in_2}{\in_1}}$$

$$\sin\theta_2 = \sin\theta_1\sqrt{\frac{\in_1}{\in_2}} = 0.5$$

$\theta_2 = 30^\circ$

Problem 4.18

Determine the critical angle for EM wave passing from glass $\in_r = 9$ to air.

Solution

$$\theta_c = \sin^{-1}\sqrt{\frac{\in_2}{\in_1}} = \sin^{-1}\sqrt{\frac{1}{9}} = 19.47^\circ$$

Problem 4.19

The dielectric constant (relative permittivity) of pure water is 80. (a) Determine the Brewster angle for parallel polarization and the corresponding angle of transmission. (b) If a plane wave of perpendicular polarization impinges at this angle, find the reflection and transmission coefficients.

Solution

(a) θ_1= Brewster angle $= \tan^{-1}\sqrt{\frac{\epsilon_2}{\epsilon_1}} = \tan^{-1}\sqrt{80} = 83.62°$

$$\sin\theta_2 = \sin\theta_1\sqrt{\frac{\epsilon_1}{\epsilon_2}} = \sin\theta_1\sqrt{\frac{1}{80}} \Rightarrow 6.379°$$

(b) $$\frac{E_r}{E_i} = \frac{\cos\theta_1 - \sqrt{\left(\frac{\epsilon_2}{\epsilon_1}\right) - \sin^2\theta_1}}{\cos\theta_1 + \sqrt{\left(\frac{\epsilon_2}{\epsilon_1}\right) - \sin^2\theta_1}} = -0.978$$

$$\frac{E_t}{E_i} = 1 + \frac{E_r}{E_i} = 1 - 0.978 = 0.02143$$

Problem 4.20

Find the critical angle for the (a) glass (ϵ_r = 4), (b) Polythene (ϵ_r = 2.25) and (c) polystyrene (ϵ_r = 2.52) to air surface.

Solution

(a) Glass to air surface

$$\theta_c = \sin^{-1}\sqrt{\frac{\epsilon_2}{\epsilon_1}} = \sin^{-1}\sqrt{\frac{1}{4}} = 30°$$

(b) $\theta_c = \sin^{-1}\sqrt{\frac{1}{2.25}} = 41.8°$

(c) $\theta_c = \sin^{-1}\sqrt{\frac{1}{2.52}} = 39.046°$

4.15 Poynting Theorem (or) Poynting Vector

4.15.1 Poynting Theorem

In order to find the power flow associated with an electromagnetic wave, it is necessary to develop a power theorem for the electromagnetic field known as the poynting theorem.

According to Maxwell's equations for time varying fields

$$\nabla \times \bar{E} = -\frac{\partial \bar{B}}{\partial t} = -\mu \frac{\partial \bar{H}}{\partial t} \qquad \text{.....(4.15.1)}$$

$$\nabla \times \bar{H} = \bar{J} + \frac{\partial \bar{D}}{\partial t} = \bar{J} + \in \frac{\partial \bar{E}}{\partial t} \qquad \text{.....(4.15.2)}$$

Let us consider the divergence of $\bar{E} \times \bar{H}$

i.e.,
$$\nabla \cdot \left(\bar{E} \times \bar{H}\right) = \bar{H}.\nabla \times \bar{E} - \bar{E}.\nabla \times \bar{H} \qquad \text{.....(4.15.3)}$$

$$= \bar{H}.\left(-\mu \frac{\partial \bar{H}}{\partial t}\right) - \bar{E}.\left(\bar{J} + \in \frac{\partial \bar{E}}{\partial t}\right)$$

$$\nabla.(\bar{E} \times \bar{H}) = -\mu \bar{H}.\frac{\partial \bar{H}}{\partial t} - \left(\bar{J}.\bar{E} + \in \bar{E}.\frac{\partial \bar{E}}{\partial t}\right)$$

$$-\nabla.(\bar{E} \times \bar{H}) = \bar{J}.\bar{E} + \in \bar{E}.\frac{\partial \bar{E}}{\partial t} + \mu \bar{H}.\frac{\partial \bar{H}}{\partial t} \qquad \text{.....(4.15.4)}$$

The two time derivatives in (4.15.4) can be written as

$$\in \bar{E}.\frac{\partial \bar{E}}{\partial t} = \frac{\partial}{\partial t}\left(\frac{1}{2}\bar{D}.\bar{E}\right) \qquad \text{.....(4.15.5)}$$

$$\mu \bar{H}.\frac{\partial \bar{H}}{\partial t} = \frac{\partial}{\partial t}\left(\frac{1}{2}\bar{B}.\bar{H}\right) \qquad \text{.....(4.15.6)}$$

Substitute equations (4.15.5) and (4.15.6) in (4.15.4)

$$-\nabla.(\bar{E} \times \bar{H}) = \bar{J}.\bar{E} + \frac{\partial}{\partial t}\left(\frac{1}{2}\bar{D}.\bar{E}\right) + \frac{\partial}{\partial t}\left(\frac{1}{2}\bar{B}.\bar{H}\right)$$

Volume integrate equation (4.15.4)

$$-\int_v \nabla.(\bar{E} \times \bar{H})dv = \int_v \bar{J}.\bar{E}dv + \int_v \frac{\partial}{\partial t}\left(\frac{1}{2}\bar{D}.\bar{E}\right)dv + \int_v \frac{\partial}{\partial t}\left(\frac{1}{2}\bar{B}.\bar{H}\right)dv \qquad \text{.....(4.15.7)}$$

Apply divergence theorem on left side and interchanging operations on right side, we get

$$-\oint_A (\bar{E}\times\bar{H}).\overline{ds} = \int_v \bar{J}.\bar{E}dv + \frac{d}{dt}\int_v \frac{1}{2}\bar{D}.\bar{E}\ dv + \frac{d}{dt}\int_v \frac{1}{2}\bar{B}.\bar{H}dv \qquad .....(4.15.8)$$

which is Poynting theorem

Right side of equation (4.15.8):

1st term is the total (but instantaneous) ohmic power dissipated within the volume.

2nd term is the total energy stored in the electric field.

3rd term is the total energy stored in the magnetic field.

The sum of expressions on right side be the total power flowing into this volume and so the total power flowing out of the volume is

$$\oint_A (\bar{E}\times\bar{H}).\overline{ds} \text{ Watts} \qquad \text{(left side of equation (4.15.8)).}$$

The cross product $\bar{E}\times\bar{H}$ is known as the Poynting vector $\bar{S}$, which is an instantaneous power density in watts/m^2

i.e., $\bar{S} = \bar{E}\times\bar{H}\ \mathrm{W/m^2}$ .....(4.15.9)

4.15.2 Power Flow through Concentric Cables or Power Flow through Coaxial Cables

Let us find the power flow through a co-axial cable by using poynting vector. Consider the power transmission to a load resistor 'R' through a co-axial cable, whose length is 'L', that contains the voltage 'V' between the conductors and the current I through the inner or outer conductors. Assume the radius of inner conductor as 'a' and outer conductor as 'b'.

Assume cable is placed along Z-axis. Since the above problem has cylindrical symmetry, we need to use cylindrical coordinate system along with pointing vector to find power flow through cable.

We know that the electric and magnetic fields are perpendicular to each other, here the magnetic field will be along ϕ-axis and electric field will be along ρ (the radial)- axis.

From the first Chapter, (co-axial capacitor topic):

the electric field $\bar{E} = \dfrac{Q}{2\pi \in \rho L}\bar{a}_\rho$

and the potential $V = \dfrac{Q}{2\pi \in L}\log\dfrac{b}{a}$

$$\Rightarrow \qquad Q = \frac{V 2\pi \in L}{\log\dfrac{b}{a}}$$

Substituting Q in the above equation

$$\bar{E} = \frac{V}{\rho\log\dfrac{b}{a}}\bar{a}_\rho \qquad \text{.....(4.15.10)}$$

From the second Chapter, (infinitely long co-axial cable topic):

$$\Rightarrow \qquad \bar{H} = \frac{I}{2\pi\rho}\bar{a}_\phi \qquad \text{.....(4.15.11)}$$

Total power flowing through the cable to the load can be obtained by integrating the poynting vector at the surface.

$$\therefore \qquad W = \int\left(\bar{E}\times\bar{H}\right)\cdot d\bar{s}$$

Here $\qquad \bar{E}\times\bar{H} = \dfrac{V}{\rho\log\dfrac{b}{a}}\bar{a}_\rho \times \dfrac{I}{2\pi\rho}\bar{a}_\phi$

$$\bar{E}\times\bar{H} = \frac{VI}{2\pi\rho^2\log\dfrac{b}{a}}\bar{a}_z$$

and $d\bar{s} = \rho d\phi d\rho\bar{a}_z$

$$\therefore \qquad W = \int_{\phi=0}^{2\pi}\int_{\rho=a}^{b}\frac{VI}{2\pi\rho^2\log\left(\dfrac{b}{a}\right)}\bar{a}_z.\rho d\phi d\rho\bar{a}_z$$

$$W = \frac{VI}{2\pi\log\left(\dfrac{b}{a}\right)}\int_{\phi=0}^{2\pi}d\phi\int_{\rho=a}^{b}\frac{1}{\rho}d\rho$$

$$W = \frac{VI}{2\pi\log\left(\dfrac{b}{a}\right)}2\pi\log\left(\frac{b}{a}\right)$$

$$W = VI \qquad \text{.....(4.15.12)}$$

From the above equation we can say that the power flow through the co-axial cable is the product of voltage and current.

4.16 Power Loss in a Conductor

By integrating the poynting vector at a point on the surface of the conductor, we get the power flowing through the surface which is known as power loss in the conductor.

The instantaneous poynting vector is $\bar{P} = \bar{E} \times \bar{H}$

The complex poynting vector $\bar{P}_{complex} = \frac{1}{2}\left(\bar{E} \times \bar{H}^*\right)$

From this average poynting vector $\bar{P}_{avg} = \frac{1}{2}\text{Real}\left(\bar{E} \times \bar{H}^*\right)$

and the reactive pointing vector $\bar{P}_{react} = \frac{1}{2}\text{Im}\left(\bar{E} \times \bar{H}^*\right)$

For a plane conductor or good conductor the intrinsic impedance $\eta = \sqrt{\frac{\omega\mu}{\sigma}}\angle 45^o$ i.e., the angle between electric and magnetic fields in a plane conductor is 45°. We know

$$\bar{E} \times \bar{H}^* = \left|\bar{E}\right| \left|\bar{H}^*\right| \sin\theta$$

Here θ = 45°

$$\therefore \qquad \bar{P}_{avg} = \frac{1}{2} EH \sin 45^o$$

$$= \frac{1}{2\sqrt{2}} \frac{E^2}{\eta} \qquad \text{.....(4.16.1)}$$

We know intrinsic impedance is equal to surface impedance for a good conductor.

∴ average poynting vector in terms of surface impedance is

$$\bar{P}_{avg} = \frac{1}{2\sqrt{2}} \frac{E^2}{Z_s} = \frac{1}{2\sqrt{2}} H^2 Z_s \text{ Watts/m}^2 \qquad \text{.....(4.16.2)}$$

Review Questions and Answers

1. **Define a wave.**

Ans. If a physical phenomenon that occurs at one place at a given time is reproduced at other places at later times, the time delay being proportional to the space separation from the first location then the group of phenomena constitutes a wave.

2. **Mention the properties of uniform plane wave.**

Ans. (i) At every point in space, the electric field E and magnetic field H are perpendicular to each other.

(ii) The fields vary harmonically with time and at the same frequency everywhere in space.

3. **Write down the wave equation for E in free space.**

Ans. $$\nabla^2 \bar{E} - \mu_0 \in_0 \frac{\partial^2 \bar{E}}{\partial t^2} = 0$$

4. **Write down the wave equation for H in free space.**

Ans. $$\nabla^2 \bar{H} - \mu_0 \in_0 \frac{\partial^2 \bar{H}}{\partial t^2} = 0$$

5. **Define intrinsic impedance or characteristic impedance.**

Ans. It is the ratio of electric field to magnetic field or it is the ratio of square root of permeability to permittivity of medium.

6. **Give the characteristic impedance of free space.**

Ans. 377 ohms

7. **Define propagation constant.**

Ans. Propagation constant is a complex number

$$\gamma = \alpha + j\beta$$

where α is attenuation constant

β is phase constant

$$\gamma = j\omega\mu(\sigma + j\omega \in)$$

8. **Define skin depth**

Ans. It is defined as that depth in which the wave has been attenuated to 1/e or approximately 37% of its original value.

$$\delta = 1/\alpha = 2 / j\omega\sigma$$

9. **Define Poynting vector.**

The pointing vector is defined as rate of flow of energy of a wave as it propagates.

$$P = E \times H$$

10. State Poynting's Theorem.

Ans. The net power flowing out of a given volume is equal to the time rate of decrease of the energy stored within the volume- conduction losses.

11. Define reflection coefficients.

Ans. Reflection coefficient is defined as the ratio of the magnitude of the reflected field to that of the incident field.

12. Define transmission coefficients.

Ans. Transmission coefficient is defined as the ratio of the magnitude of the transmitted field to that of incident field.

13. What will happen when the wave is incident obliquely over dielectric – dielectric boundary?

Ans. When a plane wave is incident obliquely on the surface of a perfect dielectric part of the energy is transmitted and part of it is reflected. But in this case the transmitted wave will be refracted, that is the direction of propagation is altered.

14. Why water has much greater dielectric constant than mica?

Ans. Water has a much greater dielectric constant than mica. Because water has a permanent dipole moment, while mica does not have.

15. What is Lorentz force?

Ans. Lorentz force is the force experienced by the test charge. It is maximum if the direction of movement of charge is perpendicular to the orientation of field lines.

16. What are uniform plane waves?

Ans. Electromagnetic waves which consist of electric and magnetic fields that are perpendicular to each other and to the direction of propagation and are uniform in plane perpendicular to the direction of propagation are known as uniform plane waves.

17. What is the significant feature of wave propagation in an imperfect dielectric?

Ans. The only significant feature of wave propagation in an imperfect dielectric compared to that in a perfect dielectric is the attenuation undergone by the wave.

18. What is the surface impedance?

Ans. The surface impedance Z_s is defined as ratio of tangential component of electric field to the conduction current density.

$$\text{i.e., } Z_s = \frac{E_{\tan}}{J_s}$$

19. What are the conditions to get Elliptically polarized wave?

Ans. (i) The phase difference between $\bar{E}_x$ and $\bar{E}_y$ must be other than zero.

(ii) Magnitudes of $\bar{E}_x$ and $\bar{E}_y$ must not be equal.

20. What is loss tangent?

Ans. The ratio of magnitudes of conduction current density and displacement current density is $\left|\frac{\sigma \bar{E}}{j\omega \in \bar{E}}\right| = \frac{\sigma}{\omega \in} = \tan\theta$ which serves a boundary between conductors and dielectrics and is called loss tangent .

Multiple Choice Questions

1. A plane wave travels in the (+z) direction in a lossy transmission line whose medium has attenuation constant $\alpha > 0$. The power carried by the wave attenuates as:
   (a) $\sim e^{-\alpha z}$
   (b) $\sim e^{\alpha z}$
   (c) $\sim e^{-2\alpha z}$
   (d) $\sim 1/z$

2. The electric field of a plane wave propagating in a nonmagnetic medium is given by $\bar{E} = 3\sin(2\pi 10^7 t - 0.4\pi x)\bar{a}_y$ V/m. Determine the wavelength making use of the phase constant (wave number).
   (a) 0.5 m
   (b) 30 m
   (c) 1 m
   (d) 5 m

3. For the wave in Problem 2, find the relative dielectric permittivity $(\in_r)$ of the medium.
   (a) 36
   (b) 1
   (c) 6
   (d) 10

4. For the wave in problem 2, find the magnitude of the magnetic field vector H_m.
   (a) 0.08 A/m
   (b) 0.024 A/m
   (c) 0.016 A/m
   (d) 0.048 A/m

5. Calculate the average power density (P_{avg}) of the traveling wave in Problem 2.
   (a) 144 mW/m^2
   (b) 72 mW/m^2
   (c) 36 mW/m^2
   (d) 10 mW/m^2

6. The electric field of a traveling wave is given as $\bar{E}(z,t) = 1\cos(\omega t - \beta z)\bar{a}_x + 1\sin(\omega t - \beta z)\bar{a}_y$ V/m. Describe the polarization of the wave.
   (a) linear, at 45° w.r.t. the x –axis
   (b) some elliptical polarization but not circular
   (c) circular, right-handed
   (d) circular, left-handed
7. For a wave normally incident from free space onto a planar interface with a perfect conductor, the reflection coefficient is
   (a) 0
   (b) 1
   (c) –1
   (d) Depends on the frequency of the wave
8. The standing wave ratio for perfect transmission (no reflection) is
   (a) 2
   (b) 0
   (c) 1
   (d) –1
9. If a wave is incident onto an interface at the Brewster's angle
   (a) It is completely transmitted
   (b) It is completely reflected
   (c) It suffers total internal reflection
   (d) The reflection coefficient is -
10. The pointing vector physically denotes the power density leaving or entering a given volume in a time varying field.
    (a) True
    (b) False
11. In a good conductor, $\bar{E}$ and $\bar{H}$ are in time phase.
    (a) True
    (b) False
12. Electromagnetic waves travel faster in conductors than in dielectrics.
    (a) True
    (b) False
13. In a travelling electromagnetic wave, E and H vector fields are
    (a) perpendicular in space
    (b) parallel in space.
    (c) E is in the direction of wave travel
    (d) H is in the direction of wave travel.
14. A wave is incident normally on a good conductor. If the frequency of a plane electromagnetic wave increases four times, the skin depth, will
    (a) increase by a factor of 2
    (b) decrease by a factor of 4.
    (c) remain the same
    (d) decrease by a factor of 2.

15. When an EM wave is incident on a dielectric, it is
    (a) fully transmitted
    (b) fully reflected
    (c) partially transmitted and partially reflected
    (d) none of these.
16. Depth of penetration in free space is
    (a) α (b) $1/\alpha$ (c) 0 (d) ∞
17. A uniform plane wave in air is incident normally on an infinitely thick slab. If the refractive index of the glass slab is 1.5, then the percentage of the incident power that is reflected from the air-glass interface is
    (a) 0% (b) 4% (c) 20% (d) 10%
18. An electromagnetic wave is incident obliquely at the surface of a dielectric medium 2 (μ_2, ϵ_2) from dielectric medium 1 (μ_1, ϵ_1). The angle of incidence and the critical angle are θ_i and θ_c respectively. The phenomenon of total reflection occurs when
    (a) $\varepsilon_1 > \varepsilon_2$ and $\theta_i < \theta_c$ (b) $\varepsilon_1 < \varepsilon_2$ and $\theta_i > \theta_c$
    (c) $\varepsilon_1 < \varepsilon_2$ and $\theta_i < \theta_c$ (d) $\varepsilon_1 > \varepsilon_2$ and $\theta_i > \theta_c$
19. Poynting vector gives
    (a) rate of energy flow (b) direction of polarization.
    (c) intensity of electric field (d) intensity of magnetic field.
20. The material is described by the following electrical parameters at a frequency of 10 GHz, $\sigma = 10^6$ mho/m, $\mu = \mu_0$ and $\dfrac{\sigma}{\sigma_c} = 10$, the material at this frequency is considered to be
    (a) a good conductor
    (b) neither a good conductor nor a good dielectric
    (c) a good dielectric
    (d) a good magnetic material

Answers

1.	(c)	2.	(d)	3.	(a)	4.	(d)	5.	(d)
6.	(c)	7.	(c)	8.	(c)	9.	(a)	10.	(a)
11.	(b)	12.	(b)	13.	(c)	14.	(d)	15.	(c)
16.	(b)	17.	(b)	18.	(d)	19.	(a)	20.	(a)

EXERCISE QUESTIONS

1. Define uniform plane wave. Prove that uniform plane wave does not have field components in the direction of the propagation. Determine the intrinsic impedance of free space.
2. What is polarization of an EM wave? Distinguish between different types of polarization? Prove that the polarization is circular when the two components of electric field are equal and are 90° apart.
3. Derive the expression for attenuation and phase constants of uniform plane wave.
4. If $\varepsilon_r = 9$, $\mu = \mu_0$ for the medium in which a wave with frequency f = 0.3 GHz is propagating, determine propagation constant and intrinsic impedance of the medium when (a) $\sigma = 0$ and (b) $\sigma = 10$ mho/m.
5. For good dielectrics derive the expressions for α, β, v and η.
6. Find α, β, v and η for Ferrite at 10 GHz $\epsilon_r = 9, \mu_r = 4, \sigma = 10\ \text{ms}/\text{m}$.
7. Determine the phase velocity of propagation, attenuation constant, phase constant and intrinsic impedance for a forward traveling wave in a large block of copper at 1 MHz $(\sigma = 5.8\times10^7,\ \epsilon_r = \mu_r = 1)$ determine the distance that the wave must travel to be attenuated by a factor of 100(40 dB).
8. The electric field intensity associated with a plane wave traveling in a perfect dielectric medium is given by $E_x(z,t) = 10\cos(2\pi\times10^7 t - 0.1\pi z)\ \text{V}/\text{m}$. (a) What is the velocity of propagation, (b) Write down an expression for the magnetic field intensity associated with the wave if $\mu = \mu_0$.
9. A large copper conductor $(\sigma = 5.8\times10^7\,\text{s}/\text{m},\ \epsilon_r = \mu_r = 1)$ support a uniform plane wave at 60 Hz. Determine the ratio of conduction current to displacement current, the attenuation constant, propagation constant, intrinsic impedance, wave length and phase velocity of propagation.
10. Explain skin depth and derive an expression for depth of penetration for good conductor
11. State and Prove Poynting Theorem.
12. A plane wave traveling in a free space has an average pointing vector of 5 watts/m^2. Find the average energy density.
13. Define surface impedance and explain how it exists.

14. Derive expression for Reflection and Transmission coefficients of an EM wave when it incidents normally on a dielectric.

15. For an incident wave under oblique incident from medium of ε_1 to medium of ε_2 with parallel polarization. (a) Define and establish the relations for the critical angle θ_c and Brewster angle θ_{Br} for non-magnetic media with neat sketches. (b) Plot θ_c and θ_{Br} versus the ratio $\varepsilon_1/\varepsilon_2$.

16. Explain the difference between the intrinsic impedance and the surface impedance of a conductor. Show that for a good conductor, the surface impedance is equal to the intrinsic impedance.

17. Define and distinguish between the terms perpendicular polarization, parallel polarization, for the case of reflection by a perfect conductor under oblique incidence.

18. Obtain an expression for the power loss in a plane conductor in terms of the surface impedance.

Chapter

5

Transmission Lines

5.1 Introduction

Energy can be transmitted either by radiating electromagnetic waves in to air as in radio and TV or it can be moved through a conducting arrangement known as transmission line.

Therefore Transmission line is a conductive method of guiding electrical energy from one place to another place. Basically transmission lines are used to link between antenna and a transmitter or receiver. There are different types of transmission lines

(i) Parallel wire type

(ii) Coaxial cable

(iii) Wave guides

(iv) Fibre optic cables

Parallel wire transmission is an open wire line. Examples for this are telephone lines and power transmission lines. In this frequencies up to 100 MHz can be transmitted. To transmit signals whose frequencies more than 100 MHz, we go for co-axial cable transmission line in which transmission is possible till 1 GHz. Above 1 GHz signals, if we want to transmit, we use wave guides. In this transmission is possible in UHF range. Further if we want to transmit the signals whose frequencies are more than UHF range, we use fibre optic cables which have got many advantages.

5.2 Primary Elements of Transmission Lines

To understand or to design a transmission line, it is necessary to know the electrical parameters that are associated with a transmission line. If a parallel wire carrying current 'I', then magnetic field will be generated around that wire and voltage is developed along the line. Due to magnetic field and current, there will be series inductance L. Due to voltage along the line there will be series resistance R. When we apply voltage between conductors, electric field will be generated, due to this there will be a shunt capacitance 'C' and shunt conductance G.

Therefore a unit of transmission line can be represented with these primary constants as shown in Fig. 5.1.

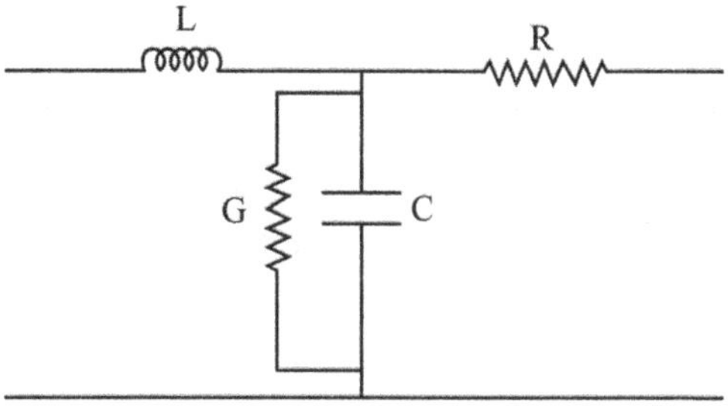

Fig. 5.1 Representation of transmission line with primary constants

As shown in the Fig.5.1, the primary elements are distributed along the transmission line uniformly. Hence the name uniform transmission line. Where 'R' is the loop resistance /unit length i.e., the sum of resistances of both the conductors per unit length. Unit for loop resistance is ohms/kilometer. 'L' is the loop inductance per unit length i.e., it is sum of the inductances of both the wires per unit length. Unit for 'L' is Henry/Km. 'C' is the shunt capacitance between the conductors per unit length. Unit for 'C' is Farad/Km. 'G' is the shunt conductance between the wires per unit length. Unit for 'G' is $\mho / \text{km}$. The series impedance of Transmission line is $Z = R + j\omega L$ and shunt admittance $Y = G + j\omega c$.

5.3 Transmission Line Equations

Let us assume a transmission line of length '*l*' and consider a part PQ i.e., length dx of transmission line which is located at a distance 'x' from sending end 'A' as shown in the Fig. 5.2.

Let the current at 'P' is I, then at Q is $I + dI$ similarly voltage at P is V then at 'Q' is $V + dV$

The series impedance between points P and Q is $(R + j\omega L)dx$ and the shunt admittance between points P and Q is $(G + j\omega C)dx$.

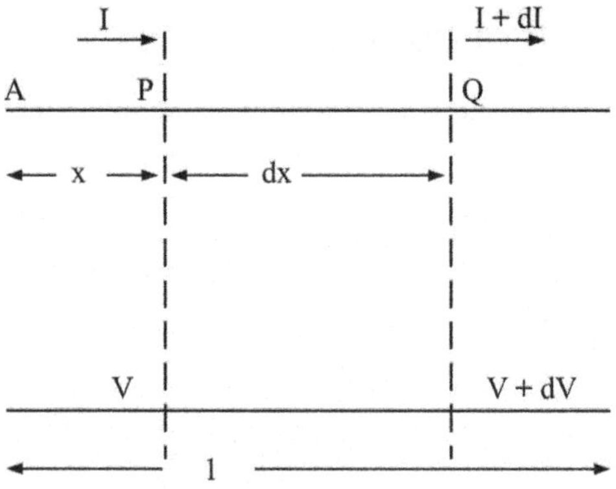

Fig. 5.2 A part of transmission line

The Potential difference between points P and Q is equal to the drop across the series impedance that exists between points P and Q.

$$\therefore \quad V - (V + dV) = I(R + j\omega L)dx$$

$$\frac{-dV}{dx} = (R + j\omega L)I \qquad .....(5.3.1)$$

The current difference between points P and Q equal to the current flowing through the shunt admittance that exist between points P and Q.

$$\therefore \quad I - (I + dI) = (G + j\omega C)dx \, . \, V$$

$$\frac{-dI}{dx} = (G + j\omega C)V \qquad .....(5.3.2)$$

Both the equations (5.3.1) and (5.3.2) contain V and I. Let us get equations which contain either V or I. Differentiate equation (5.3.1) w.r.t 'x' on both sides

$$\frac{-d^2V}{dx^2} = (R + j\omega L)\frac{dI}{dx}$$

$$= (R + j\omega L)\,[-(G + j\omega C)\,V]$$

$$\frac{1}{V}\frac{-d^2V}{dx^2} = (R + j\omega L)(G + j\omega C) \qquad .....(5.3.3)$$

Differentiate equation (5.3.2) w.r.t. 'x', we get

$$\frac{-d^2I}{dx^2} = (G + j\omega C)\frac{dV}{dx}$$

$$= -(G + j\omega C)\,[R + j\omega L]\,I$$

$$\frac{d^2I}{dx^2}=(G+j\omega C)(R+j\omega L)I \qquad \text{.....(5.3.4)}$$

Let $\gamma^2 = (G + j\omega C)\,[R + j\omega L]$

where γ is propagation constant

Then (5.3.3) and (5.3.4) become

$$\frac{d^2V}{dx^2}=\gamma^2 V \qquad \text{.....(5.3.5)}$$

$$\frac{d^2I}{dx^2}=\gamma^2 I \qquad \text{.....(5.3.6)}$$

where γ is a complex quantity = $\sqrt{(R+j\omega L)(G+j\omega C)}$

The solutions for (5.3.5) and (5.3.6) are

$$V = ae^{\gamma x} + be^{-\gamma x} \qquad \text{.....(5.3.7)}$$

$$I = ce^{\gamma x} + de^{-\gamma x} \qquad \text{.....(5.3.8)}$$

We know $e^{\gamma x} = \cosh(\gamma x) + \sinh(\gamma x)$

and $e^{-\gamma x} = \cosh(\gamma x) - \sinh(\gamma x)$

Substitute these in (5.3.7) & (5.3.8)

$$V = a[\cosh(\gamma x) + \sinh(\gamma x)] + b[\cosh(\gamma x) - \sinh(\gamma x)]$$

$$V = (a+b)\cosh(\gamma x) + (a-b)\sinh(\gamma x)$$

$$V = A\cosh(\gamma x) + B\sinh(\gamma x) \qquad \text{.....(5.3.9)}$$

Where A = a + b and B = a – b

$$I = c[\cosh(\gamma x) + \sinh(\gamma x)] + d[\cosh(\gamma x) - \sinh(\gamma x)]$$

$$I = (c+d)\cosh(\gamma x) + (c-d)\sinh(\gamma x)$$

$$I = C\cosh(\gamma x) + D\sinh(\gamma x) \qquad \text{.....(5.3.10)}$$

Where C = c + d and D = c – d

Let us get equations for V and I which contain only two unknowns A and B

Differentiate equation (5.3.9) w.r.t. 'x' on both sides and multiply with '–'

$$\frac{-dV}{dx} = -[A\sinh\gamma x + B\cosh\gamma x]\gamma = (R+j\omega L)I$$

$$\Rightarrow \qquad -\sqrt{R + j\omega L(G + j\omega C)}(\text{A sin}\gamma x + \text{B cos}\gamma x) = (\text{R} + \text{j}\omega\text{L})\text{I}$$

$$\Rightarrow \qquad -\sqrt{\frac{(G + j\omega C)}{(R + j\omega L)}}(\text{A sinh}\gamma x + \text{B cosh}\gamma x) = \text{I}$$

Let $\qquad Z_o = \sqrt{\dfrac{(R + j\omega L)}{(G + j\omega C)}}$

Where Z_o is characteristic impedance

$$I = -\frac{1}{Z_0}(\text{A sinh}\gamma x \; + \; \text{B cosh}\gamma x) \qquad \text{.....(5.3.11)}$$

$V = A\cosh(\gamma x) + B\sinh(\gamma x)$ from (5.3.9)

$$\cosh\gamma x = \frac{e^{\gamma x} + e^{-\gamma x}}{2} \qquad \sinh\gamma x = \frac{e^{\gamma x} - e^{-\gamma x}}{2}$$

$$I = -\frac{1}{2Z_o}\left[(A+B)e^{\gamma x} - (A-B)e^{-\gamma x}\right] = -\frac{1}{Z_o}\left[ae^{\gamma x} - be^{-\gamma x}\right]$$

$$I = \frac{1}{Z_o}\left[be^{-\gamma x} - ae^{\gamma x}\right] \qquad \text{.....(5.3.12)}$$

$$V = \frac{1}{2}\left[(A+B)e^{\gamma x} + (A-B)e^{-\gamma x}\right] = \left[ae^{\gamma x} + be^{-\gamma x}\right] \qquad \text{.....(5.3.13)}$$

where $\qquad a = \dfrac{A+B}{2}$

$$b = \frac{A-B}{2}$$

Let us get the equations for V and I which contain only two unknowns C and D Differentiate equation (5.3.10) w.r.t. 'x' on both sides and multiply with '–'

$$\frac{-dI}{dx} = -\gamma\left[C\sinh\gamma x + D\cosh\gamma x\right] = (G + j\omega C)V$$

$$\Rightarrow \qquad -\sqrt{(R + j\omega L)(G + j\omega C)}(\text{C sinh } \gamma x \; + \; \text{D cosh } \gamma x) = \; (\text{G} + \text{jwC})\text{V}$$

$$\Rightarrow \qquad -\sqrt{\frac{(R + j\omega L)}{(G + j\omega C)}}(\text{C sinh } \gamma \text{x} \; + \; \text{D cosh } \gamma \text{x}) \; = \; \text{V}$$

$$V = -Z_0(C \sinh \gamma x + D \cosh \gamma x) \tag{5.3.14}$$

$$I = C \cosh \gamma x + D \sinh \gamma x \qquad \text{from (5.3.10)}$$

Let us convert the above equations in exponential form

$$V = -\frac{Z_o}{2}\left[(C+D)e^{\gamma x} - (C-D)e^{-\gamma x}\right]$$

$$V = -Z_0\left[ce^{\gamma x} - de^{-\gamma x}\right]$$

Where c = (C + D)/2 and d = (C – D)/2

$$I = \frac{1}{2}\left[(C+D)e^{\gamma x} + (C-D)e^{-\gamma x}\right] = \left[ce^{\gamma x} + de^{-\gamma x}\right]$$

The pairs (5.3.9), (5.3.11) and (5.3.10), (5.3.14) are transmission line equations.

5.3.1 Determination of Constants A, B, C and D

Let us assume the voltage and currents at the sending end as V_s and I_s

At x = 0 (sending end) ; $V = V_s$

Substitute in equation (5.3.9), then

$$V_s = A\cosh\gamma(0) + B\sinh\gamma(0)a$$

$$V_s = A$$

At $x = 0, I = I_s$

Substituting in equation (5.3.11)

$$I_s = -\frac{1}{Z_0}B \Rightarrow B = -I_s Z_0$$

$$V = V_s\cosh(\gamma x) - I_s Z_0 \sinh(\gamma x) \qquad \text{.....(5.3.15)}$$

$$I = -\frac{1}{Z_0}[V_s \sinh(\gamma x) - I_s Z_0 \cosh(\gamma x)] = I_s \cosh(\gamma x) - \frac{V_s}{Z_0}\sinh(\gamma x) \qquad \text{.....(5.3.16)}$$

From equations (5.3.15) and (5.3.16), we can find the voltage and current at any point 'x' on the transmission line, if we know the voltage and current at the sending end.

If we know the voltage and current at the receiving end then let us find the general equations for voltage and current at any point 'x' on the transmission line.

Assume V_R and I_R as the voltage and current at the receiving end

At $x = l$; $V = V_R$ and $I = I_R$

Substituting in equation (5.3.14)

$$V_R = -Z_0[C\sinh(\gamma l) + D\cosh(\gamma l)] \qquad ....(5.3.17)$$

$$I_R = C\cosh(\gamma l) + D\sinh(\gamma l) \qquad .....(5.3.18)$$

From (5.3.17) & (5.3.18)

$$V_R + Z_0 C\sinh(\gamma l) = -Z_0 D\cosh(\gamma l)$$

$$I_R - C\cosh(\gamma l) = D\sinh(\gamma l)$$

Divide the above two equations

$$\frac{V_R + Z_0\, C\sinh\gamma l}{I_R - C\cosh\gamma l} = \frac{-Z_0\, D\cosh\gamma l}{D\sinh\gamma l}$$

Cross multiplying

$$\Rightarrow V_R\sinh(\gamma l) + Z_0 C\sinh^2(\gamma l) + Z_0 I_R\cosh(\gamma l) - Z_0 C\cosh^2(\gamma l) = 0$$

$$\Rightarrow V_R\sinh(\gamma l) + Z_0 I_R\cosh(\gamma l) - Z_0 C = 0$$

$$C = \frac{V_R\sinh\gamma l + I_R Z_0\cosh\gamma l}{Z_0} = X$$

Similarly

$$\frac{V_R + Z_0\, D\cosh\gamma l}{I_R - D\sinh\gamma l} = \frac{-Z_0\, C\sinh\gamma l}{C\cosh\gamma l}$$

$$\Rightarrow V_R\cosh(\gamma l) + Z_0\,\mathrm{D}\cosh^2(\gamma l) + Z_0 I_R\sinh(\gamma l) - Z_0\,\mathrm{D}\sinh^2(\gamma l) = 0$$

$$\Rightarrow V_R\cosh(\gamma l) + Z_0 I_R\sinh(\gamma l) + Z_0 D = 0$$

$$D = \frac{-(V_R\cosh\gamma l + I_R Z_o\sinh\gamma l)}{Z_o} = Y$$

Substituting in equation (5.3.14)

$$V = -Z_0[X\ sinh\ (\gamma x) + Y\ cosh\ (\gamma x)]$$

$$I = X\ cosh\ (\gamma x) + Y\ sinh\ (\gamma x)$$

$$V=-Z_0\left(\frac{V_R\sinh(\gamma l)+I_RZ_0\cosh(\gamma l)}{Z_0}\right)\sinh(\gamma x)+Z_0\left(\frac{V_R\cosh(\gamma l)+I_RZ_0\sinh(\gamma l)}{Z_0}\right)\cosh(\gamma x)$$

$$V=V_R\left(\cosh(\gamma l)\cosh(\gamma x)-\sinh(\gamma l)\sinh(\gamma x)\right)$$
$$+I_RZ_0\left(\sinh(\gamma l)\cosh(\gamma x)-\cosh(\gamma l)\sinh(\gamma x)\right)$$

$$V= V_R cosh\ \gamma(l-x)+ I_RZ_0 sinh\ \gamma(l-x) \qquad ,,,,,(5.3.19)$$

Similarly

$$I = \frac{V_R\sinh\gamma l+I_RZ_0\cosh\gamma l}{Z_0}\ cosh\ (\gamma x)\ +\ \frac{-(V_R\cosh\gamma l+I_RZ_0\sinh\gamma l)}{Z_0}\ sinh\ (\gamma x)$$

$$I = \frac{V_R}{Z_0}(\sinh(\gamma l)cosh\ (\gamma x)-\cosh(\gamma l)sinh\ (\gamma x))$$
$$+I_R\left(\cosh(\gamma l)cosh\ (\gamma x)-\sinh(\gamma l)sinh\ (\gamma x)\right)$$

$$I=\frac{V_R}{Z_0}\sinh\gamma(l-x)+I_R\cosh\gamma(l-x) \qquad .....(5.3.20)$$

5.4 Infinite Transmission Line

Consider an infinite line as shown in Fig. 5.3.

When we apply A.C. voltage at the sending end of infinite line there will be finite current due to capacitance and conductance between the lines. The ratio of V_{si} to I_{si} gives the input impedance or characteristic impedance Z_0.

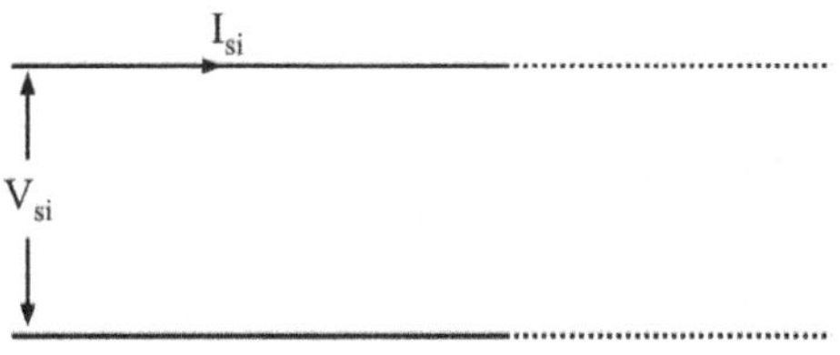

Fig. 5.3 Infinite line

$$\therefore \qquad Z_0=\frac{V_{si}}{I_{si}}$$

We know the general equation for current

$$I = c\,e^{\gamma x} + d\,e^{-\gamma x}$$

At $x = 0$, $I = I_{si}$

$$\therefore \qquad I_{si} = c + d$$

at $\qquad x = \infty \qquad I = 0$

$$\therefore \qquad 0 = ce^{\infty} + d\,e^{-\infty}$$
$$= c^{\infty} + 0$$

$\Rightarrow \quad c = 0$

$\therefore \quad d = I_{si}$

$\therefore \quad I = I_{si}\, e^{-\gamma x}$ .....(5.4.1)

which is the equation for current at any point 'x' on the infinite transmission line. Similarly we can also write the equation for voltage $V = V_{si}\, e^{-\gamma x}$ .....(5.4.2)

5.4.1 Characteristic Impedance for an Infinite Line

We know $$\frac{-dV}{dx} = (R + j\omega L)I$$

For an infinite line $V = V_{si}\, e^{-\gamma x}$

$I = I_{si}\, e^{-\gamma x}$

$$\frac{-d}{dx}\left(V_{si}e^{-\gamma x}\right) = (R + j\omega L)I_{si}e^{-\gamma x}$$

$$\Rightarrow -\left(V_{si}e^{-\gamma x}\right)(-\gamma) = (R + j\omega L)I_{si}e^{-\gamma x}$$

$$Z_0 = \frac{V_{si}}{I_{si}} = \frac{(R + j\omega L)}{\gamma}$$

$$Z_0 = \frac{(R + j\omega L)}{\sqrt{(R + j\omega L)(G + j\omega C)}} = \sqrt{\frac{(R + j\omega L)}{(G + j\omega C)}}$$

Z_0 is the characteristic impedance. For open wire transmission line the characteristic impedance is

$$Z_0 = 276 \log_{10}^{S/R}$$ .....(5.4.3)

S = spacing or distance between the wires

R = radius of either wire

Similarly Z_o for coaxial cable is

$$Z_0 = \frac{1}{2\pi}\sqrt{\frac{\mu}{\in}} \log_e (D/d)$$ .....(5.4.4)

Where D = Inner diameter of outer conductor

d = diameter of inner conductor

NOTE:

When we transmit a signal through an infinite line there will not be any reflections i.e., loss is negligible. We know that practically we can obtain an infinite line from a finite line by terminating with the characteristic impedance Z_0. If any finite line terminated perfectly with Z_0 then it is called perfectly or correctly terminated transmission line or the non resonator transmission line.

***Problem 5.1**

A coaxial line with an outer diameter of 5 mm has 50 ohm characteristic impedance. If the dielectric constant is 1.60. Calculate the inner diameter.

Solution

Given $Z_0 = 50$ ohm, D = 5 mm, $\epsilon_r = 1.6$

We have

$$Z_0 = \frac{1}{2\pi}\sqrt{\frac{\mu}{\epsilon}}\log_e(D/d)$$

$$50 = \frac{1}{2\pi}\sqrt{\frac{\mu_0\mu_r}{\epsilon_0\epsilon_r}}\log_e(D/d) = \frac{1}{2\pi}\sqrt{\frac{4\pi\times10^{-7}\times1}{8.854\times10^{-12}\times1.6}}\log_e(8mm/d)$$

$$1.0548 = \log_e(8mm/d)$$

$$e^{1.0548} = (8mm/d)$$

inner diameter d = 2.788 mm

5.5 Finite Line Terminated With Z_0

When we join a finite line with an infinite line at points A and B. The total line becomes an infinite line. We know that for an infinite line input impedance is Z_0. Instead of joining an infinite line to finite line, if we terminate the finite line with Z_0 (where Z_0 is the input impedance of an infinite line), it behaves as infinite line.

For the finite line i.e., terminated with Z_0 as shown in Fig.5.4 at the receiving end voltage and currents are V_R and I_R

$$\therefore \quad Z_0 = \frac{V_R}{I_R}$$

We know the general transmission line equations

$$V = V_s cosh\ (\gamma x) - I_s Z_0 sinh\ (\gamma x)$$

$$I = \frac{-V_s}{Z_0}\sinh(\gamma x) + I_s\cosh(\gamma x)$$

at x = l i.e., at the receiving end V = V_R and I = I_R. The above two equations become

$$V_R = V_s cosh\,(\gamma l) - I_s Z_o sinh\,(\gamma l) \quad .....(5.5.1)$$

$$I_R = I_s cosh\,(\gamma l) - \frac{V_s}{Z_0} sinh\,(\gamma l) \quad .....(5.5.2)$$

Divide equation (5.5.1) with (5.5.2)

$$Z_0 = \frac{V_R}{I_R} = \frac{V_s \cosh(\gamma l) - I_s Z_0 \sinh(\gamma l)}{I_s \cosh(\gamma l) - \frac{V_s}{Z_0} \sinh(\gamma l)}$$

$$Z_0 = \frac{Z_0 V_s \cosh(\gamma l) - Z_0 I_s Z_0 \sinh(\gamma l)}{Z_0 I_s \cosh(\gamma l) - V_s \sinh(\gamma l)}$$

$$\Rightarrow Z_0 I_s cosh\,(\gamma l) - V_s sinh\,(\gamma l) = V_s cosh\,(\gamma l) - I_s sinh\,(\gamma l)\, Z_0$$

$$Z_0 I_s [cosh\,(\gamma l) + sinh\,(\gamma l)] = V_s [cosh\,(\gamma l) + sinh\,(\gamma l)]$$

$$Z_0 = \frac{V_s}{I_s}$$

∴ The input impedance of a finite line i.e., terminated with Z_0 is also Z_0.

Conclusion: If we terminate any finite line with the characteristic impedance Z_0, it's input impedance is also Z_0 (or) if we terminate any finite line with Z_0, it behaves as an infinite line.

5.6 Secondary Constants of Transmission Line

The secondary constants of transmission line are γ and Z_o where

$$\gamma = \sqrt{(R + j\omega L)(G + j\omega C)}a \quad .....(5.6.1)$$

and

$$Z_0 = \sqrt{\frac{(R + j\omega L)}{(G + j\omega C)}} \quad .....(5.6.2)$$

From the above two equations, we can find the secondary constants by knowing primary constants or vice versa.

NOTE:

$$Z_0\, \gamma = R + j\omega L \quad \text{where } Z_0 = \sqrt{\frac{R + j\omega L}{G + j\omega C}}$$

$$\frac{\gamma}{Z_0} = G + j\omega C$$

$$\gamma = \sqrt{(R + j\omega L)(G + j\omega C)}$$

These are useful in finding R, L, G and C when Z_0 and γ are given.

5.7 Attenuation and Phase Constants

The propagation constant is $\gamma = \sqrt{(R + j\omega L)(G + j\omega C)}$

which can be represented as $\gamma = \alpha + j\beta$

where α is the attenuation constant whose unit is Nepere/km. 1 Nepere = 8.686 dB and β is the phase constant whose unit is radians/km.

π radians = 180°

1 radian = 180° /π

$$\alpha + j\beta = \sqrt{RG + j(R\omega C + G\omega L) - \omega^2 LC}$$

$$\alpha^2 - \beta^2 = RG - \omega^2 LC; \quad 2\alpha\beta = R\omega C + G\omega L$$

$$|\gamma| = \sqrt{\alpha^2 + \beta^2} = \left[\left(R^2 + \omega^2 L^2\right)\left(G^2 + \omega^2 C^2\right)\right]^{1/4}$$

$$\alpha^2 + \beta^2 = \sqrt{\left(R^2 + \omega^2 L^2\right)\left(G^2 + \omega^2 C^2\right)}$$

$$\alpha^2 + \beta^2 + \alpha^2 - \beta^2 = 2\alpha^2 = \sqrt{\left(R^2 + \omega^2 L^2\right)\left(G^2 + \omega^2 C^2\right)} + RG - \omega^2 LC$$

$$\alpha = \left[\frac{1}{2}\left[\sqrt{\left(R^2 + \omega^2 L^2\right)\left(G^2 + \omega^2 C^2\right)} + RG - \omega^2 LC\right]\right]^{1/2} \quad \text{.....(5.7.1)}$$

$$\alpha^2 + \beta^2 - \alpha^2 + \beta^2 = 2\beta^2 = \sqrt{\left(R^2 + \omega^2 L^2\right)\left(G^2 + \omega^2 C^2\right)} - RG + \omega^2 LC$$

$$\beta = \left[\frac{1}{2}\left[\sqrt{\left(R^2 + \omega^2 L^2\right)\left(G^2 + \omega^2 C^2\right)} - RG + \omega^2 LC\right]\right]^{1/2} \quad \text{......(5.7.2)}$$

5.8 Wave Length, Phase and Group Velocities

5.8.1 Wave Length

Along with the primary and secondary constants transmission lines have wave length, velocity of propagation (phase velocity or wave velocity) and group velocity.

By knowing the phase constant β and frequency we can find out the above three parameters. The wave length is defined as the distance by which wave travels in order to have phase shift of 2π.

$$\therefore \quad \beta\lambda = 2\pi$$

$$\lambda = \frac{2\pi}{\beta}$$

If we use dielectric material with the dielectric constant (K),

then the wave length $$\lambda = \frac{\lambda_f}{\sqrt{k}} \qquad .....(5.8.1)$$

Where $$\lambda_f = \frac{2\pi}{\beta} \qquad ......(5.8.2)$$

5.8.2 Velocity of Propagation or Phase Velocity

Velocity of propagation is defined as the velocity with which a signal travels or propagates along the line. The velocity of propagation is determined based on change in the phase shift i.e., β. Hence the name phase velocity.

The phase shift 2π represents one cycle in time 't' that occurs in a wave length of λ.

$\therefore \lambda = V_p t$ where V_p = phase velocity.

t = time period of one full cycle

$$\lambda = \frac{V_p}{f}$$

We know $$\lambda = \frac{2\pi}{\beta}$$

$$\therefore \quad \frac{2\pi}{\beta} = \frac{V_p}{f}$$

$$\Rightarrow \quad V_p = \frac{2\pi f}{\beta} = \frac{\omega}{\beta}$$

$$\text{Phase velocity} = \frac{\omega}{\beta} \qquad .....(5.8.3)$$

5.8.3 Group Velocity

Phase velocity is defined when the signal contains a single frequency. If a transmitted signal contains more than one frequency like in modulated signal, then it is difficult to define phase velocity. In such cases we define group velocity which is defined as the velocity with which envelope of complex signal propagates along the line.

If two angular frequencies ω_1 and ω_2 with slight difference between them, are being transmitted along the line whose phase constants are β_1 and β_2 respectively. Then the

group velocity $$V_g = \frac{\omega_2 - \omega_1}{\beta_2 - \beta_1} = \frac{d\omega}{d\beta} \quad .....(5.8.4)$$

Let us derive the relation between V_p & V_g

We know $$V_p = \frac{\omega}{\beta}$$

Differentiate w.r.t. 'ω'

$$\frac{dV_p}{d\omega} = \frac{\beta(1) - \omega\left(\frac{d\beta}{d\omega}\right)}{\beta^2}$$

$$\frac{dV_p}{d\omega} = \frac{1}{\beta} - \frac{\omega}{\beta^2}\left(\frac{d\beta}{d\omega}\right) \qquad \left(\because V_p = \frac{\omega}{\beta}\right)$$

$$\frac{dV_p}{d\omega} = \frac{1}{\beta} - \frac{V_p}{\beta}\left(\frac{d\beta}{d\omega}\right)$$

$$\beta\frac{dV_p}{d\omega} = 1 - \frac{V_p}{V_g}$$

$$\frac{V_p}{V_g} = 1 - \beta\frac{dV_p}{d\omega} \Rightarrow V_g = \frac{V_p}{1 - \beta\frac{dV_p}{d\omega}}$$

If the signal being transmitted, has single frequency then $\frac{dV_p}{d\omega} = 0$.

∴ Group velocity = phase velocity

Problem 5.2

The characteristic impedance of a certain line is 510 ∠–16° and the frequency is 1 kHz. At this frequency the attenuation is 0.01 Nepere/km and the phase function is 0.035 radians/km. Calculate the resistance, conductance, inductance and the capacitance per km and velocity of propagation.

Solution

Given $Z_0 = 510 \angle -16^\circ$; f = 1 kHz

$\alpha = 0.01$ Nepere /km $\beta = 0.035$ radians/km

$$\therefore \gamma = \alpha + j\beta$$

$$= 0.01 + j\,0.035$$

$$= 0.0364 \angle 74.055^\circ$$

We have $$\frac{\gamma}{Z_o} = G + j\omega C$$

$$\frac{0.0364\angle 74.055^o}{510\angle -16^o} = G + j\omega C$$

$$\Rightarrow \quad 7.137 \times 10^{-5} \angle 90.055^\circ = G + j\omega C$$

$$-6.851 \times 10^{-8} + j7.137 \times 10^{-5} = G + j\omega C$$

Equating real part Conductance $G = -6.851 \times 10^{-8}\ \mho / km$

Equating imaginary part $\omega C = 7.137 \times 10^{-5}$

$$\text{Capacitance } C = \frac{7.137 \times 10^{-5}}{2\pi \times 10^3} = 1.1359 \times 10^{-8} \text{ F/km}$$

We have $Z_o\ \gamma = R + j\omega L$,

$$510\angle -16^o \times 0.0364\angle 74.055^o = R + j\omega L$$

$$18.564\angle 58.055^o = R + j\omega L$$

$$9.822 + j15.753 = R + j\omega L$$

Equating real part resistance R = 9.822 Ω/km

Equating imaginary part $\omega L = 15.753$

$$L = \frac{15.753}{2\pi \times 10^3} = 2.507 \times 10^{-3} \text{ H/km}$$

and velocity o propagation is

$$V_p = \frac{\omega}{\beta} = 179519.58 \text{ km/h}$$

Problem 5.3

An open wire Telephone line has R = $10\,J/\text{km}$, L = 0.0035 H/km, C = 0.0053 × 10^{-6} F/km and G = 0.4 × $10^{-6}\,\mho/\text{km}$. Determine Z_o, α and β at 1000Hz.

Solution

We have series impedance $Z = R + j\omega L$

$$Z = 10 + j2\pi \times 1000 \times 0.0035 = 10 + j21.99$$

$$= 24.157\angle 65.54^o$$

and shunt admittance $Y = G + j\omega C$

$$Y = 0.4 \times 10^{-6} + j2\pi \times 1000 \times 0.0053 \times 10^{-6} = (0.4 + j33.3) \times 10^{-6}$$

$$= 3.33 \times 10^{-5} \angle 89.31^o$$

Propagation constant $\gamma = \sqrt{ZY}$

$$\gamma = \sqrt{8.044 \times 10^{-4} \angle 154.85^o} = \sqrt{8.044 \times 10^{-4}}\left(e^{j154.85^o}\right)^{\frac{1}{2}} = 0.02836e^{j77.425^o}$$

$$= 0.02836\angle 77.425^o$$

$$\gamma = 6.174 \times 10^{-3} + j0.0277 = \alpha + j\beta$$

$\therefore \alpha = 6.174 \times 10^{-3}$ Nepere/km and β = 0.0277 radians/km

The characteristic impedance $Z_0 = \sqrt{\dfrac{Z}{Y}}$

$$Z_0 = \sqrt{\frac{24.157\angle 65.54^o}{3.33 \times 10^{-5} \angle 89.31^o}} = \sqrt{725435.43\angle -23.77^o}$$

$$Z_0 = \sqrt{725435.43}\left(e^{-j23.77^o}\right)^{\frac{1}{2}} = 851.725e^{-j11.89^o} = 851.725\angle -11.89^o$$

$$Z_0 = 833.45 - j175.48\ \Omega/\text{km}$$

***Problem 5.4**

At 5 MHz the characteristic impedance of transmission line is (40 – j2) ohm and the propagation constant is (0.01 + j0.15) per meter. Find the primary constants.

Solution

Given f = 5 MHz, Z_0 = 40 – j^2 ohm and γ = 0.01 + j0.15 per meter

$$Z_0 = 40.05\angle -2.86^\circ \quad \text{and} \quad \gamma = 0.15\angle 86.19^o$$

We have $Z_0\ \gamma = R + j\omega L$ and $\dfrac{\gamma}{Z_0} = G + j\omega C$

$$R + j\omega L = Z_0\gamma = 40.05\angle -2.86^\circ \times 0.15\angle 86.19^\circ = 6\angle 83.33^\circ$$

$$R + j\omega L = 0.697 + j5.96$$

Equating real and imaginary parts, the primary constants are

$$R = 0.697 \text{ ohm/m and } \omega L = 5.96 \Rightarrow L = \frac{5.96}{2\pi \times 5 \times 10^6} = 0.1897\ \mu\ H/m$$

$$G + j\omega C = \frac{\gamma}{Z_o} = \frac{0.15\angle 86.19^o}{40.05\angle -2.86^o} = 3.745 \times 10^{-3}\angle 89.05^o$$

$$G + j\omega C = \frac{\gamma}{Z_o} = \frac{0.15\angle 86.19^o}{40.05\angle -2.86^o} = 3.745 \times 10^{-3}\angle 89.05^o$$

$$G + j\omega C = 6.21 \times 10^{-5} + j3.75 \times 10^{-3}$$

Equating real and imaginary parts, the primary constants are

$$G = 6.21 \times 10^{-5}\ ℧\text{/m} \quad \text{and}$$

$$\omega C = 3.75 \times 10^{-3} \Rightarrow C = \frac{3.75 \times 10^{-3}}{2\pi \times 5 \times 10^6} = 119.37p\ \text{F/m}$$

***Problem 5.5**

A telephone wire 20 m long has the following constants per loop km resistance 90 ohm, capacitance 0.062 μF, inductance 0.001 H and leakage = 1.5×10^{-6} mhos. The line is terminated in its characteristic impedance and a potential difference of 2.1 V having a frequency of 1000 Hz is applied at the sending end. Calculate:

(a) The characteristic impedance (b) wavelength and (c) The velocity of propagation

Solution

Given R = 90 ohm/km, C = 0.062 μF/km, L = 0.001 H/km, G = 1.5×10^{-6} mhos/km, V = 2.1 V and f = 1000 Hz.

$$\text{Series impedance } Z = R + j\omega L = 90 + j2\pi \times 1000 \times 0.001 = 90 + j6.253$$
$$= 90.219\angle 3.99^\circ\,\Omega$$

$$\text{And shunt admittance } Y = G + j\omega C = 1.5 \times 10^{-6} + j2\pi \times 1000 \times 0.062$$
$$= (1.5 + j359.555) \times 10^{-6} = 389.56 \times 10^{-6}\angle 89.78^\circ\ ℧$$

(a) The characteristic impedance

$$Z_0 = \sqrt{\frac{Z}{Y}} = \sqrt{\frac{90.219\angle 3.99^\circ}{389.56\times 10^{-6}\angle 89.78^\circ}}$$

$$Z_0 = \sqrt{231592.0526\angle -85.79^\circ} = \sqrt{231592.0526}\left(e^{-j85.79^\circ}\right)^{1/2}$$

$$Z_0 = 481.24e^{(-j85.79^\circ/2)} = 481.24\angle -42.895^\circ\,\Omega$$

(b) Wavelength,

We know propagation constant $\gamma = \alpha + j\beta = \sqrt{ZY}$

$$\alpha + j\beta = \sqrt{ZY} = \sqrt{(90.219\angle 3.99^\circ)(389.56\times 10^{-6}\angle 89.78^\circ)}$$

$$\alpha + j\beta = \sqrt{35145.71\angle 93.77^\circ} = \sqrt{35145.71}\left(e^{j93.77^\circ}\right)^{1/2} = 187.47e^{j(93.77^\circ/2)}$$

$$\alpha + j\beta = 187.47\angle 46.885^\circ = 128.128 + j136.849$$

$\therefore \beta = 136.849$ radians

i.e. $\beta = 0.137$ radians / km

Hence the wavelength $\lambda = \dfrac{2\pi}{\beta} = \dfrac{2\pi}{0.137} = 45.86$ km

(c) Velocity of propagation

$$v_p = \frac{\omega}{\beta} = \frac{2\pi\times 1000}{0.137} = 45.86\times 10^3 \text{ km / sec}$$

5.9 Line Distortion

Generally a modulated signal is transmitted through the transmission line which contains many frequency components. The line distortions are classified in to two types (i) frequency distortion (ii) delayed distortion.

Frequency Distortion

If all the frequency components are not attenuated by same value then frequency distortion exists i.e., one frequency component is attenuated with some value and other frequency component is attenuated with some other value. Frequency distortion will not occur, if α is not a function of 'ω'.

Delayed Distortion

If all the frequency components are not delayed by same value then delay distortion exists i.e., one frequency component is delayed by some value and other frequency component is delayed by some other value. Delayed distortion will not occur, if the phase velocity is not a function of 'ω' or phase constant β is a function of ω and ω is multiplied by constant value.

Distortion less line

We know the α and β values from equations (5.7.1) and (5.7.2)

$$\alpha = \sqrt{\frac{1}{2}\left[\left(RG - \omega^2 LC\right) + \sqrt{\left(R^2 + \omega^2 L^2\right)\left(G^2 + \omega^2 C^2\right)}\right]}$$

$$\beta = \sqrt{\frac{1}{2}\left[\left(\omega^2 LC - RG\right) + \sqrt{\left(R^2 + \omega^2 L^2\right)\left(G^2 + \omega^2 C^2\right)}\right]}$$

By observing equation for α, as 'α' is a function of 'ω', we get frequency distortion with that value of α.

Similarly from equation for β, as ω is not multiplied with constant, we get delayed distortion.

To avoid these two problems let us find out α and β values

We know $\gamma = \sqrt{(R + j\omega L)(G + j\omega C)}$

$$= \sqrt{LC}\sqrt{\left(\frac{R}{L} + j\omega\right)\left(\frac{G}{C} + j\omega\right)} \quad .....(5.9.1)$$

By making $\frac{R}{L} = \frac{G}{C}$, we get

$$\gamma = \sqrt{LC}\left(\frac{R}{L} + j\omega\right) \text{ or } \sqrt{LC}\left(\frac{G}{C} + j\omega\right)$$

$$R\sqrt{\frac{C}{L}} + j\omega\sqrt{LC} \text{ or } G\sqrt{\frac{L}{C}} + j\omega\sqrt{LC}$$

$$\therefore \alpha = R\sqrt{\frac{C}{L}} \quad \text{or} \quad \alpha = G\sqrt{\frac{L}{C}} \quad .....(5.9.2)$$

Which is independent of frequency. Thus frequency distortion can be avoided.

Similarly $\beta = \omega\sqrt{LC}$

Here ω is multiplied with constants. If we substitute this in phase velocity equation, the V_p is independent of ω which avoids delayed distortion.

Characteristics impedance $Z_0 = \sqrt{\frac{(R+j\omega L)}{(G+j\omega C)}} = \sqrt{\frac{L}{C}\frac{\left(\frac{R}{L}+j\omega\right)}{\left(\frac{G}{C}+j\omega\right)}}$

If we make $\frac{R}{L} = \frac{G}{C}$

$$Z_o = \sqrt{\frac{L}{C}} \text{ for distortion less line} \qquad .....(5.9.3)$$

Value of 'L' for Minimum Attenuation

Value of L can be obtained for minimum attenuation by differentiating α w.r.t to 'L' and keeping R, G, C and ω as constants and equating it to 0.

$$\frac{d\alpha}{dL} = \frac{1}{2}\frac{\frac{1}{2}\left\{\frac{1}{2}\frac{2\omega^2 L\left(G^2+\omega^2C^2\right)}{\sqrt{\left(R^2+\omega^2L^2\right)\left(G^2+\omega^2C^2\right)}} - \omega^2 C\right\}}{\sqrt{\frac{1}{2}\left[\left(RC-\omega^2LC\right)+\sqrt{\left(R^2+\omega^2L^2\right)\left(G^2+\omega^2C^2\right)}\right]}} = 0$$

$$\therefore \qquad \frac{\omega^2 L\left(G^2+\omega^2C^2\right)}{\sqrt{\left(R^2+\omega^2L^2\right)\left(G^2+\omega^2C^2\right)}} - \omega^2 C = 0$$

$$\Rightarrow \qquad \omega^2 L\sqrt{\frac{G^2+\omega^2C^2}{R^2+\omega^2L^2}} = \omega^2 C$$

$$\Rightarrow \qquad L^2\left(G^2+\omega^2C^2\right) = C^2\left(R^2+\omega^2L^2\right)$$

Squaring on both sides

$$L^2\left(G^2+\omega^2C^2\right) = C^2\left(R^2+\omega^2L^2\right)$$

$$L^2G^2 + \omega^2C^2L^2 = C^2R^2 + \omega^2C^2L^2$$

$$L^2G^2 = C^2R^2$$

$$\Rightarrow \qquad \frac{R}{L} = \frac{G}{C}$$

$$\Rightarrow \qquad L = \frac{RC}{G} \text{ H/km} \qquad \text{....(5.9.4)}$$

In practice the value 'L' is less than the desired value. Hence by increasing the value of 'L' distortion can be reduced further.

Value of C for Minimum Attenuation

By differentiating α w.r.t. C and keeping R, L, G, ω as constants and equating it to 0, we get value of C for minimum attenuation.

$$\frac{d\alpha}{dC} = \frac{1}{2} \frac{\frac{1}{2}\left\{\frac{1}{2}\frac{2\omega^2 C\left(R^2+\omega^2 L^2\right)}{\sqrt{\left(R^2+\omega^2 L^2\right)+\sqrt{\left(G^2+\omega^2 C^2\right)}}} - \omega^2 L\right\}}{\sqrt{\frac{1}{2}\left[\left(RG-\omega^2 LC\right)+\sqrt{\left(R^2+\omega^2 L^2\right)\left(G^2+\omega^2 C^2\right)}\right]}} = 0$$

$$\therefore \qquad \frac{\omega^2 \text{c}(\text{R}^2+\omega^2 L^2)}{\sqrt{\left(R^2+\omega^2 L^2\right)\left(G^2+\omega^2 C^2\right)}} - \omega^2 L = 0$$

$$\Rightarrow \qquad \omega^2 C\sqrt{\frac{R^2+\omega^2 L^2}{G^2+\omega^2 C^2}}$$

$$L\sqrt{G^2+\omega^2 C^2} = C\sqrt{R^2+\omega^2 L^2}$$

Squaring on both sides

$$L^2\left(G^2+\omega^2 C^2\right) = C^2\left(R^2+\omega^2 L^2\right)$$

$$L^2 G^2 + \omega^2 C^2 L^2 = C^2 R^2 + \omega^2 C^2 L^2$$

$$L^2 G^2 = C^2 R^2$$

$$\Rightarrow \qquad \frac{R}{L} = \frac{G}{C}$$

$$C = \frac{LG}{R} \text{ F/km} \qquad \text{.....(5.9.5)}$$

In practice the value 'C' is greater than the desired value. Hence by decreasing the value of 'C' distortion can be reduced further.

To decrease capacitance value the spacing between lines must be increased which is not advisable. Therefore distortion can be reduced by increasing the inductance value.

5.10 Loading

Loading is reducing the distortion by adding inductance in series with the line.

There are different types of loading

(i) Continuous loading
(ii) Patch loading
(iii) Lumped loading

With continuous loading the entire transmission line is wounded with magnetic coil. In patch loading only some parts of transmission line are wounded with magnetic coil. In lumped loading some parts of transmission line are added with toroidal coils.

5.11 Open and Short Circuited Transmission Lines

5.11.1 Open Circuited Transmission Line

In a finite transmission line terminated with characteristic impedance Z_0 at the receiving end, there will not be any reflecting waves, because it acts as an infinite transmission line. But in a finite transmission line that is either open circuited or short circuited at the receiving end, there will be reflecting waves.

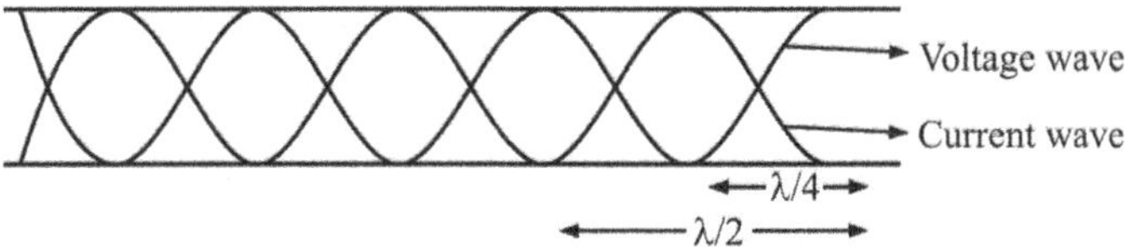

Fig. 5.4 Waves for O.C. transmission line

For open circuited (O.C.) transmission line as it is opened at the receiving end current is zero and voltage is maximum as shown in the Fig. 5.4. This is a lossless medium.

The points at which the voltage is maximum are called antinodes or maxima points. Similarly the points at which the voltage is minimum are called nodes. The nodes and antinodes occur for a half wave lengths. The first node occurs at a quarter wave length from the receiving end.

5.11.2 Short Circuited Transmission line

For a Short circuited (S.C.) transmission line since the receiving end is shorted the voltage is zero and current is maximum as shown in Fig. 5.5.

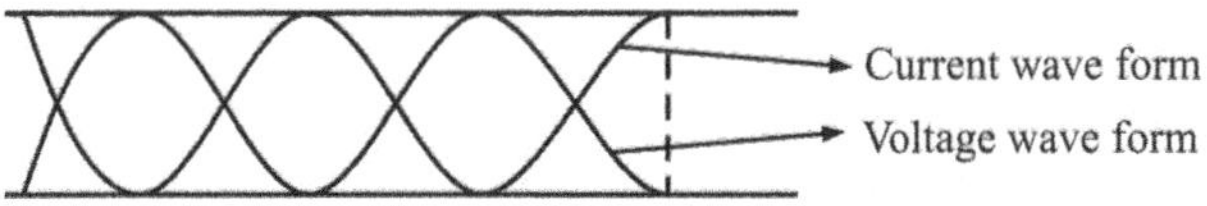

Fig. 5.5 Waves for S.C. transmission line

Z_{OC} and Z_{SC}:

Z_{OC} is the impedance seen from sending end in a O.C. transmission line as shown in Fig. 5.6.

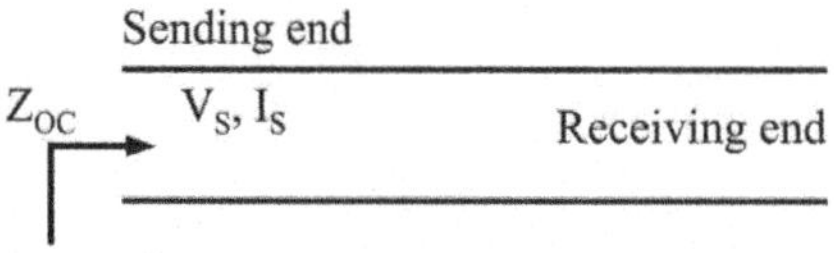

Fig. 5.6 O.C. transmission line

$$Z_{OC} = \frac{V_s}{I_s}$$

Z_{SC} is impedance seen from the sending end in a S.C. transmission line as shown in Fig. 5.7.

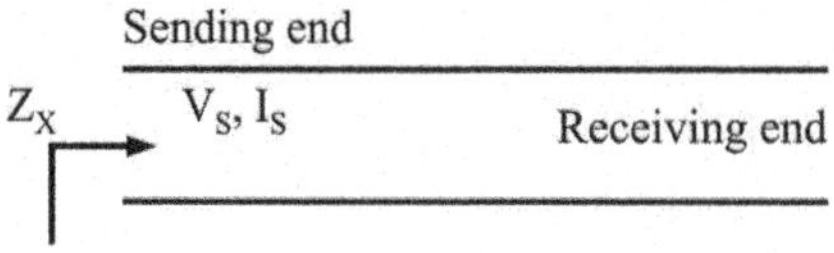

Fig. 5.7 S.C. transmission line

We know the general voltage and Current equations for finite length transmission line

$$V = V_s \cosh(\gamma x) - I_s Z_0 \sinh(\gamma x)$$

$$I = I_s \cosh(\gamma x) - \frac{V_s}{Z_0} \sinh(\gamma x)$$

Assume V_R and I_R as the voltage and currents at the receiving end and 'l' as the length of transmission line.

at $x = l$, $V = V_R$ and $I = I_R$

$$V_R = V_s \cosh(\gamma l) - I_s Z_0 \sinh(\gamma l)$$

$$I_R = I_s \cosh(\gamma l) - \frac{V_s}{Z_0} \sinh(\gamma l)$$

For O.C. transmission line $I_R = 0$

$$\therefore I_s \cosh(\gamma l) = \frac{V_s}{Z_0} \sinh(\gamma l)$$

$$Z_{OC} = \frac{V_s}{I_s} = Z_o \coth(\gamma l) \qquad .....(5.11.1)$$

Similarly for S.C. transmission line $V_R = 0$

$$\therefore\ V_s \cosh(\gamma l) = I_s Z_o \sinh(\gamma l)$$

$$Z_{SC} = \frac{V_s}{I_s} = Z_o \tanh(\gamma l)$$

$Z_{OC} \cdot Z_{SC} = Z_o^2$

$$\frac{Z_{SC}}{Z_{OC}} = \frac{\tanh(\gamma l)}{\coth(\gamma l)} = \frac{\sinh(\gamma l)}{\cosh(\gamma l)} \times \frac{\sinh(\gamma l)}{\cosh(\gamma l)} = \tanh^2(\gamma l)$$

$$\tanh(\gamma l) = \sqrt{\frac{Z_{SC}}{Z_{OC}}}$$

Problem 5.6

O.C. and S.C. impedances of a transmission line at 1.6 kHz are $900\angle{-30^\circ}\ \Omega$ and $400\ \angle{-10^\circ}\ \Omega$ respectively. Calculate it's Z_0.

Solution

$Z_{OC}\, Z_{SC} = Z_0^2$

$Z_0^2 = 300 \times 10^3\ \angle{-40^\circ}$

$Z_0 = 600\ \angle{-20^\circ}\ \Omega$

5.12 Input Impedance

Consider a finite transmission line which is terminated with general load Z_R as shown in Fig.5.8.

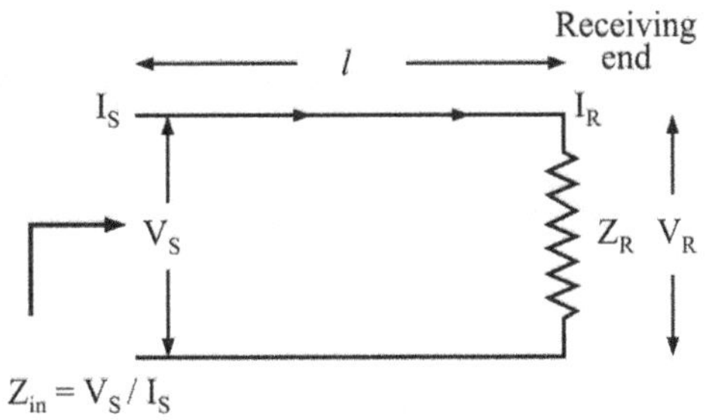

Fig. 5.8 Finite transmission line which is terminated with Z_R

Assume V_R and I_R as the voltage and current at the receiving end. Similarly V_s and I_s as the voltage and current at the sending end. We know the general transmission line equations for voltage and current in terms of V_R and I_R as

$$V = V_R \cosh \gamma(l-x) + I_R Z_0 \sinh \gamma(l-x)$$

and

$$I = \frac{V_R}{Z_o} \sinh \gamma(l-x) + I_R \cosh \gamma(l-x)$$

The input impedance seen from the sending end is $Z_{in} = \frac{V_s}{I_s}$

At x = 0, i.e., sending end V = V_s and I = I_s

$$V_s = V_R \cosh(\gamma l) + I_R Z_o \sinh(\gamma l)$$

$$I_s = \frac{V_R}{Z_o} \sinh(\gamma l) + I_R \cosh(\gamma l)$$

$$Z_{in} = \frac{V_s}{I_s} = \frac{V_R \cosh(\gamma l) + I_R Z_0 \sinh(\gamma l)}{\frac{V_R}{Z_0} \sinh(\gamma l) + I_R \cosh(\gamma l)}$$

Multiplying both numerator and denominator by $\frac{Z_0}{I_R}$

$$Z_{in} = \frac{V_s}{I_s} = \frac{Z_0\left(\frac{V_R}{I_R} \cosh(\gamma l) + Z_0 \sinh(\gamma l)\right)}{\left(\frac{V_R}{I_R} \sinh(\gamma l) + Z_0 \cosh(\gamma l)\right)}$$

$$= \frac{Z_0\left(Z_R \cosh(\gamma l) + Z_0 \sinh(\gamma l)\right)}{\left(Z_R \sinh(\gamma l) + Z_0 \cosh(\gamma l)\right)}$$

Divide both numerator & denominator with $\cosh(\gamma l)$

$$Z_{in} = Z_0 \frac{Z_R + Z_0 \tanh(\gamma l)}{Z_R \tanh(\gamma l) + Z_0} \qquad \text{.....(5.12.1)}$$

with $Z_R = Z_0$

$Z_{in} = Z_0$

With $Z_R = \infty$ for O.C.

$$Z_{in} = Z_0 \frac{1 + \frac{Z_0}{Z_R} \tanh \gamma l}{\tanh \gamma l + \frac{Z_0}{Z_R}}$$

$$= \frac{Z_0}{\tanh \gamma l} = Z_0 \cot \ \text{h} \gamma l$$

for S.C. $Z_R = 0$

$$Z_{in} = Z_0 \frac{Z_0}{Z_0} \tanh \gamma l = Z_0 \tanh \gamma l$$

Get input impedance equation in terms of exponentials

$$Z_{in} = Z_0 \frac{Z_R + Z_0 \frac{e^{\gamma l} - e^{-\gamma l}}{e^{\gamma l} + e^{-\gamma l}}}{Z_R \frac{e^{\gamma l} - e^{-\gamma l}}{e^{\gamma l} + e^{-\gamma l}} + Z_0}$$

$$= Z_0 \frac{Z_R\left(e^{\gamma l} + e^{-\gamma l}\right) + Z_0\left(e^{\gamma l} - e^{-\gamma l}\right)}{Z_R\left(e^{\gamma l} - e^{-\gamma l}\right) + Z_0\left(e^{\gamma l} + e^{-\gamma l}\right)}$$

$$= Z_0 \frac{e^{\gamma l}(Z_R + Z_0) + e^{-\gamma l}(Z_R - Z_0)}{e^{\gamma l}(Z_R + Z_0) - e^{-\gamma l}(Z_R - Z_0)} \quad \text{.....(5.12.2)}$$

5.12.1 Input Impedance for Lossless Transmission Lines

For lossless transmission line as $\alpha = 0$, γ becomes $j\beta$, $\beta = \omega\sqrt{LC}$

and $$Z_0 = \sqrt{\frac{L}{C}}$$

$\therefore$ input impedance

$$Z_{in} = \sqrt{\frac{L}{C}} \frac{Z_R + \sqrt{\frac{L}{C}} \tanh j\omega\sqrt{LC}l}{\sqrt{\frac{L}{C}} + Z_R \tanh j\omega\sqrt{LC}l}$$

Since $\tanh j\omega\sqrt{LC}l = j\tan\omega\sqrt{LC}l$

$$\therefore \quad Z_{in} = \sqrt{\frac{L}{C}} \frac{Z_R + j\sqrt{\frac{L}{C}}\tan\omega\sqrt{LC}l}{\sqrt{\frac{L}{C}} + jZ_R \tan\omega\sqrt{LC}l} \quad \text{.....(5.12.3)}$$

The variations of input impedance of lossless line when shorted and open circuited are shown in Fig.5.9.

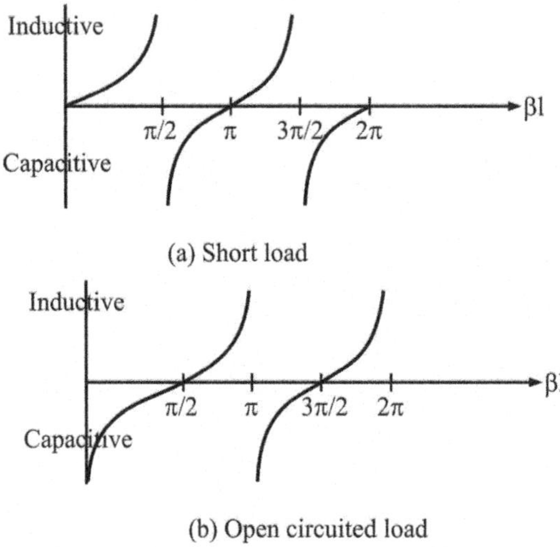

Fig. 5.9 Input impedance variation

5.13 Reflection Co-efficient

When a Transmission line is not terminated with it's characteristic impedance there will be reflected wave. We know the general equations for Transmission line

$$V = be^{-\gamma x} + ae^{\gamma x}$$

$$I = \frac{1}{Z_0}\left[be^{-\gamma x} - a\,e^{\gamma x}\right]$$

In the above two equations, the first term represents the incident wave which travels from sending end to receiving end. Second term represents reflected wave which travels from receiving end to sending end.

Reflection co-efficient is the ratio of magnitudes of the reflected wave to the incident wave. For the voltage wave this ratio is +Ve and for the current wave this ratio is –Ve i.e., there will be a phase difference of 180° in case of current. Reflection co-efficient is denoted with k.

Replace x with '–y' where 'y' is the distance from receiving end

$\therefore$ General equations for V and I becomes

$$V = be^{\gamma y} + ae^{-\gamma y}$$

$$I = \frac{1}{Z_0}\left[be^{\gamma y} - ae^{-\gamma y}\right]$$

Here also first term represents the incident wave and second term represents the reflected wave.

at y = 0 ,V = V_R and I = I_R

$\therefore$ $V_R = b + a$

$$I_R = \frac{1}{Z_0}(b-a)$$

$$\frac{1}{Z_0}V_R + I_R = \frac{2}{Z_0}b$$

$$\Rightarrow \qquad b = \frac{1}{2}V_R + \frac{Z_0}{2}I_R = \frac{1}{2}(V_R + Z_0 I_R)$$

Similarly

$$a = \frac{1}{2}(V_R - Z_o I_R)$$

Now the reflection coefficient (k) is equal to

$$\frac{\text{The second term of V}}{\text{The first term of V}} = \frac{a e^{-\gamma y}}{b e^{\gamma y}}$$

at y = 0

$$k = \frac{a}{b} = \frac{V_R - Z_0 I_R}{V_R + Z_0 I_R}$$

$$k = \frac{\dfrac{V_R}{I_R} - Z_0}{\dfrac{V_R}{I_R} + Z_0} = \frac{Z_R - Z_0}{Z_R + Z_0} \qquad \text{.....(5.13.1)}$$

Where Z_R is load impedance

5.13.1 Input Impedance in-terms of Reflection Coefficient

We have

$$Z_{in} = Z_0 \frac{e^{\gamma l}(Z_R + Z_0) + e^{-\gamma l}(Z_R - Z_0)}{e^{\gamma l}(Z_0 + Z_R) - e^{-\gamma l}(Z_R - Z_0)}$$

Divide both numerator and denominator with

$(Z_R + Z_0)e^{\gamma l}$

$$Z_{in} = z_0 \frac{1 + \left(\frac{Z_R - Z_0}{Z_R + Z_0}\right) e^{-2\gamma l}}{1 - \left(\frac{Z_R - Z_0}{Z_R + Z_0}\right) e^{-2\gamma l}}$$

$$Z_{in} = Z_0 \frac{1 + ke^{-2\gamma l}}{1 - ke^{-2\gamma l}} \qquad \text{.....(5.13.2)}$$

5.14 Standing Wave Ratio

When the transmission line is not terminated with it's characteristic impedance, we know that there will be reflected wave. The incident and reflected waves form the standing wave as shown in the Fig.5.10.

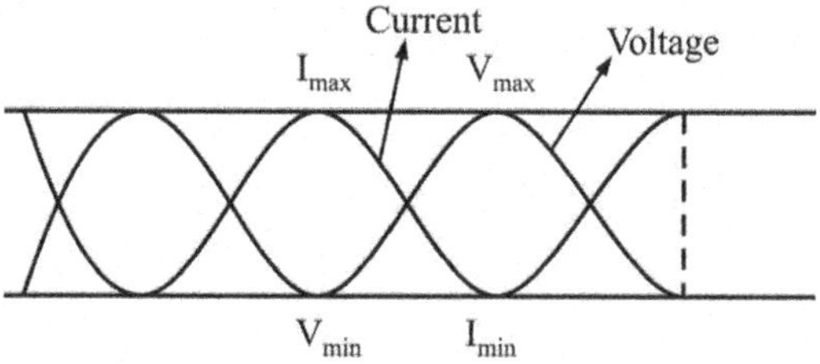

Fig. 5.10 Evaluation of SWR

Maxima points of either voltage or current are obtained due to the addition of incident and reflected waves. Similarly, the minima points of either voltage or current are obtained due to the subtraction of incident and reflected waves.

The drawback of standing wave is motion does not take place in the transmission line.

For a standing wave.

$I_{max} = |I_i| + |I_R|$

$V_{max} = |V_i| + |V_R|$

$V_{min} = |V_i| - |V_R|$

$I_{min} = |I_i| - |I_R|$

Standing wave ratio is defined as ratio of maximum and minimum of either voltage or current in a line having standing wave.

The voltage standing wave ratio $VSWR = \frac{V_{max}}{V_{min}}$.

Similarly current standing wave ratio $CSWR = \frac{I_{max}}{I_{min}}$

$$VSWR = \frac{|V_i| + |V_R|}{|V_i| - |V_R|} = \frac{\frac{|V_i|}{|V_R|} + 1}{\frac{|V_i|}{|V_R|} - 1} = \frac{\left|\frac{1}{k}\right| + 1}{\left|\frac{1}{k}\right| - 1} = \frac{1 + |k|}{1 - |k|} \quad \text{.....(5.14.1)}$$

where $\frac{|V_R|}{|V_i|} = k$

$$CSWR = \frac{|I_i| + |I_R|}{|I_i| - |I_R|} = \frac{1 + \frac{I_R}{I_i}}{1 - \frac{I_R}{I_i}} = \frac{1 - k}{1 + k} \quad \text{.....(5.14.2)}$$

where $k = \frac{-I_R}{I_i}$

***Problem 5.7**

A low transmission line of 100 ohm characteristic impedance is connected to a load of 400 ohm. Calculate the reflection coefficient and standing wave ratio. Derive the Relationships used.

Solution

Given $Z_0 = 100$ ohm and $Z_R = 400$ ohm

We have Reflection coefficient $k = \frac{Z_R - Z_0}{Z_R + Z_0}$

$$k = \frac{400 - 100}{400 + 100} = 0.6$$

Also we have voltage standing wave ratio $VSWR = \dfrac{1+|k|}{1-|k|}$

$$VSWR = \frac{1+0.6}{1-0.6} = 4$$

Derivations are as in section 5.14.

***Problem 5.8**

Explain the significance of V_{max} and V_{min} positions along the transmission line, for a complex load Z_R. Hence calculate the impedances at these positions.

Solution

Explanation is as in the section "Standing Wave Ratio".

At a voltage maximum or current minimum,

$$Z_{in} = Z_{max} = \frac{V_{max}}{I_{min}} = \frac{V_{max}}{V_{min}}\frac{V_{min}}{I_{min}} = VSWR \times Z_0$$

At a voltage minimum or current maximum,

$$Z_{in} = Z_{min} = \frac{V_{min}}{I_{max}} = \frac{V_{min}}{V_{max}}\frac{V_{max}}{I_{max}} = Z_0 \,/\, VSWR$$

5.15 Equivalent Circuits or Networks

Since the transmission line is a four terminal network. The part of transmission line can be replaced with either 'T' or 'π' networks.

Consider a transmission line as shown in the Fig.5.11.

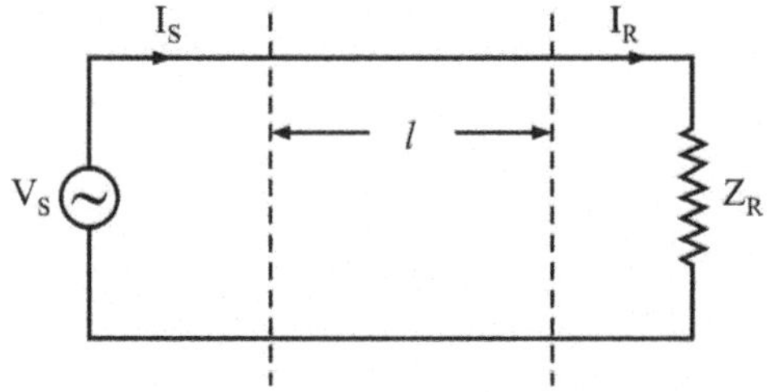

Fig. 5.11 A part of transmission line

A part of transmission line of length '*l*' can be replaced with 'T' network as shown in the Fig. 5.12.

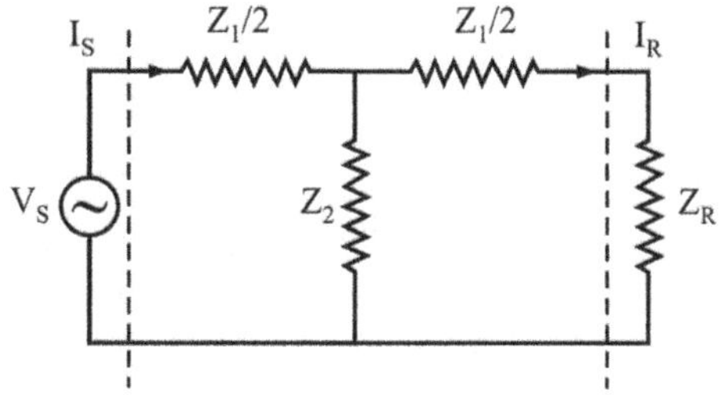

Fig. 5.12 T- Network

As the transmission line contains the secondary constants Z_0 and γ. Let us represent Z_1 and Z_2 in terms of secondary constants.

We know the general transmission line equation for current –I

$$I = I_s \cosh \gamma x - \frac{V_s}{Z_o} \sinh \gamma x$$

At $x = l$, $I = I_R$

$$I_R = I_s \cosh \gamma l - \frac{V_s}{Z_o} \sinh \gamma l \qquad \text{.....(5.15.1)}$$

By applying KVL across the input side network of Fig. 5.13

$$V_S = \frac{Z_1}{2} I_s + (I_s - I_R) Z_2$$

$$\Rightarrow \qquad I_R = I_s \left(\frac{Z_1}{2Z_2} + 1 \right) - \frac{V_s}{Z_2} \qquad \text{.....(5.15.2)}$$

Comparing (5.15.1) and (5.15.2)

$$\frac{1}{Z_2} = \frac{1}{Z_0} \sinh \gamma l$$

$$\Rightarrow \qquad Z_2 = \frac{Z_0}{\sinh \gamma l} \qquad \text{.....(5.15.3)}$$

$$\frac{Z_1}{2Z_2} + 1 = \cosh \gamma l$$

$$\frac{Z_1}{2Z_2} = -1 + \cosh \gamma l$$

$$\frac{Z_1}{2} = \frac{Z_0}{\sinh \gamma l}(\cosh \gamma l - 1)$$

$$= \frac{Z_0\left[1 + 2\sinh^2\left(\frac{\gamma l}{2}\right) - 1\right]}{2\sinh\left(\frac{\gamma l}{2}\right)\cosh\left(\frac{\gamma l}{2}\right)}$$

$$\frac{Z_1}{2} = Z_0 \tanh\left(\frac{\gamma l}{2}\right) \qquad \text{.....(5.15.4)}$$

Transmission line of length 'l' can be replaced with 'T' network as shown in Fig.5.13.

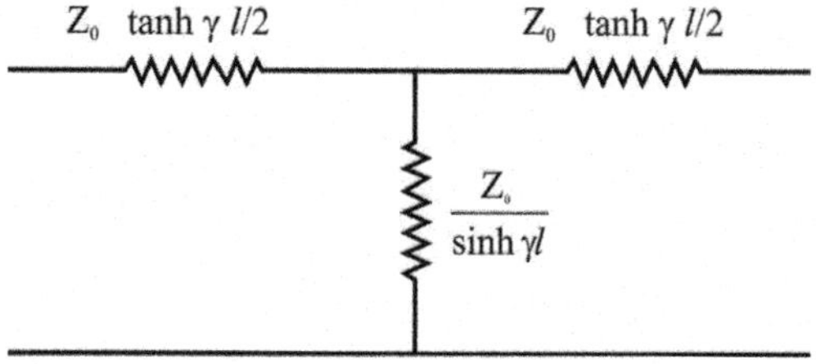

Fig. 5.13 T- Network with impedance values

For π network as shown in Fig.5.14, we can obtain Z_A, Z_B and Z_C as

$$Z_A = \frac{2Z_0 \tanh\left(\frac{\gamma l}{2}\right)\frac{Z_0}{\sinh \gamma l} + Z_0^2 \tanh^2\left(\frac{\gamma l}{2}\right)}{Z_0\, tanh\left(\frac{\gamma l}{2}\right)}$$

$$= \frac{\frac{2}{2} Z_0^{\;2} \operatorname{sech}^2\left(\frac{\gamma l}{2}\right) + Z_0^2 \tanh^2\left(\frac{\gamma l}{2}\right)}{Z_0 \tanh\left(\frac{\gamma l}{2}\right)}$$

$$Z_A = Z_C = \frac{Z_0^{\;2}}{Z_0 \tanh\left(\frac{\gamma l}{2}\right)} = Z_0 \coth\left(\frac{\gamma l}{2}\right) \qquad \text{.....(5.15.5)}$$

Similarly $Z_B = Z_0 \sinh \gamma l$

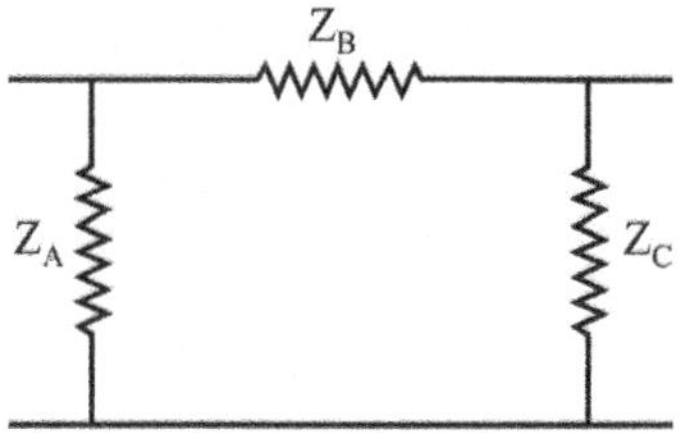

Fig. 5.14 π- Network

Conclusion: Any transmission line of length 'l' can be replaced with either 'T' network as shown in Fig. 5.13 or 'π' network as shown in Fig. 5.14.

Problem 5.9

Transmission line 10 km long is terminated properly at the far end at a frequency 1000 Hz. The attenuating and phase constants of the line are respectively 0.03 Nepere/km and 0.03 radians/km. If the far end voltage at 1000 Hz is $4\angle 0\ ^{o}$V. Calculate sending end voltage of the line.

Solution

If transmission line is terminated with Z_R then

$$V = V_R \cosh \gamma (l - x) + I_R Z_0 \sinh \gamma (l - x)$$

At $x = 0, V = V_s$

$$V_s = V_R \cosh \gamma l + I_R Z_0 \sinh \gamma l$$

Terminated properly $\therefore Z_R = Z_0$

$$I_R = \frac{V_R}{Z_R} = \frac{V_R}{Z_0}$$

$$V_s = V_R \cosh \gamma l + V_R \sinh \gamma l$$

$$= V_R (\cosh \gamma l + \sinh \gamma l)$$

$$V_s = V_R e^{\gamma l}$$

$$V_R = 4\angle 0^o$$

$$l = 10 \text{ km}$$

$$\gamma l = (\alpha + j\beta)l = (0.03 + j\,0.03)\ 10 \text{ km} = 424.26\ \angle 45^o$$

$$V_s = 4\ e^{424.26\angle 45^o}\ \text{V}$$

5.16 The Smith Chart

Engineers developed all sorts of aids such as tables, charts, graphs, etc. prior to the advent of computers and calculators to facilitate their calculations for design and analysis. A graphical means i.e., the Smith Chart has been developed to reduce the tedious manipulations involved in calculating the characteristics of transmission lines.

In this resistive component R and reactive component X are represented in circular form. The resistive components alone form the set of circles and reactive components alone form the set of circles. By superposing these two set of circles, we get the Smith chart. Using Smith chart, we can calculate input impedance, reflection coefficient, SWR etc., without using formulae.

We know the reflection coefficient

$$K = \frac{Z_R - Z_0}{Z_R + Z_0}$$

Divide numerator and denominator with Z_0

$$\therefore \quad K = \frac{\frac{Z_R}{Z_0} - 1}{\frac{Z_R}{Z_0} + 1}$$

Let $Z_r = \frac{Z_R}{Z_0}$ where Z_r is the normalized load impedance

$$\therefore \quad K = \frac{Z_r - 1}{Z_r + 1}$$

$$\therefore \quad Z_r(K-1) = -(K+1)$$

$$Z_r = \frac{1+K}{1-K}$$

Since Z_r and K are complex let us represent Z_r with $R + jX$ and K with $K_r + j K_x$

$$\therefore \quad R + jX = \frac{(K_r + jK_x) + 1}{1 - (K_r + jK_x)} = \frac{(1+K_r) + jK_x}{(1-K_r) - jK_x}$$

rationalize

$$R + jX = \frac{\left[(1+K_r) + jK_x\right]\left[(1-K_r) + jK_x\right]}{\left[(1-K_r) - jK_x\right]\left[(1-K_r) + jK_x\right]}$$

$$= \frac{1-K_r^2-K_x^2+2jK_x}{(1-K_r)^2+(K_x)^2}$$

Equating real and imaginary parts

$$R = \frac{1-K_r^2-K_x^2}{(1-K_r)^2+(K_x)^2} \; and \quad X = \frac{2K_x}{(1-K_r)^2+(K_x)^2}$$

consider

$$R = \frac{1-K_r^2-K_x^2}{(1-K_r)^2+(K_x)^2}$$

$$\Rightarrow \quad (1-K_r)^2 R+(K_x)^2 R = 1-K_r^2-K_x^2$$

$$\Rightarrow \quad (1+K_r^2-2K_r)R+(K_x)^2 R-1+K_r^2+K_x^2 = 0$$

$$\Rightarrow \quad K_r^2(1+R)+K_x^2(1+R)-2K_rR = 1-R$$

$$\Rightarrow \quad K_r^2+K_x^2-2K_r\frac{R}{1+R} = \frac{1-R}{(1+R)}$$

Add $\frac{R^2}{(1+R)^2}$ on both sides

$$\Rightarrow \quad K_r^2+K_x^2-2K_r\frac{R}{1+R}+\frac{R^2}{(1+R)^2} = \frac{1-R}{(1+R)}+\frac{R^2}{(1+R)^2}$$

$$\Rightarrow \quad \left(K_r-\frac{R}{1+R}\right)^2+K_x^2 = \frac{1}{(1+R)^2} \qquad \text{....(5.16.1)}$$

which is equation for circle with radius $\frac{1}{1+R}$ and centre $\left(\frac{R}{1+R}, 0\right)$

For R = 0, the radius is ‘1’ and center is (0, 0). The circle is as shown in the Fig. 5.15.

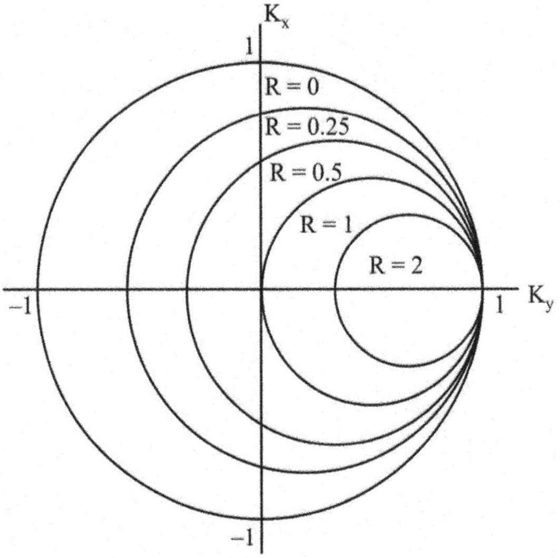

Fig. 5.15 R-circles

Similarly for R = 1, the radius is 0.5 centre is (0.5, 0) the circle is as shown in the Fig. 5.15. By observing the Fig.5.15, we can say that for any value of R the circles touch at (1, 0).

Now take
$$X = \frac{2K_x}{(1-K_r)^2 + (K_x)^2}$$

$$\Rightarrow \quad \left(1+K_r^2 - 2K_r\right)X + (K_x)^2 X = 2K_x$$

$$\Rightarrow \quad K_r^2 X + K_x^2 X + X - 2K_x - 2K_r X = 0$$

Divide with X

$$\Rightarrow \quad K_r^2 - 2K_r + 1 + K_x^2 - 2\frac{K_x}{X} = 0$$

$$\Rightarrow \quad (K_r - 1)^2 + K_x^2 - \frac{2K_x}{X} = 0$$

Add $\frac{1}{X^2}$ on both sides

$$\Rightarrow \quad (K_r - 1)^2 + \left(K_x - \frac{1}{X}\right)^2 = \frac{1}{X^2}$$

which also form a circle with radius $\frac{1}{X}$ and center $(1, \frac{1}{X})$ For X = ∞, the radius of the circle is ‘0’ and centre of the circle is (1, 0), The circle is as shown in Fig. 5.16.

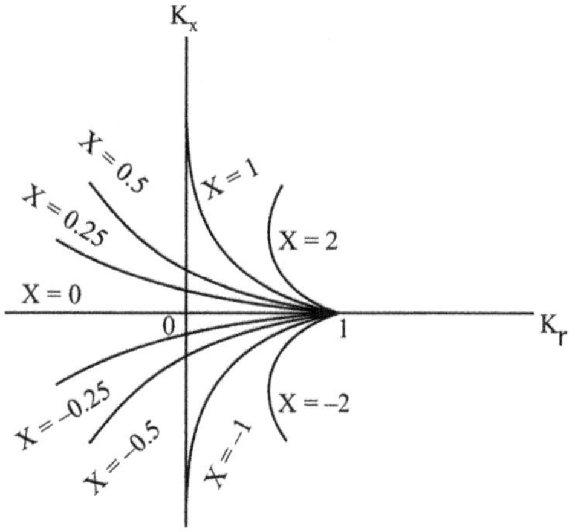

Fig. 5.16 X-circles

For X = 0, the radius of the circle is ∞ and center is (1, ∞) which is a straight line i.e., K_r line. By superposing Fig. 5.15 and 5.16, we get the smith chart as sown in Fig. 5.17.

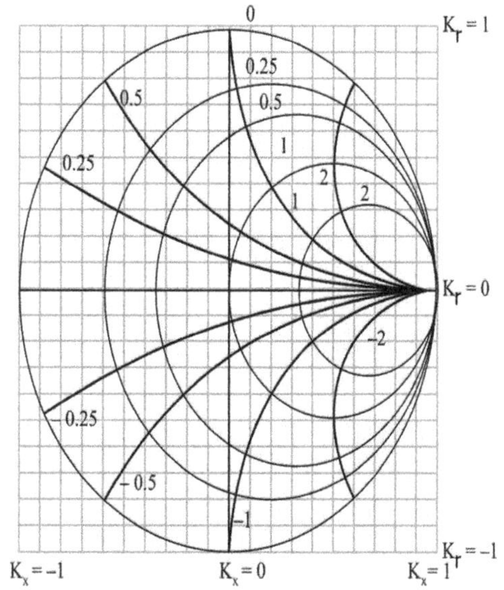

Fig. 5.17 R and X-circles

The complete Smith chart is shown in Fig. 5.18.

The following points are to be noted about the Smith chart

(a) At point P_{sc} on the chart R = 0, X = 0; that is $Z_R = 0 + j0$ showing that P_{sc} represents a short circuit on the transmission line. At point P_{oc}, R = ∞ and X = ∞;

that is $Z_R = \infty + j\infty$, which implies that P_{oc} corresponds to an open circuit on the line.

(b) Moving along the circle in clockwise direction indicates that moving in a transmission line from load to generator side (sending side). Moving in anti clockwise direction indicates in transmission line moving from generator to the load side. The outer circle in smith chart represents the distance in terms of wave length from the load. A complete revolution (360°) around the Smith chart represents a distance of $\lambda/2$ on the line. Therefore λ distance on the line corresponds to a 520° movement on the chart. i.e., $\lambda \rightarrow 720^0$.

(c) V_{max} occurs where $Z_{in,max}$ is located on the chart and that is on the positive K_r axis or on OP_{oc}. V_{min} is located at the same point where we have $Z_{in,min}$ on the chart; that is, on the negative K_r axis or on OP_{sc}. The V_{max} and V_{min} are $\lambda/4$(150°) apart.

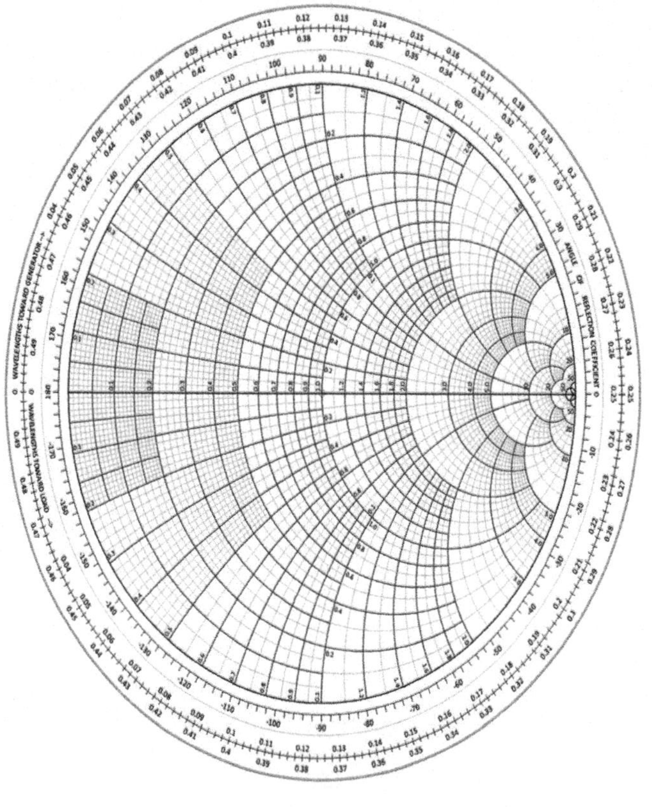

Fig. 5.18 The complete Smith chart

Problem 5.10

A 30 m long lossless transmission line with $Z_0 = 50\ \Omega$ operating at 2 MHz is terminated with a load $Z_R = 60 + j\,40\ \Omega$. If $v_p = 0.6v_0$ on the line. Find

(a) The reflection coefficient 'K'

(b) The SWR 'S'

(c) The input impedance

Solution

Method: 1 (without smith chart)

(a) Reflection coefficient $K = \dfrac{Z_R - Z_0}{Z_R + Z_0}$

$$= 0.3523\ \angle 56^{\circ}$$

(b) $\text{SWR} = S = \dfrac{1+|K|}{1-|K|} = \dfrac{1+0.3523}{1-0.3523} = 2.08$

(c) Z_{in} for lossless transmission line is

$$Z_{in} = Z_0 \frac{Z_R + jZ_0 \tan\beta l}{Z_0 + jZ_R \tan\beta l}$$

βl = Electrical length of transmission line

We have

$$v_p = \frac{\omega}{\beta}$$

$$\beta = \frac{\omega}{v_p} = \frac{2\pi f}{v_p}$$

$$\beta l = \frac{2\pi f l}{v_p} = \frac{2\pi \times 2\times 10^6 \times 30}{0.6\times 3\times 10^8} = 2.0944 \times \frac{180^o}{\pi} = 120^o$$

$$Z_{in} = 50\frac{60 + j40 + j50(-1.732)}{50 + j(60 + j40)(-1.732)}$$

$$= 50\frac{73.14\angle 34.9^o}{48.59\angle 26^o} = 75.26\angle 8.9^o$$

$$= 23.95 + j1.35\ \Omega$$

Method: 2 (with smith chart)

(a) Calculate the normalized load impedance $Z_r = \frac{Z_R}{Z_0} = 1.2 + j0.8$

Locate Z_r on the Smith chart of Fig.5.19 at point A where the R = 1.2 circle and the X = 0.8 circle meet. To get K at Z_r, extend OA to meet the R = 0 circle at B and measure OA and OB. Since OB corresponds to |K| = 1, then at A,

$|K| = \frac{OA}{OB} = \frac{3.5\text{ cm}}{10\text{ cm}} = 0.350$, here OA and OB may vary depending on the size of the Smith chart used, but the ratio remains same.

Angle θ_K is read directly on the chart as the angle between OC and OA;

i.e. $\theta_K = 56^\circ$

$\therefore \quad K = 0.350\angle 56^\circ$

(b) To obtain the standing wave ratio S, draw a circle with radius OA and center at O.

This is the constant S or |K| circle. Locate point C where the S-circle meets the K_r axis. The value of R on the Smith chart is S,

i.e., S = R = 2.1

(c) To obtain Z_{in}, first express l in terms of λ or in degrees.

$$\lambda = \frac{V_p}{f} = \frac{0.6 \times 3 \times 10^8}{2 \times 10^6} = 90\text{ m}$$

$$l = 30\text{ m} = \frac{30}{90}\lambda = \frac{\lambda}{3}$$

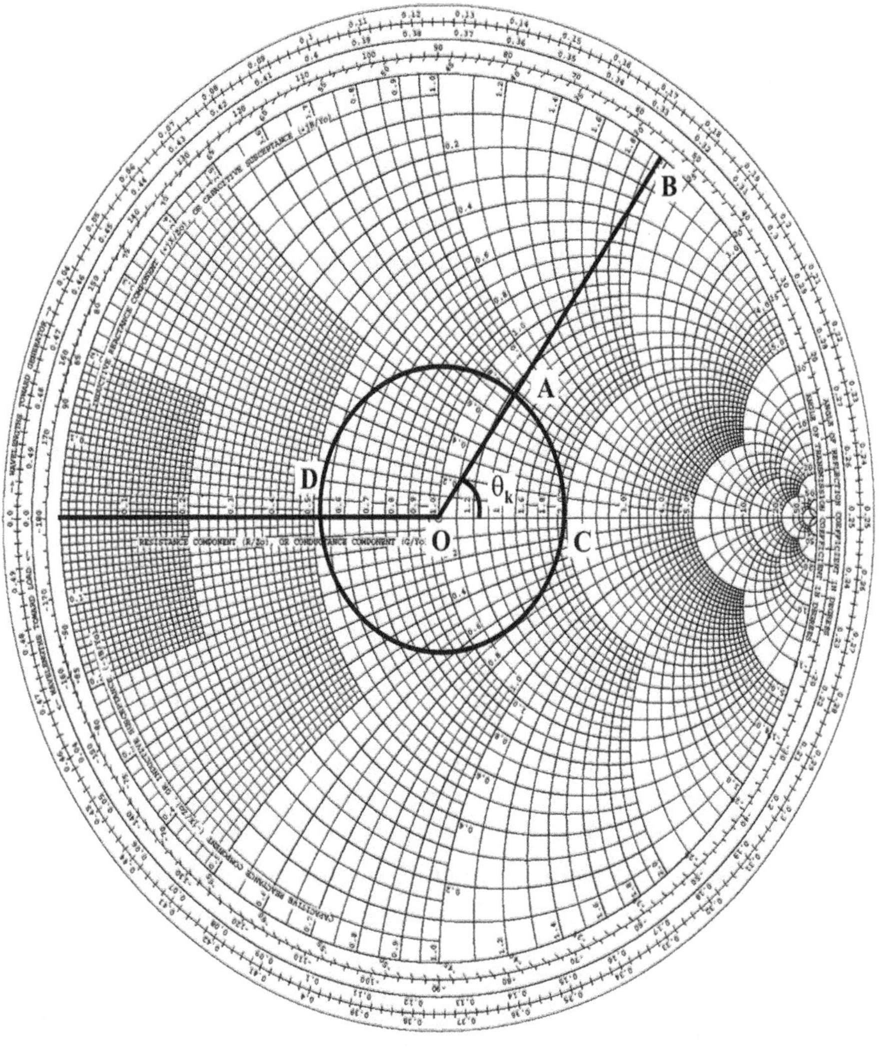

Fig. 5.19

Since $\lambda \rightarrow 720^{\circ}$; $\frac{\lambda}{3} \rightarrow \frac{720^{\circ}}{3} = 240^{\circ}$

To get input impedance, we need to move towards the generator i.e., move in the clockwise direction by 240° on the S circle from point A to point D. At D, we get normalized $z_{in} = 0.47 + j0.035$

$\therefore$ actual $Z_{in} = Z_0 z_{in} = 50(0.47 + j0.035) = 23.5 + j1.75\Omega$

5.17 Applications of Transmission Line

Transmission lines are used to serve different purposes. The following are the few applications where transmission lines are used.

5.17.1 $\lambda/4, \lambda/2, \lambda/8$ Lines

When Z_R is not equal to Z_0 for main transmission line, then there will be reflections due to mismatch. The matching is achieved by using shorted sections of transmission lines.

For the transmission line of length $l = \frac{\lambda}{4}$

$$\beta l = \frac{2\pi}{\lambda}\frac{\lambda}{4} = \frac{\pi}{2}$$

$$Z_{in} = Z_0 \frac{Z_R \cos\beta l + jZ_0 \sin\beta l}{Z_0 \cos\beta l + jZ_R \sin\beta l}$$

$$Z_{in} = \frac{Z_0^2}{Z_R}$$

$$\frac{Z_{in}}{Z_0} = \frac{Z_0}{Z_R} \qquad \text{.....(5.17.1)}$$

A mismatched load Z_R can be properly matched to a line(with characteristic impedance Z_0) by inserting prior to the load a transmission $\lambda/4$ long(with characteristic impedance Z_0') as shown in Fig.5.20. The $\lambda/4$ section of the transmission line is called a quarter-wave transformer because it is used for impedance matching like an ordinary transformer. From equation (5.17.1), Z_0' is selected such that $Z_0' = \sqrt{Z_0 Z_R}$ which is obtained by replacing Z_0 with Z_0' and Z_{in} with Z_0 in equation (5.17.1).

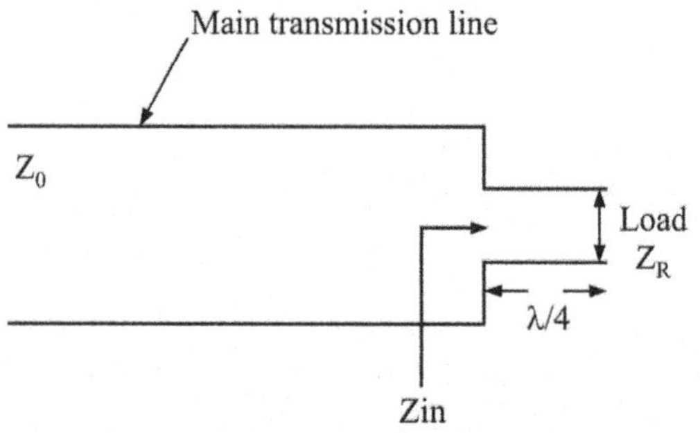

Fig. 5.20 Transmission line terminated with quarter wave transformer

By selecting a terminated transmission line of suitable length, it is possible to produce the equivalent of a pure resistance, inductance and capacitance or any desired combination of the above.

Equivalent circuits for shorted and open lines are shown in Table 5.1

Table 5.1 Equivalent circuits

Transmission line	Equivalent circuit	Input impedance
A shorted Transmission line of length $l < \lambda/4$ (ex: $\lambda/8$ line)	Inductor	$Z_{in} = -jZ_0 \tan \beta l$ For $\lambda/8$ line $Z_{in} = -jZ_0$
An open Transmission line of length $l < \lambda/4$ (ex: $\lambda/8$ line)	Capacitor	$Z_{in} = jZ_0 \cot \beta l$ For $\lambda/8$ line $Z_{in} = jZ_o$
A shorted Transmission line whose length is $\lambda/4 < l < \lambda/2$	Capacitor	$Z_{in} = -jZ_0 \tan \beta l$
An open Transmission line whose length is $\lambda/4 < l < \lambda/2$	Inductor	$Z_{in} = jZ_0 \cot \beta l$
A shorted Transmission line of length $l = \lambda/4$ (Quarter wave transformer)	Tank circuit	$Z_{in} = \dfrac{Z_0^2}{\tanh \alpha l}$
An open Transmission line whose length is $l = \lambda/2$	Tank circuit	$Z_{in} = \dfrac{Z_0^2}{\tanh \alpha l}$

5.17.2 Impedance Matching

Consider a Transmission line terminated with load impedance Z_R , which has input impedance Z_{in} and Characteristic impedance Z_o as shown in Fig.5.21. If $Z_{in} = Z_0 = Z_R$, then line is said to be matched.

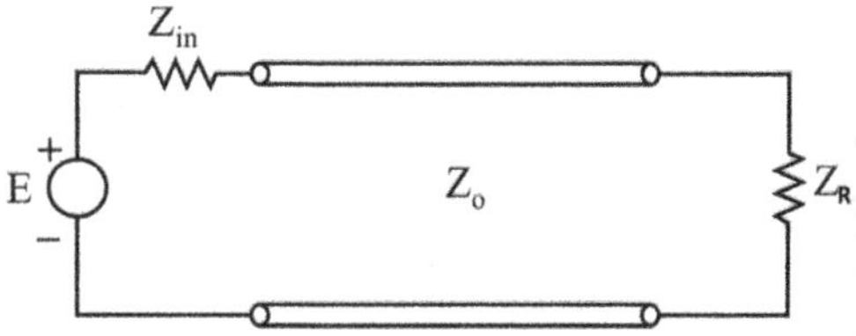

Fig. 5.21 Transmission line terminated with load Z_R

We know the reflection coefficient as

$$k = \frac{Z_R - Z_0}{Z_R + Z_0}$$

k ranges as $-1 \leq k \leq 1$ and is a complex quantity i.e. $k = |k| \angle \theta_R$

(i) If $Z_R = \alpha$ i.e., $k = 1$ (O.C)

(ii) If $Z_R = 0$ i.e., $k = -1 = 1\angle 180^0$ (S.C)

(iii) If $Z_R = Z_0$ i.e., $k = 0$

Based on the values of 'k' impedance and admittance circles are as shown in Figs.5.22 and 5.23.

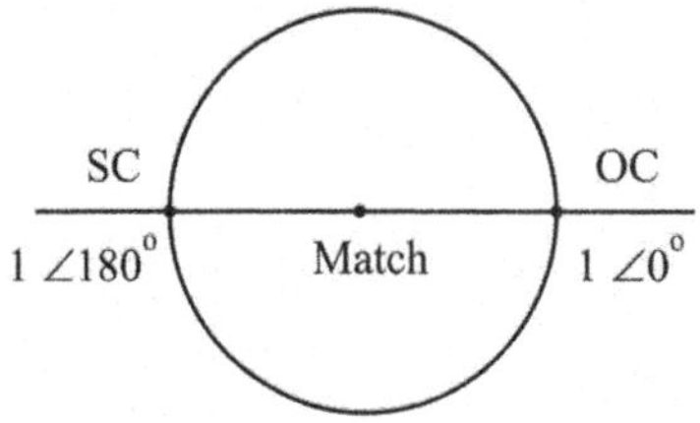

Fig. 5.22 Impedance circle

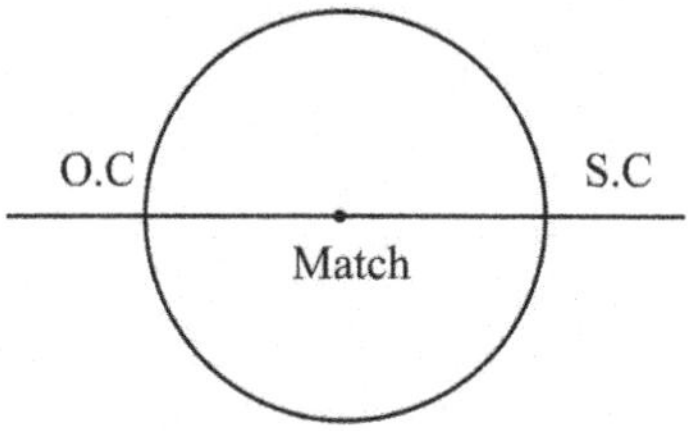

Fig. 5.23 Admittance circle

Stub is a short length transmission line. The single stub and double stub matching is used to achieve perfect impedance matching in a main transmission line.

Single Stub (shunt connected short circuited stub):

If any transmission line i.e., not perfectly matched can be made to have perfect matching transmission line by connecting single stub in shunt to the main transmission line as shown in Fig. 5.24. These stubs can be either O.C or S.C. Generally S.C. transmission lines are used for perfect load matching.

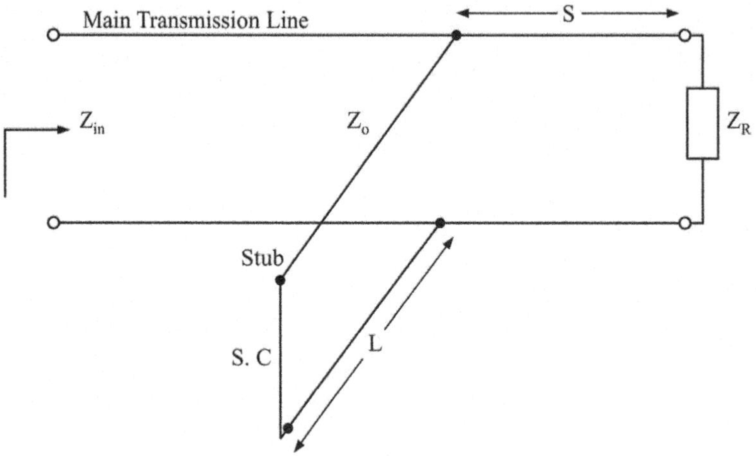

Fig. 5.24 Single stub matching

A single stub placed at a suitable point from the load side provides a susceptance which is equal but opposite in sign to the input susceptance, i.e., at the point of attachment susceptance is zero.

Thus the combination of the stub and line will present a conductance which is equal to the characteristic conductance of the line.

The stub should be placed as nearer to load as possible so that the prevailing mismatch if any is for the minimum possible length.

To get the length(L) and position(S) of stub(short-circuit), let us consider

$$Z_{in} = Z_0 \frac{Z_R + Z_0 \tanh \gamma l}{Z_R \tanh \gamma l + Z_0}$$

$$Y_{in} = \frac{1}{Z_{in}} = \frac{Z_R \tanh \gamma l + Z_o}{Z_0(Z_R + Z_0 \tanh \gamma l)} = Y_0 \frac{Y_0 \tanh \gamma l + Y_R}{Y_0 + Y_R \tanh \gamma l}$$

Let $\frac{Y_{in}}{Y_0} = Y_s$ = Normalized input admittance

$$Y_s = \frac{\frac{Y_R}{Y_0} + \tanh \gamma l}{1 + \frac{Y_R}{Y_0} \tanh \gamma l}$$

And let $\frac{Y_R}{Y_0} = Y_L$ = Normalized load admittance

For lossless Transmission line $\gamma = j\beta$

$$Y_s = \frac{Y_L + j \tan \beta l}{1 + jY_L \tan \beta l}$$

Rationalizing, we get

$$Y_s = \frac{Y_L(1 + \tan^2 \beta l) + j \tan \beta l(1 - Y_L^2)}{1 + (Y_L \tan \beta l)^2} \quad \text{.....(5.17.2)}$$

The length where the imaginary part should go to zero is where the stub has to be positioned(S). Also, normalized admittance has to be unity i.e., real part of equation (5.17.2) should be unity.

$$\frac{Y_L(1 + \tan^2 \beta S)}{1 + (Y_L \tan \beta S)^2} = 1$$

$$\Rightarrow \quad Y_L + Y_L \tan^2 \beta S = 1 + Y_L^2 \tan^2 \beta S$$

$$\Rightarrow \quad Y_L - 1 = Y_L \tan^2 \beta S(Y_L - 1)$$

$$\Rightarrow \quad \tan^2 \beta S = \frac{1}{Y_L}$$

The stub is positioned at $S = \frac{1}{\beta}\tan^{-1}\left(\frac{1}{\sqrt{Y_L}}\right)$

$$S = \frac{1}{\beta}\tan^{-1}\left(\sqrt{\frac{Y_0}{Y_R}}\right) = \frac{1}{\beta}\tan^{-1}\left(\sqrt{\frac{Z_R}{Z_0}}\right) \qquad \text{.....(5.17.3)}$$

At this point the susceptance of the transmission line is the imaginary part of equation (5.17.2)

i.e., Normalized susceptance $= \frac{j\tan\beta S(1-Y_L^2)}{1+(Y_L\tan\beta S)^2}$

we have $\tan\beta S = \sqrt{\frac{1}{Y_L}} = \sqrt{\frac{Y_0}{Y_R}}$

$$\therefore \quad \text{Normalized susceptance} = j\frac{\sqrt{\frac{Y_R}{Y_0}}\left(1-\frac{Y_R^2}{Y_0^2}\right)}{1+\frac{Y_R^2}{Y_0^2}\frac{Y_0}{Y_R}} = j\sqrt{\frac{Y_R}{Y_0}}\left(1-\frac{Y_R}{Y_0}\right)$$

$$\text{The susceptance} = j\sqrt{\frac{Y_R}{Y_0}}\left(1-\frac{Y_R}{Y_0}\right)Y_0 = j\sqrt{\frac{Y_R}{Y_0}}\left(Y_0 - Y_R\right) \qquad \text{.....(5.17.4)}$$

To have no reflections, this susceptance should be equal but opposite in sign to the susceptance of stub.

We know that the input impedance of a short circuited transmission line as $Z_{in} = jZ_0\tan\beta l$

The stub input impedance $Z_{stub} = jZ_0\tan\beta L$

Where L is length of the stub.

The susceptance of stub $= Y_{stub} = -jY_0\cot\beta L$ .....(5.17.5)

Equating equations (5.17.4) and (5.17.5) with –ve sign, we get

$$jY_0\cot\beta L = j\sqrt{\frac{Y_R}{Y_0}}\left(Y_0 - Y_R\right)$$

$$\cot\beta L = \sqrt{\frac{Y_R}{Y_0}}\left(1-\frac{Y_R}{Y_0}\right)$$

$$\Rightarrow \qquad L = \frac{1}{\beta}\cot^{-1}\left(\sqrt{\frac{Y_R}{Y_0}}\left(1-\frac{Y_R}{Y_0}\right)\right)$$

$$L = \frac{1}{\beta}\cot^{-1}\left(\sqrt{\frac{Z_0}{Z_R}}\left(1-\frac{Z_0}{Z_R}\right)\right) \qquad \text{.....(5.17.6)}$$

In terms of reflection coefficient these expressions are obtained as

We know $Z_{in} = Z_0 \dfrac{1+ke^{-2\gamma l}}{1-ke^{-2\gamma l}}$

If $\qquad k = |k|e^{j\theta}$, then

$$Z_{in} = Z_o \frac{1+|k|e^{j(\theta-2\beta l)}}{1-|k|e^{j(\theta-2\beta l)}} \quad \text{where } \gamma = j\beta \text{ for lossless line}$$

$$Y_s = \frac{Y_{in}}{Y_0} = \frac{1-|k|e^{j(\theta-2\beta l)}}{1+|k|e^{j(\theta-2\beta l)}} = \frac{1-|k|\cos\psi - j|k|\sin\psi}{1+|k|\cos\psi + j|k|\sin\psi}, \text{ Where } \psi = \theta - 2\beta l$$

$$Y_s = \frac{(1-|k|\cos\psi - j|k|\sin\psi)(1+|k|\cos\psi - j|k|\sin\psi)}{(1+|k|\cos\psi + j|k|\sin\psi)(1+|k|\cos\psi - j|k|\sin\psi)}$$

$$Y_s = \frac{1-|k|^2 - 2j|k|\sin\psi}{1+|k|^2 + 2|k|\cos\psi}$$

Equating the real part to unity

$$\frac{1-|k|^2}{1+|k|^2+2|k|\cos\psi} = 1$$

$$\Rightarrow \qquad 1-|k|^2 = 1+|k|^2 + 2|k|\cos\psi$$

$$\Rightarrow \qquad \cos\psi = -|k| \Rightarrow \psi = \cos^{-1}(-|k|)$$

$$\theta - 2\beta S = \cos^{-1}(-|k|) = \cos^{-1}(|k|) - \pi$$

Position of the stub is $S = \dfrac{\theta + \pi - \cos^{-1}(|k|)}{2\beta}$ .....(5.17.7)

The shunt susceptance should be equal to the susceptance of the stub with negative sign, the length of the stub can be obtained as

$$L = \beta \tan^{-1}\left(\frac{\sqrt{1-|k|^2}}{2|k|}\right) \qquad .....(5.12.8)$$

and in terms of VSWR

$$L = \beta \tan^{-1}\left(\frac{\sqrt{VSWR}}{VSWR-1}\right) \qquad .....(5.17.9)$$

Advantages

(a) The length and characteristic impedance need not be changed.

(b) Mechanical adjustment of the length of the stub and position of the stub for matching is possible.

Disadvantages

(a) For variable frequency applications as the frequency changes, the location of the stub has to be changed, which is not possible in single stub.

(b) For the applications (coaxial lines) in which terminating impedances are to be changed, the length and position of the stub must be changed, which is not possible in single stub.

*Problem 5.11

An aerial of (200 – j300) ohm is to be matched with 500 ohm lines. The matching is to be done by means of low loss 600 ohm stub line. Find the position and length of the stub line used if the operating wave length is 2 meters.

Solution

Given $Z_R = 200 - j300$ ohm $= 360.55 \angle -56.3^0$, $Z_0 = 500$ ohm and $\lambda = 20$ m

$$k = \frac{Z_R - Z_o}{Z_R + Z_o} = \frac{200 - j300 - 500}{200 - j300 + 500} = -0.2068 - j0.5172 = 0.557\angle -111.8^0 \;\therefore\; |k| = 0.555$$

and $\theta = -111.5^0 = -0.6211\pi$

$$\text{Position of the stub } S = \frac{\theta + \pi - \cos^{-1}(|k|)}{2\beta} = \lambda\frac{\theta + \pi - \cos^{-1}(|k|)}{4\pi}$$

$$S = 20\frac{-0.6211\pi + \pi - \cos^{-1}(0.557)}{4\pi} = 0.33465 \text{ m}$$

Length of the stub $L = \beta \tan^{-1}\left(\frac{\sqrt{1-|k|^2}}{2|k|}\right)$

$$L = \frac{2\pi}{\lambda}\tan^{-1}\left(\frac{\sqrt{1-|k|^2}}{2|k|}\right) = \frac{2\pi}{20}\tan^{-1}\left(\frac{\sqrt{1-(0.557)^2}}{2(0.557)}\right) = 0.201 \text{ m}$$

Problem 5.12

A 100 Ω transmission line is connected to a cellular phone antenna with load impedance $Z_R = 26 - j16$ Ω. Find the position and the length of a shunt short-circuit stub required to match the 100 Ω line.

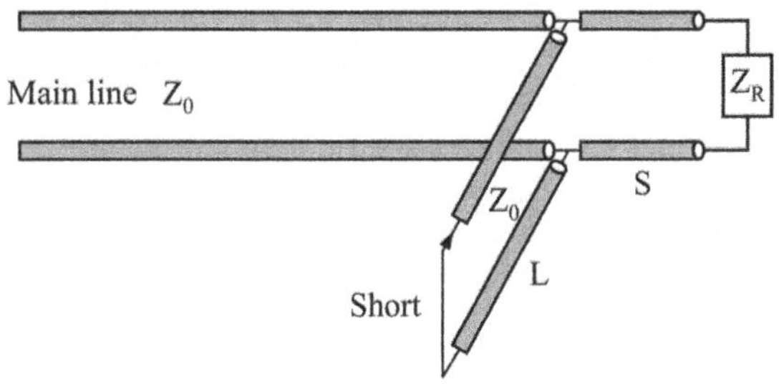

Fig. 5.25

Given $Z_R = 26 - j16$ Ω, $Z_o = 100$ Ω, find S and L for the diagram shown in Fig.5.25

- The normalized impedance $Z_R / Z_o = 0.26 - j0.16$

  Plot this normalized impedance on smith chart as shown in Fig.5.26.
- Let O be the centre of the smith chart and P be the normalized impedance point on the smith chart. Draw a circle with O as centre and OP as radius, which gives the constant VSWR circle.
- Extend the line from P to O and it cuts the constant VSWR circle at point Q as shown in smith chart. Where point Q is normalized admittance. $Y_L/Y_O = 3 + j1.9$

  (The stub, which is given is the shunt stub so prefer output admittance instead of output impedance).
- Rotate the line OQ towards generator such that the line cuts the intersection of VSWR circle and R = 1 circle. Let the intersection point is S.
- From the OQ line to OS line measure the distance on 'wavelength towards generator' scale, which will give the position 'S' of the line.

  The OQ line cuts the 'wavelength towards generator' scale at 0.223λ

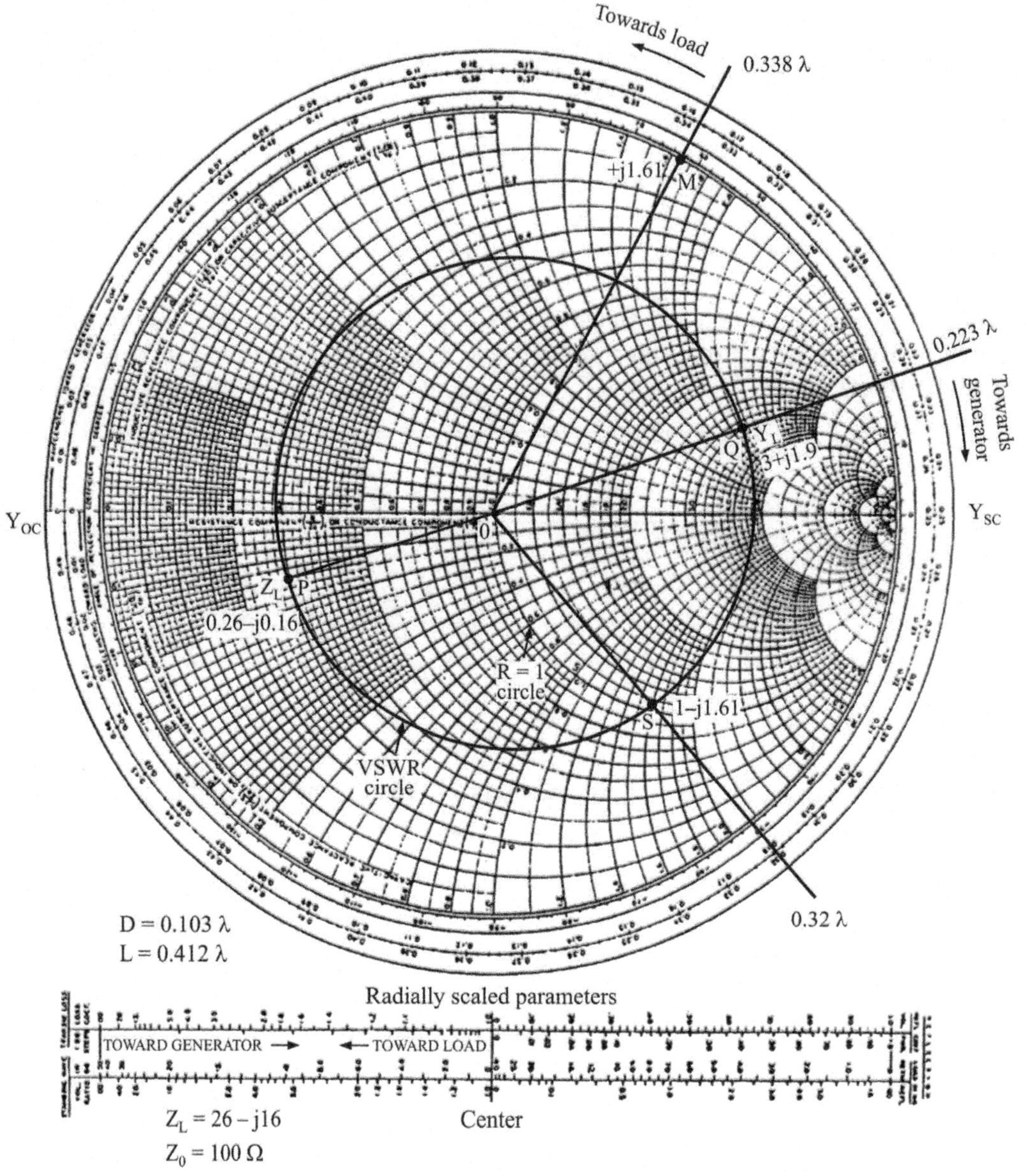

Fig. 5.26 Smith chart for Problem 5.12

The OS line cuts the 'wavelength towards generator' scale at $0.32\ \lambda$

So, position'S'= $0.32\ \lambda - 0.223\ \lambda$

$$= 0.103\ \lambda$$

- From the Fig.5.26, the admittance at point S is $1 - j\ 1.61$

- In order to make $Y_L/Y_O = 1$, we must cancel the imaginary part in the existing admittance at point S. so +j1.61 is used to cancel the imaginary part of admittance

  Draw a point (0, + j1.61) in the smith chart as shown in Fig.5.26, let this point be M.

  The line OM cuts the 'wavelength towards load' scale at 0.338 λ

  From the OM line to the OY_{sc} (as shown in Fig.5.26) measure the distance on a 'wavelength towards load' scale, which will give the length of the stub L.

$$\text{So, } L = 0.25\,\lambda + (0.5\,\lambda - 0.338\,\lambda)$$
$$= 0.412\,\lambda.$$

Problem 5.13

A 50 Ω transmission line is connected to a cellular phone antenna with load impedance Z_R =100+j100 Ω. Find the position and the length of a shunt open-circuit stub required to match the 50 Ω line

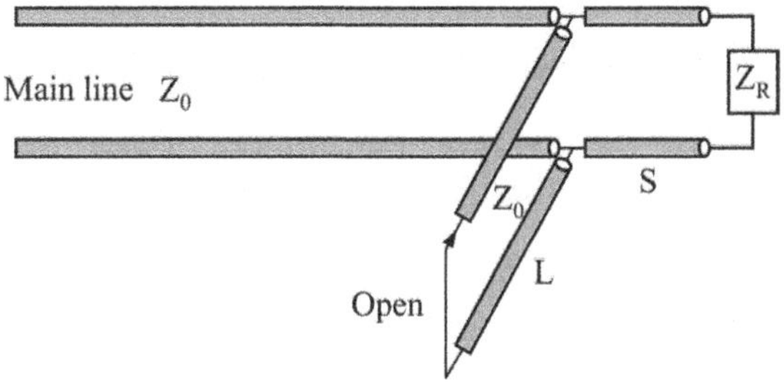

Fig. 5.27

Given $Z_R = 100 + j100\ \Omega$, $Z_o = 50\ \Omega$, find S and L for the diagram as shown in Fig.5.27

- The normalized impedance $Z_R / Z_o = 2 + j2$

  Plot this normalized impedance on smith chart as shown in Fig.5.28.

- Let O be the centre of the smith chart and P be the normalized impedance point on the smith chart. Draw a circle with O as centre and OP as radius, which gives the constant VSWR circle.

- Extend the line from P to O and it cuts the constant VSWR circle at point Q as shown in Fig.5.28. Where point Q is normalized admittance. $Y_L/Y_O = 0.28 - j0.26$

  (The stub, which is given is the shunt stub so prefer output admittance instead of output impedance).

- Rotate the line OQ towards generator such that the line cuts the intersection of VSWR circle and $R = 1$ circle. Let the intersection point is S.
- From the OQ line to OS line measure the distance on a 'wavelength towards generator' scale, which will give position 'S'

The OQ line cuts the 'wavelength towards generator' scale at 0.458λ

The OS line cuts the 'wavelength towards generator' scale at 0.177λ

$$\text{So, } S = 0.177\lambda + 0.5\lambda - 0.458\lambda$$

$$= 0.219\lambda$$

- From the Fig.5.29, the admittance at point S is $1 + j1.56$
- In order to make $Y_L/Y_O = 1$, we must cancel the imaginary part in the existing admittance at point S. so $-j1.56$ is used to cancel the imaginary part of admittance

Draw a point $(0, -j1.56)$ in the smith chart as shown in Fig.5.28, let this point be M

The line OM cuts the 'wavelength towards load' scale at 0.16λ

From the OM line to the OY_{oc} (as shown in smith chart) measure the distance on a 'wavelength towards load' scale, which gives the length L.

$$\text{So, } L = 0.5\lambda - 0.16\lambda$$

$$= 0.34\lambda.$$

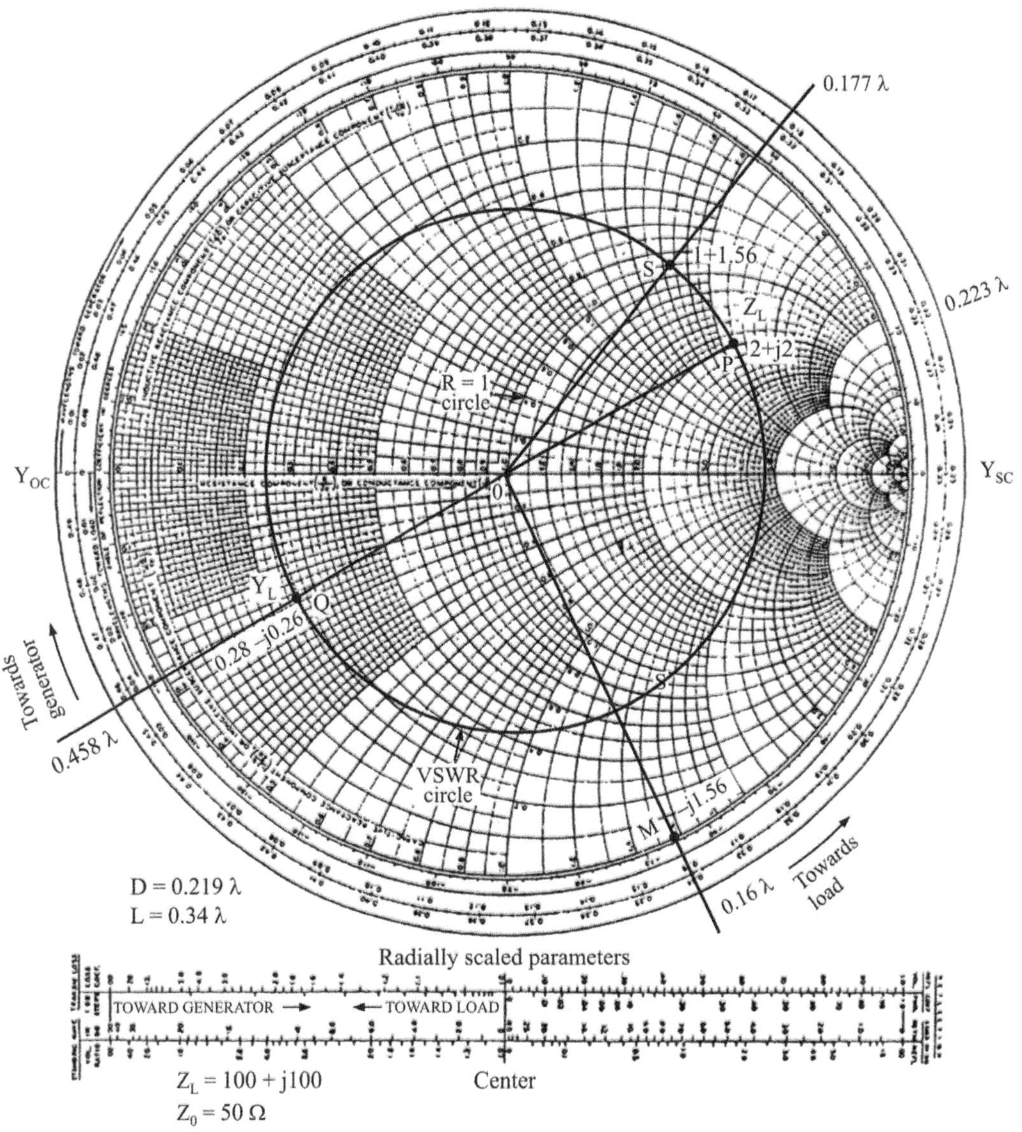

Fig. 5.28 Smith Chart for Problem 5.13

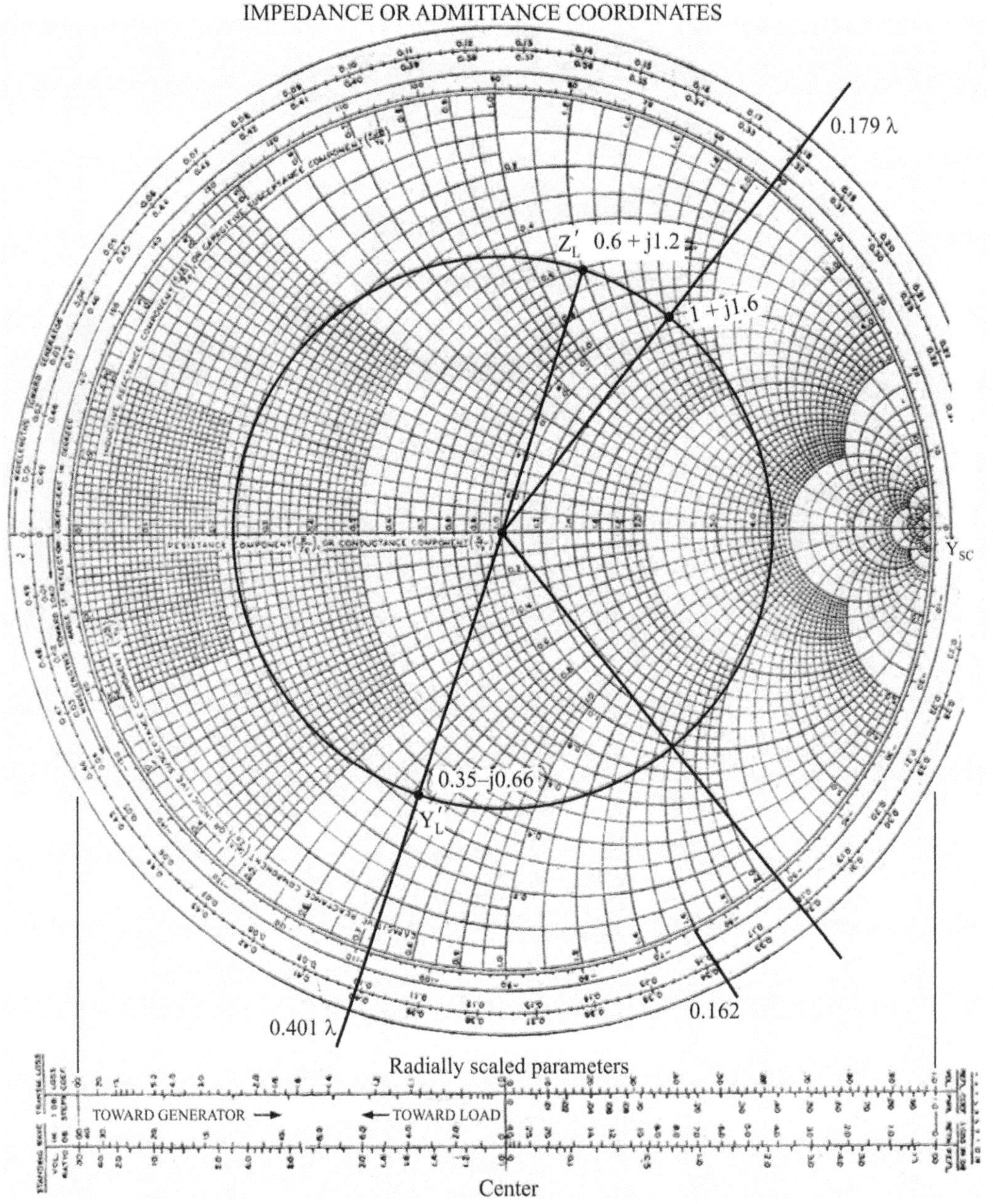

Fig. 5.29 Smith Chart for problem 5.14

Problem 5.14

Consider the shunt connected open circuited stub as shown in Fig.5.27

Given $Z_R = 30 + j\,60\ \Omega$, $Z_o = 50\ \Omega$, find L and S

Solution:

1. Normalized impedance is $Z'_L = \frac{Z_R}{Z_o} = \frac{30 + j60\ \Omega}{50} = 0.6 + j\,1.5\ \Omega$
2. From the smith chart in Fig.5.28, Normalized admittance is

   $$Y'_L = \frac{Y_L}{Y_o} = 0.35 - j\,0.66$$

3. Draw constant S-circle, which intersects r = 1 circle at 1 + j1.6 (i.e., nearer to load)
4. So, $S = 0.5\,\lambda - 0.401\lambda + 0.179\,\lambda$ TWG

   $= 0.278\,\lambda$

5. –j 1.6 is used to match the load

   So, $L = 0.162\,\lambda$

Problem 5.15

A 50 Ω transmission line is connected to a cellular phone antenna with load impedance $Z_R = 100 + j80\ \Omega$. Find the position and the length of a series open-circuit stub required to match the 50 Ω line

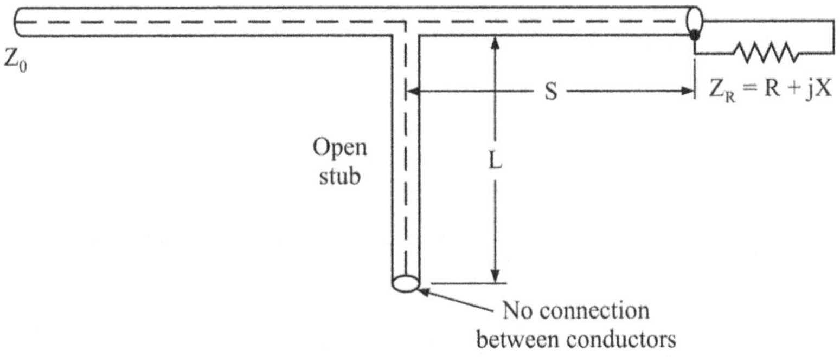

Fig. 5.30

Given Z_L =100+j80 Ω, Z_o=50 Ω, find S and L for the diagram as shown in Fig.5.30.

- The normalized impedance $Z_R / Z_o = 2 + j1.6$

  Plot this normalized impedance on smith chart as shown in Fig. 5.31.
- Let O be the centre of the smith chart and P be the normalized impedance point on the smith chart. Draw a circle with O as centre and OP as radius, which gives the constant VSWR circle.

- Rotate the line OP towards generator such that the line cuts the intersection of VSWR circle and R = 1 circle. Let the intersection point is S.

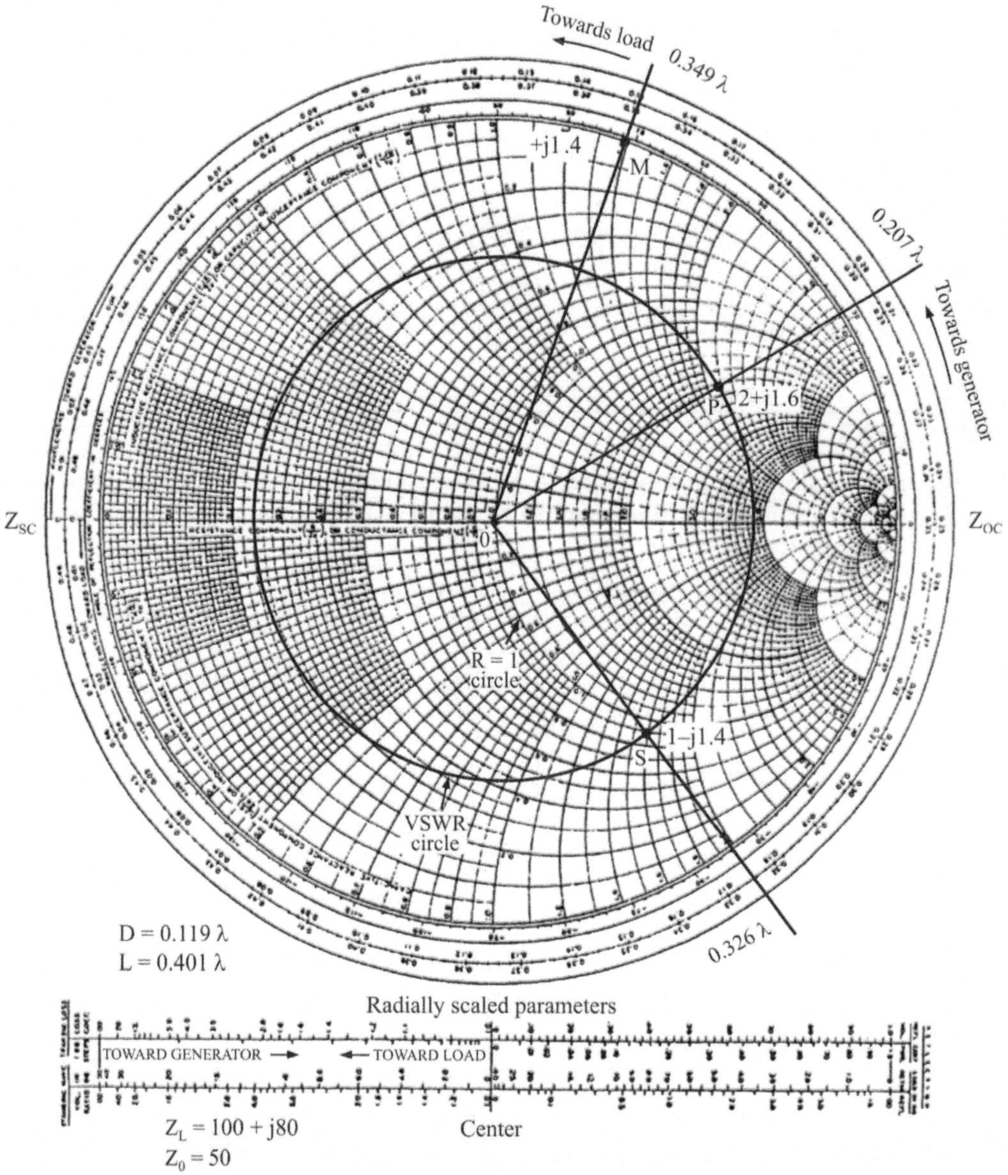

Fig. 5.31 Smith chart for problem 5.15

- From the OP line to OS line measure the distance on a 'wavelength towards generator' scale, which will give the position 'S'
  The OP line cuts the 'wavelength towards generator' scale at 0.207λ
  The OS line cuts the 'wavelength towards generator' scale at 0.326 λ
  $$\text{So, } S = 0.326\,\lambda - 0.207\,\lambda$$
  $$= 0.119\,\lambda$$
- From the smith chart shown in Fig.5.32, the impedance at point S is $1 - j1.4$
- In order to make $Z_R/Z_O = 1$, we must cancel the imaginary part in the existing impedance at point S. so +j1.4 is used to cancel the imaginary part of impedance
  Draw a point (0, + j1.4) in smith chart shown in Fig.5.32, let this point be M
  The line OM cuts the 'wavelength towards load' scale at 0.349 λ
  From the OM line to the OZ_{oc} (as shown in smith chart) measure the distance on a 'wavelength towards load' scale, which gives the L.
  $$\text{So, } L = 0.25\,\lambda + 0.5\,\lambda - 0.349\,\lambda$$
  $$= 0.401\,\lambda.$$

Problem 5.16

A 50 Ω transmission line is connected to a cellular phone antenna with load impedance Z_R =100+j200 Ω. Find the position and the length of a series short-circuit stub required to match the 50 Ω line

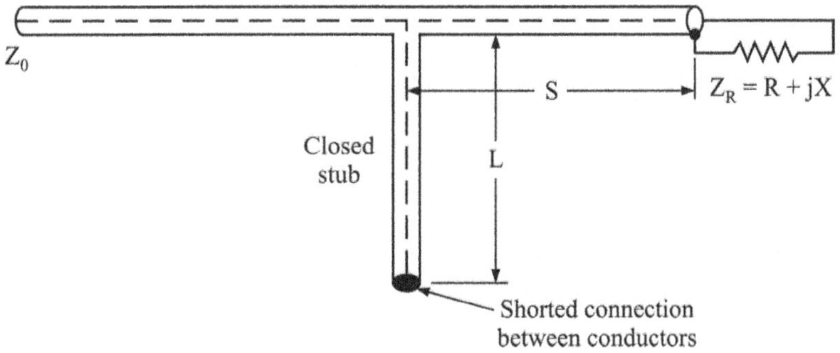

Fig. 5.32

Given $Z_L = 100 + j200\ \Omega$, Z_o=50 Ω, find S and L for the diagram as shown in Fig.5.32

- The normalized impedance $Z_R / Z_o = 2 + j4$

  Plot this normalized impedance on smith chart as shown in Fig.5.33.

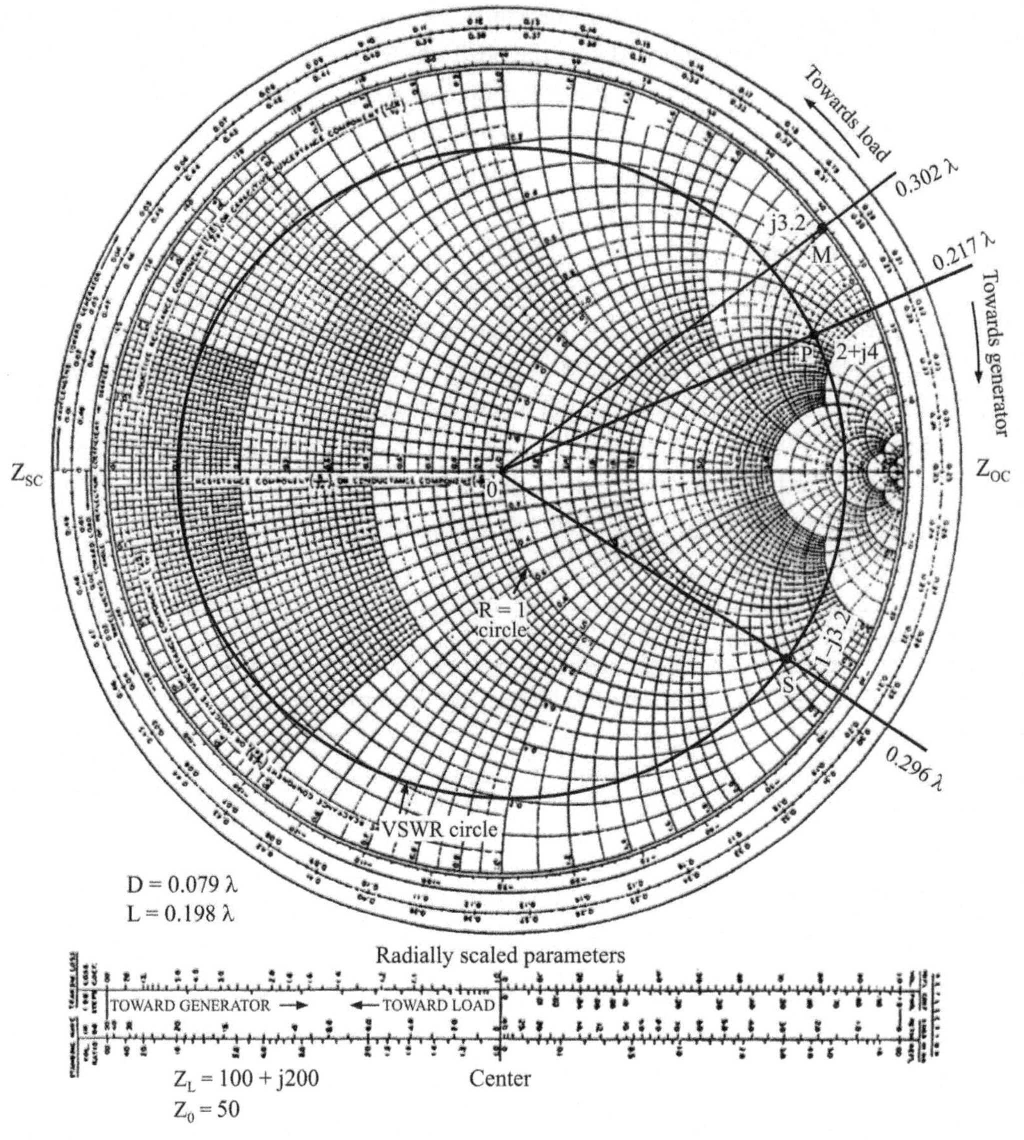

Fig. 5.33 Smith chart for problem 5.16

- Let O be the centre of the smith chart and P be the normalized impedance point on the smith chart. Draw a circle with O as centre and OP as radius, which gives the constant VSWR circle.

- Rotate the line OP towards generator such that the line cuts the intersection of VSWR circle and R = 1 circle. Let the intersection point is S.

- From the OP line to OS line measure the distance on a 'wavelength towards generator' scale, which will give the position 'S'

  The OP line cuts the 'wavelength towards generator' scale at 0.217λ

  The OS line cuts the 'wavelength towards generator' scale at 0.296 λ

  So, $S = 0.296\,\lambda - 0.217\,\lambda$

  $= 0.079\,\lambda$

- From the smith chart shown in Fig:5.33, the admittance at point S is $1 - j\,3.2$
- In order to make $Z_R/Z_O = 1$, we must cancel the imaginary part in the existing impedance at point S. so $+\,j3.2$ is used to cancel the imaginary part of impedance.

  Draw a point (0, + j3.2) in smith chart shown in Fig.5.33, let this point be M

  The line OM cuts the 'wavelength towards load' scale at 0.302 λ

  From the OM line to the OZ_{sc} line (as shown in smith chart) measure the distance on a 'wavelength towards load' scale, which gives the L.

  So, $L = 0.5\,\lambda - 0.302\,\lambda$

  $= 0.198\,\lambda.$

Double Stub Matching

The disadvantages of single stub matching can be reduced by using double stub matching as shown in Fig.5.34, in which two short-circuited stubs whose lengths are adjustable independently but whose positions are fixed.

Usually, these stubs are separated by a length λ/4 or 0.375λ.

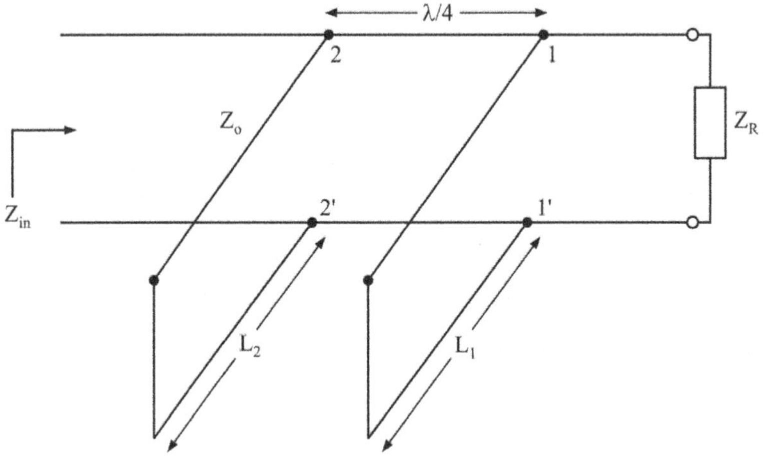

Fig. 5.34 Double stub matching

Problem 5.17

A 50 Ω transmission line is connected to a cellular phone antenna with load impedance $Z_R = 100 + j100\ \Omega$. Find the position and the lengths of a shunt short-circuit double stub required to match the 50 Ω line.

Given $Z_R = 100 + j100\ \Omega$, $Z_o = 50\ \Omega$ find L_1 and L_2.

- The normalized impedance $Z_R / Z_o = 2 + j2$

  Plot this normalized impedance on smith chart as shown in Fig.5.35.

- Let O be the centre of the smith chart and P be the normalized impedance point on the smith chart. Draw a circle with O as centre and OP as radius, which gives the constant VSWR circle.

- Extend the line from P to O and it cuts the constant VSWR circle at point Q as shown in smith chart. Where point Q is normalized admittance. $Y_L/Y_O = 0.25 - j0.26$

  (The stub, which is given is the shunt stub so prefer output admittance instead of output impedance).

- Rotate R = 1 circle with a distance λ/4 (given in problem) in anti-clockwise direction. Let this circle be R' circle. (as shown in smith chart)

- Given D = 0.03 λ, rotate the line OQ towards generator on 'wavelength towards generator' scale with a distance of 0.03 λ then it cuts the VSWR circle at $0.24 - j0.072$. let this point be C

- Let the intersection of R = 0.24 circle and R' circle is K, k = $0.24 + j0.37$ (here there is a possibility of two intersection points but take the intersection point which is nearer.)

- Draw a circle path in clockwise direction with O as centre and OK as radius, which cuts the R = 1 circle at $1 + j1.7$, let this point be S

- The difference between the points K and C, which gives $K - C = j0.442$

  So, +j0.442 is used to find out the L_1

  Draw a point (0, + j0.442) in smith chart as shown in Fig.5.35, let this point be N

  The line ON cuts the 'wavelength towards load' scale at 0.434 λ

  From the ON line to the OY_{sc} (as shown in smith chart) measure the distance on a 'wavelength towards load' scale, which will gives the L_1.

  $$\text{So, } L_1 = 0.25\,\lambda + (0.5\,\lambda - 0.434\,\lambda)$$
  $$= 0.316\,\lambda.$$

- From the smith chart, the admittance at point S is $1 + j1.7$

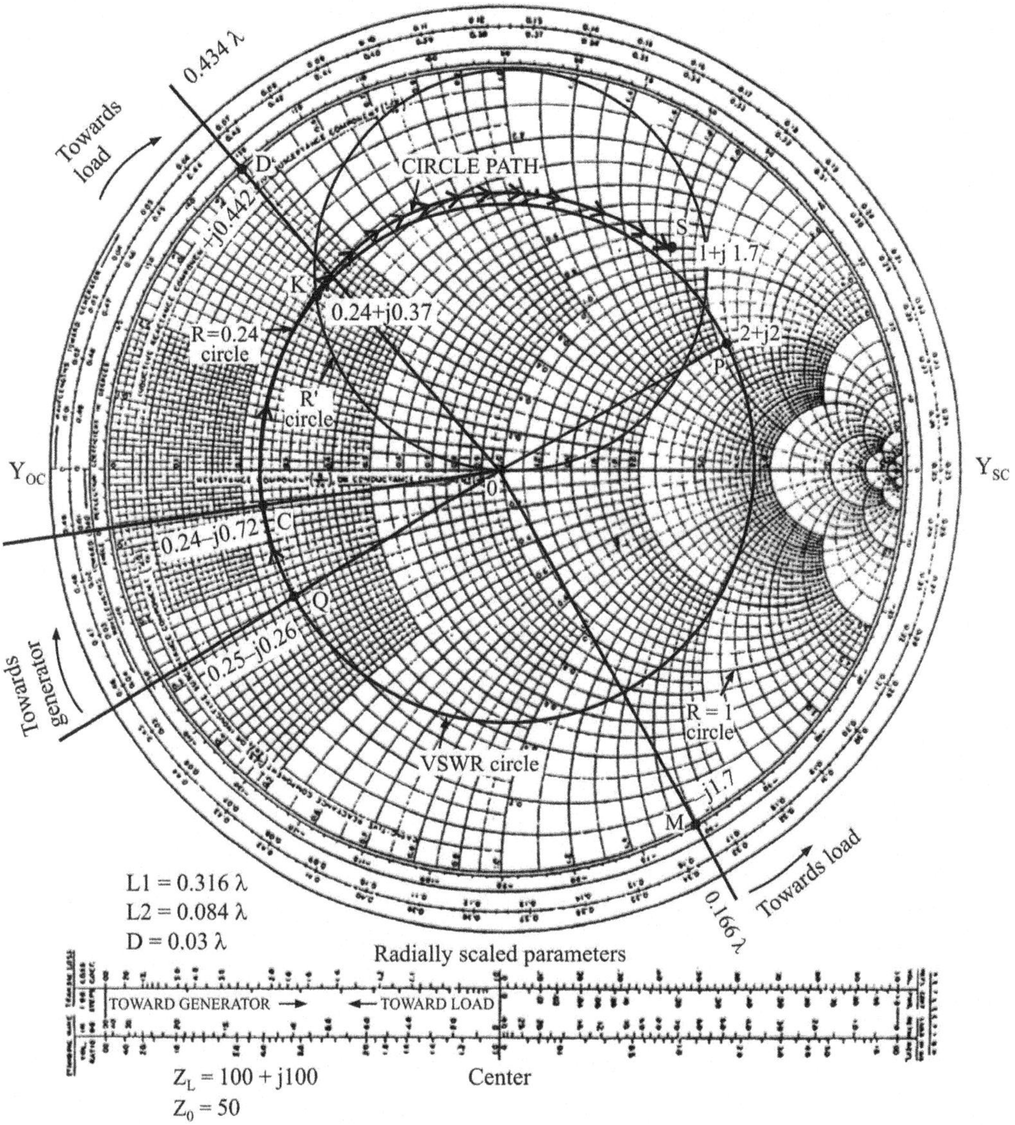

Fig. 5.35 Smith chart for problem 5.17

In order to make $Y_L/Y_O = 1$, we must cancel the imaginary part in the existing admittance at point S. so – j1.7 is used to cancel the imaginary part of admittance

Draw a point (0, – j1.7) in smith chart, let this point be M

The line OM cuts the 'wavelength towards load' scale at 0.166 λ

From the OM line to the OY_{sc} (as shown in smith chart) measure the distance on a 'wavelength towards load' scale, which will gives the L_2.

$$\text{So, } L_2 = 0.25\,\lambda - 0.166\,\lambda$$

$$= 0.084\,\lambda.$$

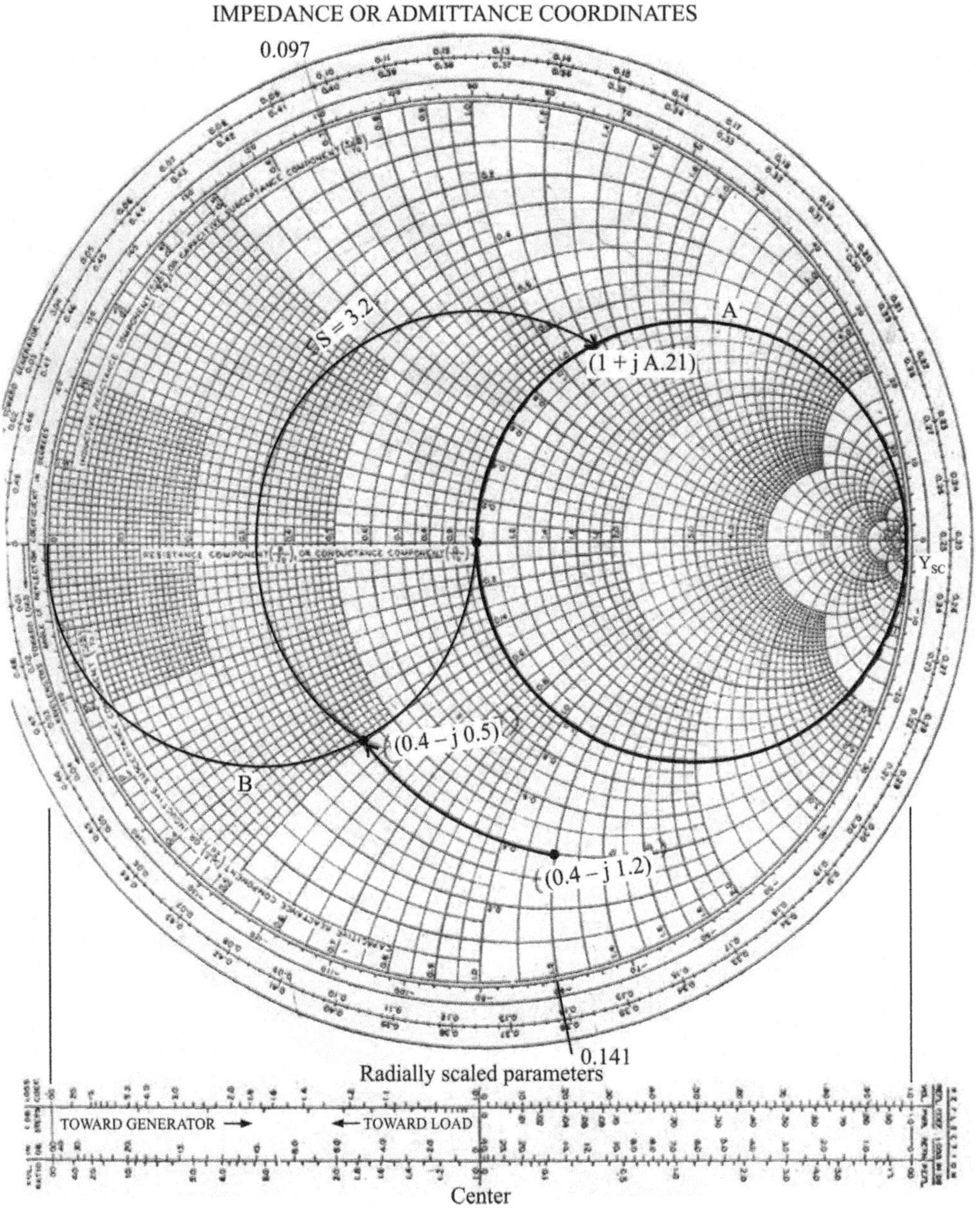

Fig. 5.36 Smith chart for problem 5.18

Problem 5.18

Given $Y_L = 20 - j60\ \mho$, $G_o = 50\ \mho$ find L_1 and L_2. (distance between two stubs are equal to $\lambda/4$).

Solution:

1. $\dfrac{Y_L}{G_o} = 0.4 - j1.2$, (plot this point)

2. Rotate $g_a = 1$ circle Anti-clockwise by $\lambda/4$ distance i.e., circle B as shown in Fig.5.36.

3. Move the Admittance $\dfrac{Y_L}{G_o}$ towards circle B and the intersection point is

   $\dfrac{Y_1}{G_o} = 0.4 - j0.5$ (i.e., conductance value is not changed).

4. Transform the point $\dfrac{Y_1}{G_o}$ with VSWR = 3.2 on to the circle A i.e., Now $\dfrac{Y_2}{G_o} = 1.0 + j1.21$.

5. So, stub 2 having inductive susceptance −1.21 and stub 1 having susceptance = 0.7 $\left(\because \text{from} \dfrac{Y_2}{G_o} \text{to} \dfrac{Y_1}{G_o}\right)$

   $\therefore\ L_1 = 0.25 + 0.097 = 0.348\ \lambda$

   $L_2 = 0.25 - 0.141 = 0.109\ \lambda$

5.18 Microstrip Transmission Line

Microstrip is a type of electrical transmission line which can be fabricated using printed circuit board technology, whose cross-sectional view is as shown in Fig.5.37 and is used to convey microwave-frequency signals. It consists of a conducting strip separated from a ground plane by a dielectric layer known as the substrate. Microwave components such as antennas, couplers, filters, power dividers etc., can be formed from microstrip, the entire device existing as the pattern of metallization on the substrate. Microstrip is thus much less expensive than traditional waveguide technology, as well as being far lighter and more compact.

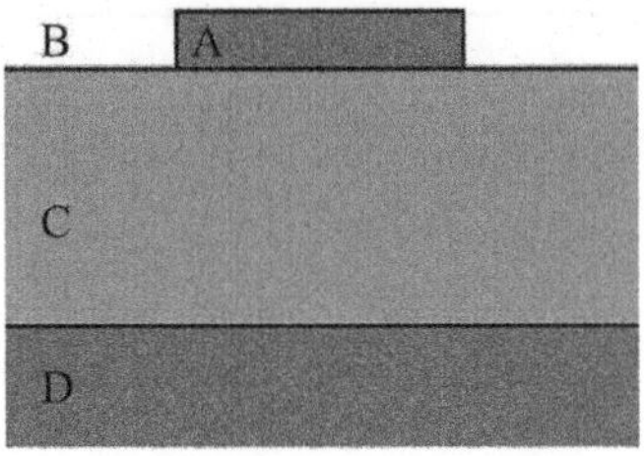

Fig. 5.37 Cross-section of microstrip geometry. Conductor 'A' is separated from ground plane 'D' by dielectric substrate 'C'. Upper dielectric 'B' is typically air.

The disadvantages of microstrip compared with waveguide are the generally lower power handling capacity, and higher losses. Also, unlike waveguide, microstrip is not enclosed, and is therefore susceptible to cross-talk and unintentional radiation.

For lowest cost, microstrip devices may be built on an ordinary substrate. However it is often found that the dielectric losses in ordinary substrate are too high at microwave frequencies, and that the dielectric constant is not sufficiently tightly controlled. For these reasons, an alumina substrate is commonly used.

On a smaller scale, microstrip transmission lines are also built into monolithic microwave integrated circuits.

Microstrip lines are also used in high-speed digital PCB designs, where signals need to be routed from one part of the assembly to another with minimal distortion, and avoiding high cross-talk and radiation.

Microstrip is very similar to stripline and coplanar waveguide and it is possible to integrate all three on the same substrate.

Review Questions and Answers

1. **Define the line parameters?**

   (i) The parameters of a transmission line are:

   (ii) Resistance (R)

   (iii) Inductance (L)

   (iv) Capacitance (C)

   (v) Conductance (G)

   (vi) Resistance (R) is defined as the loop resistance per unit length of the wire. Its unit is ohm/km

   (vii) Inductance (L) is defined as the loop inductance per unit length of the wire. Its unit is Henry/km

(viii) Capacitance (C) is defined as the loop capacitance per unit length of the wire. Its unit is Farad/km

(ix) Conductance (G) is defined as the loop conductance per unit length of the wire. Its unit is mho/km

2. What are the secondary constants of a line? Why the line parameters are called distributed elements?

The secondary constants of a line are:

(i) Characteristic Impedance

(ii) Propagation Constant

Since the line constants R, L, C, G are distributed through the entire length of the line, they are called as distributed elements. They are also called as primary constants.

3. Define Characteristic impedance

Characteristic impedance is the impedance measured at the sending end of the line.

It is given by $Z_0 = \sqrt{\frac{(R + j\omega L)}{(G + j\omega C)}}$ where

$R + j\omega L$ is the series impedance

$G + j\omega C$ is the shunt admittance

4. Define Propagation constant

Propagation constant is defined as the natural logarithm of the ratio of the sending end current or voltage to the receiving end current or voltage of the line. It gives the manner in the wave is propagated along a line and specifies the variation of voltage and current in the line as a function of distance. Propagation constant is a complex quantity and is expressed as $\gamma = \alpha + j\ \beta$. The real part is called the attenuation constant whereas the imaginary part of propagation constant is called the phase constant.

5. What is a finite line? Write down the significance of this line?

A finite line is a line having a finite length on the line. It is a line, which is terminated, in its characteristic impedance ($Z_R = Z_0$), so the input impedance of the finite line is equal to the characteristic impedance ($Z_i = Z_0$).

6. What is an infinite line?

An infinite line is a line in which the length of the transmission line is infinite. A finite line, which is terminated in its characteristic impedance, is termed as infinite line. So for an infinite line, the input impedance is equivalent to the characteristic impedance.

7. What is wavelength of a line?

The distance the wave travels along the line while the phase angle is changing through 2π radians is called a wavelength.

8. What are the types of line distortions?

The distortions occurring in the transmission line are called waveform distortion or line distortion. Waveform distortion is of two types:

(a) Frequency distortion

(b) Phase or Delay Distortion.

9. How frequency distortion occurs in a line?

When a signal having many frequency components are transmitted along the line, all the frequencies will not have equal attenuation and hence the received end waveform will not be identical with the input waveform at the sending end because each frequency is having different attenuation. This type of distortion is called frequency distortion.

10. How to avoid the frequency distortion that occurs in the line?

In order to reduce frequency distortion occurring in the line,

(a) The attenuation constant α should be made independent of frequency.

(b) By using equalizers at the line terminals which minimize the frequency distortion. Equalisers are networks whose frequency and phase characteristics are adjusted to be inverse to those of the lines, which result in a uniform frequency response over the desired frequency band, and hence the attenuation is equal for all the frequencies.

11. What is delay distortion?

When a signal having many frequency components are transmitted along the line, all the frequencies will not have same time of transmission, some frequencies being delayed more than others. So the received end waveform will not be identical with the input waveform at the sending end because some frequency components will be delayed more than those of other frequencies. This type of distortion is called phase or delay distortion.

12. How to avoid the frequency distortion that occurs in the line?

In order to reduce frequency distortion occurring in the line,

(a) The phase constant β should be made dependent of frequency.

(b) The velocity of propagation is independent of frequency.

(c) By using equalizers at the line terminals which minimize the frequency distortion. Equalizers are networks whose frequency and phase characteristics are adjusted to be inverse to those of the lines, which result in a uniform frequency response over the desired frequency band, and hence the phase is equal for all the frequencies.

13. What is a distortion less line? What is the condition for a distortion less line?

A line, which has neither frequency distortion nor phase distortion is called a distortion less line. The condition for a distortion less line is RC = LG. Also,

(a) The attenuation constant should be made independent of frequency.

(b) The phase constant should be made dependent of frequency.

(c) The velocity of propagation is independent of frequency.

14. What is the drawback of using ordinary telephone cables?

In ordinary telephone cables, the wires are insulated with paper and twisted in pairs, therefore there will not be flux linkage between the wires, which results in negligible inductance, and conductance. If this is the case, then there occurs frequency and phase distortion in the line.

15. How the telephone line can be made a distortion less line?

For the telephone cable to be distortion less line, the inductance value should be increased by placing lumped inductors along the line.

16. What is Loading?

Loading is the process of increasing the inductance value by placing lumped inductors at specific intervals along the line, which avoids the distortion.

17. What are the types of loading?

(a) Continuous loading

(b) Patch loading

(c) Lumped loading

18. Define reflection loss

Reflection loss is defined as the number of nepers or decibels by which the current in the load under image matched conditions would exceed the current actually flowing in the load.

19. What is Impedance matching?

If the load impedance is not equal to the source impedance, then all the power that are transmitted from the source will not reach the load end and hence some power is wasted. This is called impedance mismatch condition. So for proper maximum power transfer, the impedances in the sending and receiving end are matched. This is called impedance matching.

20. Define the term insertion loss

The insertion loss of a line or network is defined as the number of nepers or decibels by which the current in the load is changed by the insertion. Insertion loss = Current flowing in the load without insertion of the network/Current flowing in the load with insertion of the network.

21. When reflection occurs in a line?

Reflection occurs because of the following cases:

1. when the load end is open circuited
2. when the load end is short-circuited
3. when the line is not terminated in its characteristic impedance

When the line is either open or short circuited, then there is not resistance at the receiving end to absorb all the power transmitted from the source end. Hence all the power incident on the load gets completely reflected back to the source causing reflections in the line. When the line is terminated in its characteristic impedance, the load will absorb some power and some will be reflected back thus producing reflections.

22. What are the conditions for a perfect line? What is a smooth line?

For a perfect line, the resistance and the leakage conductance value were neglected. The conditions for a perfect line are $R = G = 0$. A smooth line is one in which the load is terminated by its characteristic impedance and no reflections occur in such a line. It is also called as flat line.

23. List the applications of the smith chart.

The applications of the smith chart are,

(i) It is used to find the input impendence and input admittance of the line.

(ii) The smith chart may also be used for lossy lines and the locus of points on a line then follows a spiral path towards the chart center, due to attenuation.

(iii) In single stub matching

24. What are the difficulties in single stub matching?

The difficulties of the smith chart are

(i) Single stub impedance matching requires the stub to be located at a definite point on the line. This requirement frequently calls for placement of the stub at an undesirable place from a mechanical view point.

(ii) For a coaxial line, it is not possible to determine the location of a voltage minimum without a slotted line section, so that placement of a stub at the exact required point is difficult.

(iii) In the case of the single stub it was mentioned that two adjustments were required, these being location and length of the stub.

Multiple Choice Questions

1. Transmission line equations assume the following propagation along the line direction
   (a) TM (b) TEM
   (c) TE (d) All of the above
2. For a lossless line (R,G = 0) the characteristic impedance depends on
   (a) The ratio L/C
   (b) The angular frequency ω
   (c) The ratio L/C and the angular frequency ω
   (d) None
3. For a coaxial line, the conductivity of the dielectric between the two conductors leads to the following parameter being non-zero
   (a) R (b) G (c) L (d) C
4. Which of the following statements are not true of the line parameters R, L, G, and C?
   (a) R and L are series elements.
   (b) G and C are shunt elements.
   (c) G = 1/R.
   (d) The parameters are not lumped but distributed.
5. For a lossy transmission line, the characteristic impedance does not depend on
   (a) The operating frequency of the line
   (b) The length of the line
   (c) The conductivity of the conductors
   (d) The conductivity of the dielectric separating the conductors
6. The origin of the Smith chart corresponds to a reflection coefficient of value
   (a) 1 (b) 0
   (c) –1 (d) Any of these

7. The input impedance of a transmission line depends on
   (a) The reflection coefficient at the load end
   (b) The frequency of operation
   (c) The length of the line
   (d) All of these

8. A 50 ohm transmission line is connected to a load impedance yielding a VSWR of unity. The load impedance is
   (a) 50 ohm (b) 100 ohm
   (c) 1 ohm (d) 0 ohm

9. If the reflection coefficient at a point on a transmission line is –0.3, the transmission coefficient is
   (a) 0.3 (b) –0.3
   (c) 1 (d) 0

10. If maximum and minimum currents in a transmission line are 4 A and 2 A respectively, CSWR is
    (a) 0.5 (b) 2 (c) 1 (d) 4

11. For a transmission line terminated in its characteristic impedance, which of the following statement is incorrect:
    (a) It is a smooth line.
    (b) The energy distribution between magnetic and electric field is not equal.
    (c) Standing wave does not exist.
    (d) Efficiency of transmission of power is maximum.

12. For a line of characteristic impedance, Z_0 terminated in a load, Z_R such that $Z_R > Z_0$, the Voltage Standing Wave Ratio (VSWR) is given by
    (a) $\frac{Z_R}{Z_0}$ (b) Z_R
    (c) Z_0 (d) $\frac{Z_0}{Z_R}$

13. The intrinsic impedance of free space is
    (a) 75 ohm (b) 73 ohm
    (c) 120 π ohm (d) 377 ohm.

14. The characteristic impedance is given by

(a) $Z_0 = \frac{\sqrt{Z_{oc}}}{Z_{sc}}$ (b) $Z_0 = \frac{\sqrt{Z_{sc}}}{Z_{oc}}$

(c) $Z_0 = \sqrt{Z_{oc}Z_{sc}}$ (d) $Z_0 = Z_{oc}Z_{sc}$

15. Consider a transmission line of characteristic impedance 50 ohms and the line is terminated at one end by + j50 ohms, the VSWR produced in the transmission line will be

(a) +1 (b) Zero (c) Infinity (d) –1

16. Which one of the following conditions will not gurantee a distortionless transmission line

(a) R = 0 = G

(b) RC = LG

(c) very low frequency range (R >>ωL, G >> ωC)

(d) very high frequency range (R<< ωL, G << ωC)

17. For a transmission line terminated by a load, the reflection co-efficient magnitude $|\Gamma|$ and the voltage standing wave ration S are related as:

(a) $S = \frac{1}{(1+|\Gamma|)}$ (b) $S = \frac{1}{(1-|\Gamma|)}$

(c) $S = \frac{(1-|\Gamma|)}{(1+|\Gamma|)}$ (d) $S = \frac{(1+|\Gamma|)}{(1-|\Gamma|)}$

18. For a line of characteristic impedance, Z0 terminated in a load ZR such that

$Z_R = \frac{Z_0}{3}$, the reflection coefficient is

(a) 1/3 (b) 2/3 (c) –1/3 (d) –1/2

19. If a line is terminated in an open circuit, the VSWR is

(a) 0 (b) 1 (c) ∞ (d) –1

20. A transmission line with a characteristic impedance Z_1 is connected to a transmission line with characteristic impedance Z_2. If the system is being driven by

a generator connected to the first line, then the overall transmission coefficient will be

(a) $\dfrac{2Z_1}{Z_1+Z_2}$ (b) $\dfrac{Z_1}{Z_1+Z_2}$

(c) $\dfrac{2Z_2}{Z_1+Z_2}$ (d) $\dfrac{Z_2}{Z_1+Z_2}$

Answers

1	(b)	2	(a)	3	B	4	(c)	5	(b)
6	(b)	7	(d)	8	(a)	9	(a)	10	(b)
11	(b)	12	(a)	13	(d)	14	(c)	15	(c)
16	(c)	17	D	18	(d)	19	(c)	20	(a)

Exercise Questions

1. Determine the reflection coefficients when
   (a) $Z_R = Z_0$
   (b) Z_R = short circuit
   (c) Z_R = open circuit
   (d) Also find out the magnitude of reflection coefficient when Z_R is purely reactive
2. A transmission line in which no distortion is present has the following parameters $Z_0 = 50$ ohm, $\alpha = 20$ mNP/m, $v = 0.6\ v_0$. Determine R, L, G, C and wavelength at 0.1 GHz.
3. What are the salient aspects of primary constants of a two wire transmission line.
4. A lossless transmission line used in a TV receiver has a capacitance of 50 PF/m and an inductance of 200 nH/m. Find out the characteristic impedance for 10 meter long section of the line and 500 meter section.
5. Derive the characteristic impedance of a transmission line in terms of its line constar
6. Explain the meaning of the terms characteristic impedance and propagation constant of a uniform transmission line and obtain the expressions for them in terms of parameters of line.
7. Starting from the equivalent circuit, derive the transmission line equations for V and I in terms of the source parameters.

8. A two wire line has a characteristic impedance of 600 ohm and is to feed a 150 ohm resistor at 200 MHz. A half wave line is to be used as a tube, 1.2 cm in diameter. Find centre to centre spacing in air.
9. Explain how UHF lines can be treated as circuit elements, giving the necessary equivalent circuits.
10. A loss less line of 100 ohm is terminated by a load which produces SWR = 3. The first Maxima is found to be occurring at 320 cm. If f = 300 MHz, determine load impedance.
11. Draw the equivalent circuits of a transmission lines when (a) length of the transmission line $l < \lambda/4$, with shorted load (b) when $l < \lambda/4$, with open end (c) $l = \lambda/4$.
12. Find out SWR if (a) Z_0 = 100 ohm, R_L = 80 ohm (b) when Z_0 = 80 ohm, R_L = 100 ohm.
13. Explain the principle of Impedance matching with Quarter wave Transformer?
14. A 100 ohm loss less line connects a signal of 100 kHz to a load of 140 ohm. The load power is 100 mW. Calculate (a) Voltage Reflection coefficient (b) VSWR (c) Position of V_{Max}, I_{Max}, V_{min} and I_{min}.
15. Define the reflection coefficient and derive the expression for i/p impedance in terms of reflection coefficient.
16. Explain how the i/p impedance varies with the frequency with sketches.
17. Explain the significance and utility of $\lambda/8$, $\lambda/4$ and $\lambda/2$ lines.
18. Explain the significance and design of single stub impedance Matching.

    Discuss the factors on which stub length depends.